Science, Conservation, and National Parks

Science, Conservation, and National Parks

EDITED BY STEVEN R. BEISSINGER,
DAVID D. ACKERLY, HOLLY DOREMUS,
AND GARY E. MACHLIS

The University of Chicago Press
Chicago and London

The University of Chicago Press, Chicago 60637
The University of Chicago Press, Ltd., London
© 2017 by The University of Chicago
All rights reserved. Published 2017.
Printed in the United States of America

26 25 24 23 22 21 20 19 18 17   1 2 3 4 5

ISBN-13: 978-0-226-42295-4 (cloth)
ISBN-13: 978-0-226-42300-5 (paper)
ISBN-13: 978-0-226-42314-2 (e-book)
DOI: 10.7208/chicago/9780226423142.001.0001

Library of Congress Cataloging-in-Publication Data

Names: Beissinger, Steven R., editor. | Ackerly, David D., editor. | Doremus,
    Holly D., editor. | Machlis, Gary E., editor. | University of California,
    Berkeley, organizer, host institution. | National Geographic Society (U.S.),
    organizer. | United States. National Park Service, organizer.
Title: Science, conservation, and national parks / edited by Steven R.
    Beissinger, David D. Ackerly, Holly Doremus, and Gary E. Machlis.
Description: Chicago ; London : The University of Chicago Press, 2017.
    | Papers from a summit, "Science for Parks, Parks for Science: the next
    century", organized by University of California, Berkeley, in partnership
    with the National Geographic Society and the National Park Service and
    held 25–27 March 2015 at the University of California, Berkeley. | Includes
    bibliographical references and index.
Identifiers: LCCN 2016020068 | ISBN 9780226422954 (cloth : alk. paper) |
    ISBN 9780226423005 (pbk. : alk. paper) | ISBN 9780226423142 (e-book)
Subjects: LCSH: National parks and reserves—United States—Congresses. |
    National parks and reserves—Management—Congresses. | Environmental
    sciences—Congresses. | Environmental sciences—Social aspects—Con-
    gresses. | Nature conservation—Congresses.
Classification: LCC SB482.A4 S36 2016 | DDC 363.6/8—dc23 LC record
    available at https://lccn.loc.gov/2016020068

♾ This paper meets the requirements of ANSI/NISO Z39.48-1992
    (Permanence of Paper).

*For the parks of the world, and to the pioneers of park science—
past, present, and future*

George Melendez Wright (*left*) and Ben H. Thompson (*right*)
in a snowdrift in Yellowstone National Park, May 1932. Photo taken
by Joseph S. Dixon and provided by Pamela Wright Lloyd.

CONTENTS

I think of this book as a time capsule that has been assembled at a critical moment for humanity and its relationship with the rest of life on Earth. It captures the current state of knowledge, the challenges, and the controversies that embody conservation at the beginning of the 21st century. Parks and protected areas—national, regional, and local—play key roles both in conserving biological diversity at a time when species extinctions are accelerating and in engaging people with nature at a moment when much of humanity lives apart from most other life forms. Now, more than a century after the founding of the US National Park Service (NPS) in 1916, those who are entrusted with the care of the parks face unprecedented challenges to sustain their ecological integrity and their facilities.

Our national, state, and regional parks are under increasing threat from a changing climate, from storms and fires of greater severity, from urban encroachment and pollution, from invasions of nonnative species, from plant and animal extinctions, from the changing attitudes of a public that has become more urbanized, and from the political pressures of narrow interest groups that have sometimes led to benign neglect of parks. These challenges will continue to grow over the coming decades.

This book, and the summit at the University of California, Berkeley, from 25 to 27 March 2015 that spawned it, builds on the historic linkage between UC Berkeley and the NPS. National parks and public education are arguably America's "two best ideas," and they grew up together at UC Berkeley. Much of the major inspiration for, and the perspiration that produced, the NPS came from UC Berkeley and its graduates over a century ago. Moreover, much of the early and influential research in national parks was done by Berkeley faculty and graduates. This remarkable history is revisited at the end of this book in the appendix (as it set the stage for the summit and for this book).

UC Berkeley, in partnership with the National Geographic Society and the NPS, convened the summit entitled "Science for Parks, Parks for Science: The Next Century" to celebrate the NPS centennial and to focus on science that is relevant to parks and protected areas in the United States and worldwide. Why science for parks and parks for science? The mission of national parks, in the United States and globally, implicates science in two complementary, often interwoven, ways that are addressed in this book. First, science plays important roles in determining the size and location of lands needed to conserve as parks, in identifying threats to parks, in developing and evaluating management solutions, in translating abstract conservation goals into concrete results to inform decisions, and in understanding how people interact with and benefit from parks. Second, parks and protected areas provide unique sites for scientific study of environmental processes that are important to sustain both life and humanity. They can act as "control sites" for understanding human impacts on species and ecosystems, or on cultural resources. National parks may be essential for some kinds of studies to the extent that these parks are more strongly shielded from human impacts than other lands, protect larger areas than other types of parks, or protect resources not found elsewhere. Finally, scientific studies that include citizen participation in parks may also serve to create human connections with parks and nature, which may be essential for the long-term maintenance of protected land systems and for biodiversity conservation.

The summit organizers chose to focus on science writ large—biological science, physical science, and social science. But what makes science so important anyway? Strong inference from science brings data and theory together to make a formal model, or hypothesis, about how systems work. But science doesn't end there. From a model, we create predictions, and then we collect new data to test whether those predictions are supported. When science is done less rigorously or produces results that are less demonstrative, other scientists often challenge and improve the process. By repeating these tests and amending our models, a self-correcting system of understanding emerges from science that can produce important and often unbiased insights. In the absence of data, theory, and models (i.e., science), managers and politicians are left with opinions and perceptions—rather than evidence—to guide decision making.

A goal of the summit was to secure a future for parks by enabling and catalyzing a community of scholars and practitioners to push forward the

frontiers of science for and in parks and protected areas. The summit organizers invited 30 plenary speakers and discussants—led by E. O. Wilson, Jane Lubchenco, and academics from other institutions—to engage
participants on key subjects. Why emphasize the voices of academics? Because the academy is a place of free discourse and because conversations
about difficult subjects often begin here first. The summit encouraged the
exchange of ideas, convening three "strategic conversations" that featured
differing viewpoints on themes critical to parks and protected areas. The
plenary sessions and strategic conversations were complemented by over
200 contributed oral and poster presentations that were attended by more
than 550 participants. Another 1,000+ viewers from around the world
watched the talks as they were live streamed. This book captures most of
the contributions of the plenary speakers and the strategic conversations.

It does not, however, capture all the challenges facing parks, all the ways
that science can contribute, or all the voices involved with or affected by
parks. There was a tension in putting together the summit and this volume; both take their inspiration from the NPS centennial, but also seek
to address a broader vision about science for parks and parks for science.
The volume editors tried to balance a treatment that emphasized national
parks, especially in the United States, but provided connections to parks
around the world and to parks under other administrative jurisdictions. In
comparison with other park missions, this book achieves its strongest coverage on issues affecting biological resources in parks and the interactions
of people and parks. For example, there are no chapters devoted to the science of conserving cultural artifacts. The editors felt that the challenges facing managers of historic buildings, sites, and battlefields, while significant
and challenging, were too narrow for the general readership of this book.
Moreover, while cultural parks are numerically important in the US National Park System, which includes over 400 units, their acreage is dwarfed
by the 59 national parks devoted to biological or scenic resources. Instead,
the editors encouraged authors to include consideration of cultural features in their chapters where appropriate, including coverage of cultural
and spiritual connections to nature. It was also the editors' intention to feature the voices of Native Americans and Native Hawaiians in the chapters
and strategic discussions; multiple individuals were invited to participate,
but none were willing or able to step forward.

I would like to thank the many people and organizations that made
important contributions to this book and to the success of the summit,
which was the result of two years of planning. The UC Berkeley College
of Natural Resources (CNR), under the leadership of Dean Keith Gilless,

provided financial and logistical support for the summit and the production of this book. CNR's Jennifer Brand and Bernadette Powell worked tirelessly to make all aspects of the summit a success and, along with Maya Goehring-Harris, were largely responsible for its flawless execution. The Summit Program Committee, which I chaired, consisted of coeditors David Ackerly and Holly Doremus of UC Berkeley, along with ex officio members Gary Machlis (also a coeditor), Angela Evenden, and Raymond Sauvajot of the NPS. Dick Beahrs and Linda Schacht provided important leadership as cochairs of the Summit Planning Committee. Kelly Iknayan and Sarah MacLean assisted with summit promotion. Abstracts submitted for summit presentations were reviewed by a committee of UC Berkeley faculty and graduate students and NPS scientists, coordinated by Todd Dawson and Angela Evenden. UC Berkeley undergraduate, graduate, and postdoctoral student volunteers helped shepherd all the moving parts of the summit. UC Berkeley graduate students who participated in the "Science for Parks" seminar held in spring 2015 did an outstanding job of generating and fielding questions for plenary speakers and discussants. They also edited the strategic discussions that appear in this book. Tierne Nickel provided editorial support and coordinated the production of the entire book. Financial support was received from the NPS for live streaming the summit presentations, which was accomplished flawlessly by UC Berkeley's Jon Schainker and his video production staff. Major financial support for the summit was provided by the National Geographic Society and Save the Redwoods League. Additional financial sponsorship was provided by the California State Parks Foundation, East Bay Regional Park District, Golden Gate National Parks Conservancy, LSA Associates, The Nature Conservancy, and the Yosemite Conservancy.

Finally, I thank the members of the Beissinger Lab who helped pick up the slack during the preparation and execution of the summit and this book, and most importantly my family for their unflagging support during this endeavor.

—Steven R. Beissinger
  Berkeley, California

Mission and Relevance of National Parks

We begin this volume with several chapters focused on the role of parks within their larger geographic and policy contexts. Although this book was inspired by the centennial of the US National Park System, it is intended to be attentive to and provide lessons for the use and understanding of parks and protected areas globally. This section throws us immediately into that larger world, with three chapters that ask from a global perspective what goals protected areas should serve, what sorts of protected areas might qualify as "parks," and how those areas can balance the potentially conflicting goals of human use and resource protection. All are questions that the US National Park Service has grappled with over the past 100 years and continues to address as it enters its second century.

In the United States, there are a wide variety of protected lands, with different historical origins, managed by different entities, and serving different purposes. At the federal level, in addition to the national parks, there are the national forests, national wildlife refuges, and Bureau of Land Management lands. Each system has its own set of goals, although some overlap in mission and potential for conflict exist, and occasional turf wars between agencies occur. The national forests and Bureau of Land Management lands are governed by multiple use mandates; they are supposed to serve a variety of interests over time and space, including resource extraction as well as wildlife protection and provision of recreation opportunities. The national wildlife refuges, originally founded primarily to enhance production of game birds, now also serve the larger goal of protecting the biological integrity of ecosystems as well as conserving individual species. The national parks, historically focused on the protection of spectacular scenery, have the joint mission of conserving natural and historic resources while providing for human enjoyment of those resources now and in the future. Many states have one or more systems of protected lands as well, serving conservation and recreational needs. Local governments may also hold lands for these purposes. Management of state and local lands is governed by state and local law, which may apply generally to an entire system or specifically to individual land units. Finally, there are privately owned protected lands. Like other private lands, those serve whatever goal their owner has in mind. Individual owners make conservation choices that are constrained primarily by their own values, which may be idiosyncratic. Privately protected lands also include those owned and managed by small and large nonprofit entities, ranging from local land trusts to The Nature Conservancy. They must deal with additional constraints imposed by their charters and fundraising needs.

Over time, lands are added to and removed from federal, state, or local ownership. The choice to add lands to (or subtract lands from) a national park system requires decisions about both what is optimal and what is possible. The optimal decision puts lands into public ownership if the private market will not adequately serve the goals to which they are dedicated, places public lands under the control of the level of government best suited to effectuating those goals, and matches the managing agency to those goals. To further complicate matters, the optimal decision may seem unattainable, at least in the short run, because of political or budgetary constraints. Judgments may also be needed, therefore, about trade-offs between what is desirable and what is possible, and between short- and long-term goals.

Once a decision has been made to incorporate lands into a national park system, another set of daunting potential trade-offs may need to be confronted. A conspicuous feature of the US national parks is the mandate to achieve both resource protection and resource use. The Organic Act of 1916 declared the purpose of the US National Park System to be "to conserve the scenery and the natural and historic objects and the wild life therein and to provide for the enjoyment of the same in such manner and by such means as will leave them unimpaired for the enjoyment of future generations."[1] That dual purpose may be essential to building a long-term constituency for protected lands, but it creates challenges for managers because this mandate harbors both tension and ambiguity. The Organic Act does not explicitly acknowledge the potential for conflict between conservation and enjoyment of park resources that might occur, much less offer principles for striking a balance between the two. Nor does it address

1. Formerly 16 USC § 1, now codified at 54 USC § 100101(a).

what it means to keep park resources "unimpaired" for the future, which is where the divide between acceptable and unacceptable change to a park might lie. Nor, finally, does it define or limit the universe of activities that constitute legitimate "enjoyment" of park resources. Those are questions that cannot be answered in the abstract, or forever. They are necessarily context specific. Moreover, we should not think of the mission of national parks as static or rigid. The principles articulated in the Organic Act are unchanging, but the way those principles apply to specific facts is necessarily a function of the times.

How we interpret the mission of the US National Park System has implications that go well beyond the resolution of specific management conflicts. That interpretation, for instance, is crucial to determining whether the park system in the United States should add more land or more units, and if so, what acquisitions should have the highest priority. In addition, the mission of the US National Park System speaks to global issues. As the first nation to establish national parks and a national park system, the United States historically has been highly influential in the development and spread of the national park idea around the world. Today, the United States can also learn from the extensive experience and diverse models of parks elsewhere.

The three chapters in this section all approach the question of the mission of parks from a large-scale perspective, looking globally at what resources should be protected and how strongly. All focus not just on conservation of the biota, but also on the connections of people to the biota and to protected areas. They differ in their focal points and emphases but have much common ground.

Edward O. Wilson begins with the need to slow the accelerating species extinction rate, bringing it down to a rate near the prehuman baseline. Success, he argues in his essay and an accompanying discussion with Steven R. Beissinger, will require the protection of roughly half of Earth's surface as "inviolable habitat," protected from intensive human activity. Implicit in Wilson's discussion is a dual view of the mission of parks, and of protected areas more generally. From Wilson's perspective, their primary purpose is to conserve biodiversity. But an important secondary purpose, and one that in the end helps serve the first, is to introduce new generations of people to science, and particularly to natural history. To that end, the parks should be research and education centers. Although he does not make this connection explicitly, Wilson's view of the purposes of protected reserves fits comfortably with the Organic Act's mission statement, filtered through a natural historian's lens. That lens elevates conservation above use, and endorses scientific study as the best way to enjoy, and to connect with, the parks.

Kirsten Grorud-Colvert, Jane Lubchenco, and Allison K. Barner follow with a chapter that echoes Wilson's dual themes, but casts them in a more humanistic light and focuses specifically on protection of marine areas. The establishment of marine parks has lagged greatly behind parks on lands through the 20th century, but has been an area of tremendous conservation activity in recent years. Grorud-Colvert, Lubchenco, and Barner argue forcefully for making protection of special places in the ocean as important as it has been on land. A primary mission of parks for these authors is not merely conservation, although that is important, but also inspiration: national parks help people understand themselves and their place in nature. Like Wilson, these authors note that parks are generally the most protected category of reserves, with strong limits on resource extraction and other ecosystem-disturbing activities. They cite data showing that strong protection of seascapes can quickly and positively affect degraded ecosystems, with impacts that extend beyond the boundaries of marine reserves. Like Wilson, these authors endorse strong limits on extractive activities in parks, but they see certain human uses of parks as highly beneficial. They emphasize the formation of deep human bonds to nature. Implicit in Wilson's chapter is that scientific engagement can nourish such bonds. Grorud-Colvert, Lubchenco, and Barner agree, and add that art and tourism can have similar effects. They offer an inspirational vision of ocean parks as important locations for conservation of and connection to nature.

In the third chapter in this section, Ernesto C. Enkerlin-Hoeflich and Steven R. Beissinger examine the role of protected areas, both terrestrial and marine, worldwide. They trace the values that protected areas serve around the world, and how they have expanded over the past century from conservation to include ecosystem services, poverty reduction, climate change mitigation, and human health benefits. They also discuss global targets for protected area coverage, which have grown from 10% of the world's ecosystems at the 1992 World Parks Congress, to the 2020 goal of 17% of terrestrial and inland water areas and 10% of coastal and marine areas set by the Aichi Biodiversity Target in 2010, and finally to the "nature needs half" campaign promoted by E. O. Wilson and others to protect half the planet in undisturbed ecosystems. Currently, protected areas compose about 15.4% of the planet's terrestrial areas and inland water areas and 3.4% of the oceans, a level of protection that the Promise of Sydney emerging from the 2014 World Parks Congress pledges to expand. Enkerlin-Hoeflich and Beissinger argue that, in practice, to set aside half of Earth's surface for nature will require careful integration of highly protected areas into national and international planning, and the thoughtful and pragmatic integration of human use, and even resource extraction, in some protected areas. Like the two preceding chapters, Enkerlin-Hoeflich and Beissinger acknowledge the importance of building public support for the long-term sustainability of protected areas. They add a new perspective to this discussion: the role of cooperation—from governments to religious organizations—for building that kind of support.

The section concludes with a strategic conversation looking both to the past and to the future of the mission of the US national parks. Discussants included Denis Galvin, a career National Park Service employee, now retired; George Miller, former US congressman; and Frances Roberts-Gregory, a graduate student at the University of California, Berkeley. The panelists discussed the complexity and fluidity of the mission of the US national parks, the need for cooperation and communication with neighboring communities, and what makes national parks distinct among protected landscapes. They also addressed the core challenges for US national parks in the next 100 years, focusing on public engagement with the parks and with science, and on climate change, which affects everything the parks are and aspire to be.

Those challenges bring us back to the Organic Act. In the face of an uncertain future, its lack of specificity is an opportunity, not just a challenge. The Organic Act wisely leaves room for our collective understanding of the purposes of the parks to evolve over time. It won't be easy to decide what the parks of the future should be, or how they should be managed to achieve their purposes. Should backcountry fires and invasive species be fought aggressively or allowed to take their course? Should roads, parking lots, and campgrounds be added or removed? Where within the parks, if anywhere, should motorized off-road vehicles, rock climbing, base jumping, and other forms of recreation be permitted? In the face of global climate change, what should we want the landscapes of our national parks to look like 100 years from now, and what biota should we want them to support?

We, our children, and our grandchildren will continue to struggle with those and similar questions about the mission and purposes of parks and other protected lands, from the most local level to the global. That struggle will be worth it if it helps keep the parks as important, inspiring, and connected to the present as they are today.

ONE

Parks, Biodiversity, and Education: An Essay and Discussion

EDWARD O. WILSON

This is a very important meeting and book, and I'm grateful to be part of it. First, I'll summarize what scientists have learned about biodiversity and extinction, especially during the past 20 years. Then I'll suggest what I believe is the only viable solution to stanch the continuing high and growing rate of species extinction. Then, finally, I'll make the point already obvious to many of you, that our national parks are logical centers for both scientific research and education for many domains of science, but especially and critically biodiversity and conservation of the living part of the environment.

The world is turning green, albeit pastel green, but humanity's focus remains on the physical environment—on pollution, the shortage of fresh water, the shrinkage of arable land, and on that great, wrathful demon, climate change. In contrast, Earth's biodiversity, and the wildlands on which biodiversity is concentrated, have continued to receive relatively little attention. This is a huge strategic mistake. Consider the following rule of our environmental responsibility: If we save the living environment of Earth, we will also save the physical nonliving environment, because each depends intimately on the other. But if we save only the physical environment, as we seem bent on doing, we will lose them both.

So, what is the condition of the living environment, in particular its diversity and stability? How are we handling this critical element of Earth's sustainability? Let me summarize the basic information that scientists have assembled up to the present time, most of it during the last decade.

First, what is biodiversity? It's the collectivity of all inherited variation in any given place, whether a vacant lot in a city, an island in the Pacific, or the entire planet. Biodiversity consists of three levels: an ecosystem such as a pond, a forest patch, or coral reef; then the species composing the ecosystem; and finally at the base, the genes that prescribe the traits that distinguish the species that compose the living part of the ecosystem.

How many species are known in the whole world? At the present time, almost exactly two million. How many are there actually, both known and unknown? Excluding bacteria and the archaea, which I like to call the dark matter of biology because so little is known of their biodiversity, the best estimate of the diversity of the remainder (that is, the fungi, algae, plants, and animals) is nine million species, give or take a million. Except for the big animals, the vertebrates, comprising 63,000 known species collectively of mammals, birds, reptiles, amphibians, and fishes, and 270,000 species of flowering plants, very little to nothing is known of the remaining millions of kinds of fungi and invertebrates. These are the foundation of the biosphere, the mostly neglected little things that run the planet.

To put the whole matter in a nutshell, we live on a little-known planet. At the present rate of elementary exploration, in which about 18,000 additional new species are described and given a Latinized name each year, biologists will complete a census of Earth's biodiversity only sometime in the 23rd century.

I'm aware of only three national parks in the world at the present time in which complete censuses have been undertaken: the Great Smoky Mountains National Park, the Boston Harbor National Park and Recreation Area, and the Gorongosa National Park of Mozambique. The Great Smoky Mountains National Park is the most advanced, with 50,000 hours of fieldwork by experts and assistants completed, about 18,000 species recorded, and a rough estimate of 40,000 to 60,000 species considered likely when all transient, rare, or undescribed species have been registered. Fewer than 1%, let me repeat, 1%, of the species have been studied beyond this first roll call. (Incidentally, the largest biodiversity in a North American park would be the one under consideration for the Mobile Alabama Delta and Red Hills immediately to its north.)

Next, what is the extinction rate? With the data sets of the best-known vertebrate animal species, and additional information from paleontology and genetics, we can put the extinction rate, to the closest power of 10, at 1,000 times greater than the extinction rate that existed before the coming of humans. For example, from 1895 to 2006, 57 species and distinct geographic subspecies of freshwater fishes were driven to extinction in the United States by human activity. These losses have removed roughly 10% of the total previously alive. The extinction rate is estimated to be just under 900 times the level thought to have existed before the coming of humans.

This brings us to the effectiveness of the global conservation moment, a contribution to world culture pioneered by the United States. It has raised public awareness and stimulated a great deal of research. But what has it accomplished in saving species, hence biodiversity? The answer is that it has slowed the rate of species extinction but is still nowhere close to stopping it. A study made by experts on different groups of land vertebrates, species by species, found that the rate in these most favored groups has been cut by about 20% worldwide. Furthermore, the Endangered Species Act of 1973, by focusing on recognized endangered vertebrates in the United States, with legal process and processes designed for each species in turn, has brought 10 times more species back to health than have been lost to extinction.

Nevertheless, the species, and with them the whole of biodiversity, thus continue to hemorrhage. The prospects for the rest of the century are grim. All have heard of the 2C threshold, 2°C (or 3.6°F), the increase in the ground average temperature above which the planet will enter a regime of dangerous climate changes. A parallel circumstance exists in the living world.

Earth is at or approaching an extinction rate of 1,000 times above prehuman levels, and the rate is accelerating. Somewhere between a rate of 1,000 times and 10,000 times, Earth's natural ecosystems will reach the equivalent of the 2C global warming threshold and begin to disintegrate as half or more of the species pass into extinction.

We're in the situation of surgeons in an emergency room who've brilliantly slowed the bleeding of an accident patient to 50%. You can say, "Congratulations! The patient will be dead by morning."

There is a momentous moral decision confronting us here today. It can be put in the form of a question: What kind of a species, what kind of an *entity*, are we to treat the rest of life so cheaply? What will future generations think of those now alive who are making an irreversible decision of this magnitude so carelessly? The five previous such mass extinctions, the last one 65 million years ago that ended the Age of Reptiles, required variously 5–40 million years to recover.

Does any serious person really believe that we can just let the other eight million species drain away, and our descendants will be smart enough to take over the planet and ride it like the crew of a real space ship? That they will find the way to equilibrate the land, sea, and air in the biosphere, on which we absolutely depend, in the absence of most of the biosphere?

Many of us, I believe, here present understand that only by taking global conservation to a new level can the hemorrhaging of species be brought down to near the original baseline rate, which in prehuman times was one species extinction per 1–10 million species per year. Loss of natural habitat is the primary cause of biodiversity extinction—ecosystem, species, and genes, all of it. Only by the preservation of much more natural habitat than hitherto envisioned can extinction be brought close to a sustainable level. The number of species sustainable in a habitat increases somewhere between the third and fifth root of the area of the habitat, in most cases close to the fourth root. At the fourth root, a 90% loss in area, which has frequently occurred in present-day practice, will be accompanied by an automatic loss of one-half of the number of species.

At the present time, about 15% of the global land surface and 3% of the global ocean surface are protected in nature reserves. Not only will most of them continue to suffer diminishment of their faunas and floras, but extinction will accelerate overall as the remaining wildlands and marine habitats shrink because of human activity.

The only way to save the rest of life is to increase the area of protected, inviolable habitat to a safe level. The safe level that can be managed with a stabilized global population of about 10 billion people is approximately half of Earth's land surface plus half of the surface of the sea. Before you start making a list of why it can't be done, that half can't be set aside for the other 10 million or so species sharing the planet with us, let me explain why I believe it most certainly can be done—if enough people wish it to be so.

Think of humanity in this century, if you will, as passing through a bottleneck of overpopulation and environmental destruction. At the other end, if we pass through safely and take most of the rest of the life forms with us, human existence could be a paradise compared to today, and virtual immortality of our species could be ensured—again, if enough wish it to be.

The reason for using the metaphor of a bottleneck instead of a precipice is that four unintended consequences of human behavior provide an opening for the rest of the century. The first unintended consequence is the dramatic drop in fertility at or below zero population growth whenever women gain a modicum of social and economic independence. Population growth is slowing worldwide, and the world population has been predicted most recently by the United Nations to reach between 9.6 billion and 12.3 billion by the end of the century. This assumes that the peoples of sub-Saharan Africa will pass through the demographic transition and fertility rates there will drop to levels consistent with the rest of the world.

The second unintended consequence is from the ongoing abandonment of rural, primitive agricultural economies by the implosion of people into cities, freeing land for both better agriculture and the conservation of natural environments by restoration. It's worth noting also that the present daily production of food globally is 2,800 calories per person. The problem is not food production but transportation and the poor quality of artisanal agriculture. We can fix that. Present-day agriculture is still primitive, with a lot of wiggle room.

The third unintended consequence is the reduction of the human ecological footprint by the evolution of the economy itself. The ecological footprint is the amount of space required for all the needs of each person on average. The idea that the planet can safely support only two to three billion people overlooks the circumstance that the global economy is evolving during the digital revolution, and at a fast rate. The trend is overwhelmingly toward manufacture of products that use less materials and energy, and require less to use and repair. Information technology is improving at almost warp speed. The result is a shrinkage of the ecological footprint. We need an analysis of the trend and its impact. If economists have thought of analyzing this effect and its meaning for the environment, instead of stumbling around in the fever swamp parameters of the early 21st century, I haven't seen it.

The fourth unintended consequence is the easing of demand on the natural environment inherent in the evolutionary shift now occurring from an extensive economy to an intensive economy, one that focuses—in the manner of Moore's law—on improvements of existing classes of products instead of acquisition of new and bigger projects, expanding physical development, and promotion of capital growth based on land acquisition. Humanity may be shifting toward a nongrowth economy focused on quality of life instead of capital and economic power as the premier measure of success.

This brings me to the focal issue of the conference. Inevitably, biodiversity and ecosystem science will move toward parity with molecular, cell, and brain science among the biological disciplines. They have equal challenges. They have equal importance to our daily lives. As this expansion occurs, national parks and other reserves will be the logical centers for fundamental research. They are our ready-made laboratories, in which the experiments have been largely performed. They will also be among the best places to introduce students at all levels to science. We already know that is the case for geology, earth chemistry, and water systems studies. Soon it will be obviously true also for studies of the living environment. Students and teachers alike will have the advantage of hands-on science at all levels. Even at the most elementary level, they are soon caught up in original discoveries of citizen science. After 42 years of teaching experience at Harvard, I believe that natural ecosystems are by far the most open and effective door to science education.

The databases alluded to in this essay are drawn from among those in my book *Half Earth* (New York: W. W. Norton, 2016). The need for continent-size reserves, in particular those built from wildlife corridors, has been argued by others, including Harvey Locke, Michael Soulé, and fellow participants of the Yukon to Yellowstone Conservation Initiative and Wildlands Network. More recently, in 2011, R. F. Noss et al. have added arguments in "Bolder Thinking for Conservation" (*Conservation Biology*, vol. 26, pp. 1–4). I argued the case for half the planetary surface as a reserve in *A Window on Eternity: A Biologist's Walk through Gorongosa National Park* (New York: Simon & Schuster, 2014), and my friend and colleague Tony Hiss accordingly coined the term "Half Earth" in "The Wildest Idea on Earth" (*Smithsonian*, November 2014, pp. 66–73).

Moderated Discussion at the Berkeley Summit "Science for Parks, Parks for Science" on 26 March 2015

STEVEN R. BEISSINGER: Let's talk about some of the interesting issues that you've raised. *One that you alluded to is that the attention focused in our society right now is on climate change, not biodiversity loss. What do you think needs to be done to change that?*

EDWARD O. WILSON: What obviously needs to be done is that we need not just larger parks and more of them, but we need them connected. There's a movement that is taking place in the conservation community, globally. Here it is the Wilderness Society and the Wildlands Network that are promoting the idea, not only of larger parks, but also of joining them to make corridors.

I think one of the things that we could accomplish in this country, at the present time, is what I like to call "Boxing America." We have often discussed, and it's been mapped very well, "Y2Y," or the Yellowstone to Yukon corridor. That can be extended to the Rockies, to the mountains of southern Arizona, to the Sierra Madre Occidental, and then farther south. It can also cross the Taiga, the great coniferous wilderness area across Canada, through areas that are surprisingly sparsely occupied by people to the Adirondacks, and continue south to the end of the Appalachians. Then we have a corridor already mostly put together through the length of Florida, from the Everglades to Okefenokee. I've been very actively concerned in building rapidly a corridor that goes from close to Tallahassee all the way to Mississippi, thus "Boxing America."

Then, as the climate changes—dries, heats, and so on—this will allow species of plants and animals to migrate, that is, to breed and expand their population and remain in existence. That should be a worldwide way of planning expansion of land.

BEISSINGER: *What would you see as the role of the US National Park Service in that plan and in your plan to reach your 50% goal?*

WILSON: More, bigger, and taking central place in America's strategic planning right alongside defense.

BEISSINGER: Great! There's been a growing recognition of the potential of working landscapes to conserve biodiversity. *Do working landscapes contribute to acreage goals that you need to achieve the 50%?*

WILSON: I think it's not just a stupid but a dangerous way of looking at it. What kind of landscapes have we got now if they're not working landscapes? If they are not working landscapes, are they lazy landscapes?

This is a wrongheaded and quite dangerous worldview, that somehow our national parks, our park systems, our reserves should be valued in terms of their importance for humanity. Then if you can't get some product out of them without extinguishing birds, at least you would be able to measure their value by the aesthetic and psychological benefits that people receive from them. That's completely wrong. To start with, we do not even know what biodiversity is in our parks for the most part. We really need to have an ethic that recognizes the importance of the natural world in its own right, at least until such time that we can begin to half understand it, where it came from, and what it all means. This is unmitigated arrogance to think of nature as in some way fungible.

I've noticed that people who have written most prominently on this worldview are also those with the least experience in studying ecosystems, species, and other levels of biodiversity.

I'm sorry for the heat I'm putting into this, but this is something that should be countered immediately because it's dangerous. I would like to quote Alexander Humboldt, who encountered some resistance of this sort 200 years ago when he said, "The most dangerous worldview is the worldview of those who have not viewed the world."

BEISSINGER: And he viewed the world, didn't he? Humboldt went everywhere! *Does that mean, then, that we have to have areas devoid of people to make this goal work for you? Protected areas devoid of people, without people living in them?*

WILSON: Oh, of course not, I think that's a misconception promoted by the new conservationists. "Oh, this means we're going to clear everybody out and not let people in." Not in the least. There are indigenous people; there are people and their families who have occupied the areas that can be included in the expanded parks who receive easements, as was done for the Great Smoky Mountains National Park. It is just the land that is conserved. It's conserved in a way that the fauna and the flora are protected and allowed to evolve and maintain whatever equilibrium they had prior to occupation.

BEISSINGER: I have a question here from the audience: *Do you believe classical natural history is on the decline, and if so, what can be done about it?*

WILSON: It is. It has been on the decline ever since the molecular revolution, which of course gave us the golden age of modern biology. I say that frankly. I remember one day when, as an assistant professor at Harvard in the 1950s, my archrival Jim Watson came. I remember the time I suggested that no ecologist had been thought of to add to the faculty. I said, "Well so-and-so is over in this graduate school of design. Might he be given a courtesy or associate membership in this department?" Jim, who is a good friend of mine now, said, "Are you out of your mind?" I said, "What do you mean out of my mind?" He said, "Anybody that would suggest bringing an ecologist to a biology department must be out of their mind."

Well, that was the attitude of so many when we saw the birth of the great developments in molecular, then cell, then developmental, and then neurobiological (or brain) science. But now we have to understand that the organization of ecosystems from an infinitude of biodiversity is one of the great challenges of modern science.

And that is the point I'm trying to make—that we are going to see a rebirth of what I like to call scientific natural history. I want to see, furthermore, eventually a reinstatement of the "logos" or the "ologies." I want to see in places like Berkeley and Harvard a return of herpetology, ichthyology, entomology, and so on, with full departments and majors, in which the students enter to study biodiversity and bring in the armamentarium of modern biology to enrich their studies, but whose central interest is the taxonomic group. I want to see people who are students, fellow professionals, and biologists, who are in love with the group they are studying. They want to know everything about it, and they want to make discoveries based upon it. That's what we need desperately and I think that will happen.

Maybe the "ologies" won't come back, but they should. We shouldn't be justifying studies on biodiversity by saying, "Oh they add a lot to our understanding of evolutionary biology, or developmental biology, and so on." That's pathetic, that's a beggar's recommendation. We need to make vertebrate zoology, ornithology, and so on, equal in emphasis to neurobiology, for example.

BEISSINGER: Thank you and we're lucky at Berkeley. We've managed to maintain and conserve those "ology" courses.

WILSON: You're like one of these big parks with a lot of what we call "relict species."

BEISSINGER: Present company excluded, right?

WILSON: Yes! But now, we want to see them grow and flourish and speciate and come back.

BEISSINGER: There you go! Lineages you don't want to lose, right? Thinking about losing lineages, some scientists have advocated triage, or letting some species perish if the cost of conserving them would be better spent maximizing the benefits for a greater number of species. *Do you think triage is necessary for conservation of biodiversity to be ultimately successful?*

WILSON: That's ridiculous. The idea that we can have some knuckleheaded engineer or biologist come in and look over the endangered species and say, as we have done under the Endangered Species Act, "Well, we're just going to have to let that one go, we can't spend a million dollars to bring back that warbler, or so on." In real life, we can bring them back with knowledge and effort. It's not that expensive and it's also synergistic with other human endeavors. As you start to implement larger reserves, better knowledge, and the techniques of sustaining ecosystems, it will become less expensive, just like everything else.

The idea has been floated that some species are just destined to die, their time has come, and that's basically it, so why should we be spending taxpayer dollars trying to save some species that's going to die anyway? That is absolute nonsense. The reason that a species becomes rare and extinct is usually because its support system in the natural environment has been taken away, or species have been allowed to come in that are invasive and are pressing it out. Its members are not old and senile and decrepit, but quite the contrary. The young are just as vital and reproductive, unless the population gets too small, when there is too much inbreeding. But even that can be fixed. Okay, I'll sum it up with just one phrase, because we're running out of time: save them all.

BEISSINGER: Let's switch gears and talk about some questions I've received online about curiosity and engaging people in our parks. *How can the national parks best shape biological curiosity for future generations of Americans?*

WILSON: Well, that's precisely relevant to what we are here for at this meeting. We're talking about the idea that seems to me to be gaining momentum—that without undue disturbance, we can have research and educational centers within a park. When students are brought in contact with the rich natural environment (which they are innately prone to be interested in) and allowed to learn and also do research of their own (such as help in the search for a rare species, help identify or collect unknown forms for identification, and so on), they are engaged and are much more likely to move directly into science.

BEISSINGER: In your own life, you've spoken about how your childhood curiosity drove you as a biologist. *Are national parks and wilderness necessary to inspire that awe of nature? Can we find it in smaller places?*

WILSON: Just woods! Any fragment of a wild or semiwild environment will do. I grew up in the richest part of the United States per unit area and was delighted to collect all sorts of insects, including butterflies. When I was a little kid, I was called by my colleagues "Bugs Wilson," and then I went into a snake phase. Forty species of snakes are in the Central Gulf Plain. I was catching them. I even took a strike by a rattlesnake on my left finger, and I was then called "Snake Wilson."

BEISSINGER: *How did your parents feel about that?*

WILSON: I lucked out by having parents who didn't question me too closely when I went out the door. We could have another whole hour on the best way to raise children vis-à-vis the natural world, and letting kids explore and find things on their own. It's a very stimulating kind of environment for kids.

BEISSINGER: Thank you so much Professor Wilson. It's been a remarkable discussion.

TWO

Seas the Day: A Bluer, Saltier Second Century for American Parks

KIRSTEN GRORUD-COLVERT, JANE LUBCHENCO, AND ALLISON K. BARNER

Introduction

National parks, famously labeled "America's best idea" by Wallace Stegner (Benson 1996), have been heralded across the decades as our collective windows into the past, inspiration in the present, and hope for the future. Championed by men and women who led the way in creating refuges for people in nature, national parks reflect a bold idea whose value, though initially unappreciated, increased rapidly through time. We look to these wild islands to understand ourselves and our place in nature, to play and learn, to dream. Through time, the system of national parks has been complemented by the designation of wilderness areas, monuments, state parks, wildlife refuges, sanctuaries, and more. The same philosophy of protecting special places on behalf of everyone unifies these places despite their different management authorities. Collectively, they were and are a brilliant idea.

However, unforeseen challenges have arisen as millions of visitors have embraced the notion of parks, conflicts over values and funding have emerged, and human influence on the planet has grown. Through time, scientific knowledge and practical knowledge have surfaced to inform decisions about ways to keep parks, people, and the ecosystems in which they are embedded vibrant. The importance of protected areas will continue to grow with time, but only if we use the available knowledge and take responsibility for ensuring their vitality. As David Quammen (2006) wrote, "Our national parks are as good, only as good, as the intensity with which we treasure them."

As we look toward the next century of protected places, we have the benefit of over 100 years' experience with national parks. But this experience, and the understanding of the importance of wild places, did not always exist. The national parks were a daring idea, championed by visionaries who changed our national understanding of wild places. John Muir, who believed that "one who gains the blessings of one mountain day . . . is rich forever," inspired first a president and then a nation to preserve the wonder and value of nature in its truest form. Thomas Moran's breathtaking paintings of Yellowstone allowed those in Washington, DC, a glimpse of these treasures and fostered a willingness to protect them. Theodore Roosevelt established the National Park System and believed it to be an "essential democracy" of places, preserved for all Americans to enjoy. Stephen Mather's ideas to make the national parks accessible to all led to an expanded system of roads and trails, management choices that now challenge our current efforts to protect our "loved to death" parks. George Melendez Wright believed that these parks should not be treated as zoos, but that the value of their animals "lies in their wildness." Marjory Stoneman Douglas believed that the Everglades' river of grass should be protected solely to preserve its animals and plants even though others dismissed it as a "snake swamp." Adolphe Murie carefully observed the wolves at Mount McKinley and illustrated how understanding animals' interactions with their environment can show us the best way to protect our natural systems (Duncan and Burns 2009). These visionaries—politicians, artists, scientists, and civil leaders—defined the key ingredients for the parks we know today: protecting wilderness for its own sake and making it available to everyone. An original audacious idea took root, was embraced by Americans and foreign visitors alike, and blossomed into a defining feature of the American landscape and psyche.

National parks inspire us. We like the very idea that special places have been recognized as treasures for the whole nation, and indeed the world, to share. Parks protect spectacular scenery, amazing wildlife, essential habitat, and, in principle, the continued functioning of core elements that make up these different ecosystems. They provide knowledge, recreation, inspiration, education, economic benefits, and hedges against overuse in the areas outside the park. These experiences and benefits have inspired comparable national parks around the world.

Now, in celebration of the 100th birthday of the National Park Service, we pause to reflect on these successes and also on the challenges for parks and people in the second century. How, for example, do we keep parks healthy in the face of myriad threats ranging from too many people to climate change? How do we resolve existing conflicts with adjacent land uses?

How might we introduce parks to new and more diverse generations of Americans? Recommitting to and updating the original vision for existing parks is timely and essential to the future of parks.

This chapter envisions a parallel and complementary focus for the second century of America's parks, one that is inspired by the audacity of the original vision. The centennial offers a unique opportunity to be as bold as the initial vision, to extend the now well understood concept of special places on land (green parks) to a proposed equivalent for the largely ignored, special places in the ocean (blue parks). To be sure, some blue parks exist today, but the primary focus of protecting special places has been on land. To date, we've only dipped our toes in the water of ocean conservation. Now, to honor the foresight of Muir, Roosevelt, and Stoneman Douglas, this chapter proposes taking the plunge into a new public vision—one that embraces the ocean under US jurisdiction as equally worthy of attention, one that extends the spectrum of protection from green through blue.

The Insufficiently Protected Sea

Before the idea of national parks became a reality, few Americans were familiar with the concept of formally protecting a place—setting aside an area for its intrinsic value. Most nonindigenous Americans had never seen, nor given much thought to, now-iconic places such as Yellowstone, Yosemite Valley, or the Grand Canyon. Today, the same is true for most of the ocean—out of sight, out of mind. Most people don't know if or where ocean areas are protected, or why protecting them is important. For example, although in 2010 some 62% of US poll respondents supported efforts to protect the environment (Pew Research Center 2010), a 2014 poll by The Ocean Project showed that Americans were largely unaware of issues affecting the ocean, such as overfishing, ocean acidification, climate change, and pollution (Meyer, Isakower, and Mott 2015). These findings echo responses in an earlier study, in which only 4.3% of a random sampling of 1,233 Americans considered themselves very well informed about ocean issues (Steel et al. 2005), although 71.3% knew the term "marine protected area" (MPA). Polls from California and New England have shown that people tend to significantly overestimate the fraction of the ocean that is protected. In 2002, interviews with 750 residents of New England and Atlantic Canada showed that many believed that 20%–23% of their regional ocean waters were already fully protected. In fact, less than 1% was protected (Conservation Law Foundation 2002). In a separate poll in 2002,

16%–24% of 1,000 Californians mistakenly believed 22% of their state waters were fully protected (Edge Research 2002), whereas less than 3% was actually protected at that time (Gleason et al. 2013).

On the global stage, although governments have recognized the need to protect special places in the ocean, their aspirations for ocean protection are less ambitious than those for land. Moreover, there is demonstrably more progress in meeting land protection targets compared with those for the oceans. For example, protected areas currently cover 14% of the world's terrestrial areas (Deguignet et al. 2014), and efforts are on target to meet the United Nations (UN) Convention on Biological Diversity's goal of having 17% of terrestrial and inland water areas protected by 2020 (Secretariat of the Convention on Biological Diversity 2014). In stark contrast, protected areas of any kind today cover only 3.7% of the ocean, and only a paltry 1.9% is fully or strongly protected (Lubchenco and Grorud-Colvert 2015).[1] The UN has set a less ambitious goal for the ocean than for the land, calling for protection of only 10% of coastal and marine areas by 2020, with even minimally protected areas counted in this goal. Even that modest goal is not likely to be met (Secretariat of the Convention on Biological Diversity 2014).

Why Ocean Parks?

Note that the vast majority of the ocean used to be a de facto fully protected area because most of it was inaccessible to extractive uses; technology has now made most of the ocean accessible and used (Roberts 2007). Now countries large and small are considering how to reestablish protection and how to meet the internationally agreed-on targets. Each country has the authority to determine the location and degree of protection for places within its own exclusive economic zone (EEZ), which encompasses an area up to 200 miles from its shores. In the ocean as on land, the choice of protection level is important. Different types of designation connote different levels of protection. US national forests, for example, are open to commercial activities such as logging, livestock grazing, hunting, and fishing. US national parks on land generally forbid hunting, commercial fishing, livestock grazing, mining, and logging, but they allow recreational fishing. US wilderness areas disallow roads and structures, as well as motorized equipment and transport, and are managed to keep an area "wild," but they can allow hunting and fishing (Public Broadcasting System 2012).

---

1. See also www.MPAtlas.org (accessed 16 March 2016).

Levels of protection and managing authorities vary for protected areas in the ocean as well. In the United States, for example, both states and the federal government can create protected areas, but the type of protection is not consistent within or across different management agencies. Moreover, the language describing level of protection in the ocean is somewhat confusing, within the United States and internationally. The commonly used term "marine protected area" (MPA) says little about the level of protection. "MPA" simply means an area that is intended to achieve some conservation purpose (Day et al. 2012). MPAs include the full spectrum, from prohibiting the collection of a single species to prohibiting all activities.

Because the level of protection is so important, the following language is often used to describe degree of protection (Lubchenco and Grorud-Colvert 2015). *Fully protected* means that no extractive or destructive activities are allowed. The terms "marine reserves" and "no-take" are synonymous with "fully protected." A fully protected area can be a single area, such as the Pitcairn Islands Marine Reserve, or a "network" of fully protected areas that are designed to be connected by the movement of juveniles or adults. Example networks of marine reserves are found in the US Channel Islands National Marine Sanctuary, West Hawai'i, Papua New Guinea, the Gulf of California, Moorea in French Polynesia, and the state marine reserves of California (Grorud-Colvert et al. 2014), where "it is unlawful to injure, damage, take, or possess any living, geological, or cultural marine resource . . . and the area shall be maintained to the extent practicable in an undisturbed and unpolluted state."[2]

Strongly protected areas are very close to fully protected areas in their level of protection, but they permit a very small amount of subsistence or recreational fishing, or both. Strictly speaking, they are not fully protected, but from the standpoint of the human impact on the area, they are likely equivalent to fully protected areas. The newly expanded Pacific Remote Islands Marine National Monument is an example. Subsistence and recreational fishing is allowed, but because population density and tourism are exceedingly low, the impact of such activities is similarly small.

Partially protected areas (sometimes called "lightly protected" areas) allow some significant extractive activities but prohibit others. For example, Gray's Reef National Marine Sanctuary in the United States allows some forms of fishing, including rod-and-reel and handlines, but prohibits dredging and drilling. The Monterey Bay National Marine Sanctuary allows fishing but prohibits drilling and extraction of gas and oil.

2. California Fish and Game Code § 2850–2863.

Multiple use areas permit different types of use in different areas within a single MPA. For example, the Channel Islands National Marine Sanctuary, the Florida Keys National Marine Sanctuary, and Dry Tortugas National Park are all zoned for different activities; these zones range from partially to fully protected.

MPAs can serve multiple management and conservation goals that scale with varying levels of protection, all involving trade-offs between the short and long term and among the social, ecological, and economic implications of protecting an ocean area (Lubchenco and Grorud-Colvert 2015). In this chapter, we use "MPA" as a generic term and specify specific types of MPAs: "fully protected" (all extractive activities prohibited; also called "marine reserves"), "strongly protected" (almost all extractive activities prohibited), and "partially protected" (considerable extractive activities allowed). We use the term "ocean parks" to mean either fully or strongly protected areas in the ocean.

There are important differences among these levels of protection. A global scientific comparison of fully protected areas versus partially protected areas concludes that fully protected areas tend to have substantially greater ecological benefits, including two times greater total biomass of fishes and other organisms by number and weight than areas where some forms of fishing are allowed (Lester and Halpern 2008; Sciberras et al. 2015). Extensive research on fully protected marine reserves demonstrates their utility for increasing density and biomass of target and nontarget species and preserving biodiversity (fig. 2.1) (Lester et al. 2009; Claudet et al. 2010; Ballantine 2014; Edgar et al. 2014). The benefits of fully protected areas can spill over into surrounding waters (e.g., Johnson, Funicelli, and Bohnsack 1999; Kaunda-Arara and Rose 2004; Starr, O'Connell, and Ralston 2004), and can improve fish catch levels (e.g., Harmelin-Vivien et al. 2008; Vandeperre et al. 2011). These benefits increase when networks are established with reserve sizes and spacing that are designed to facilitate connections for fish and other organisms as they disperse across different scales (Grorud-Colvert et al. 2014). Analyses of commercial fishing, recreational fishing, and tourism show that, together, the economic and biological benefits of well-managed MPAs can far outweigh their costs (Sala et al. 2013). Further, strongly protected MPAs can produce a net increase in local economic activity, improving food security, employment, social surplus value, and overall welfare (Reithe, Armstrong, and Flaaten 2014). We emphasize that designation alone accomplishes little without enforcement of prohibitions, just as the designation of terrestrial parks would be of little

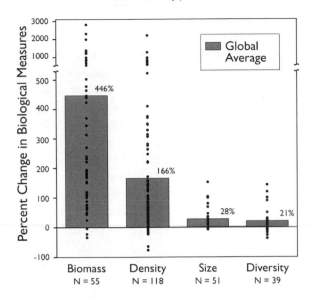

2.1 Average changes (*bars*) in fishes, invertebrates, and seaweeds within marine reserves around the world measured before and after protection was implemented or by comparing areas with and without protection. Although changes varied among reserves (*dots*), most reserves had positive changes. Data from Lester et al. (2009); figure used with permission from Lubchenco et al. (2007).

value without reliable follow-through. Moreover, if restrictions are not durable, conservation benefits are lost.

To be sure, there are other management tools that can help protect marine biodiversity, such as the Endangered Species Act or the Marine Mammal Protection Act, but neither provides protection for the full suite of microscopic to macroscopic species in a place. Effective fishery management is also a critically important tool for areas outside protected areas, complementing the place-based management inside. None of these tools address other threats to marine biodiversity such as pollution, climate change, or ocean acidification. Parallel actions are needed to address those other threats.

A growing body of scientific data indicates that fully protected marine reserves and other types of MPAs should cover multiple spatial scales to achieve many of the common management goals, including the protection and restoration of an ocean ecosystem and its services, the maintenance of fishing lifestyles and incomes, and the preservation of recreational and cultural opportunities. For example, if a goal is protection of biodiversity in general or of target species in particular, a variety of relevant habitats will need to be protected. During their lifetimes, many marine organisms move among adjacent habitats along coastlines, or between deeper and shallower waters. Some species straddle the boundary between state-managed coastal waters and federally managed waters more than three miles from shore. For example, the bocaccio (*Sebastes paucispinis*)—a long-lived commercially and recreationally important fish found along the US West Coast—spans both state and federal waters, taking refuge in drifting kelp mats, sandy areas, eelgrass beds, boulder fields, and deepwater caves during different times in its life (Love, Yoklavich, and Thorsteinson 2002).

Many large pelagic species such as tuna (Scombridae), swordfish (*Xiphias gladius*), marlin (Istiophoridae), oceanic sharks, sea turtles (Chelonioidea), and marine mammals travel even farther afield into the high seas beyond national jurisdiction, making it difficult to establish a single marine reserve that would encompass their entire life cycle. However, contrary to many assumptions, research shows that highly migratory species can benefit from protection of key portions of their habitats, in particular places and times where they are vulnerable to capture. These can include migration corridors, nurseries, and spawning or feeding sites (e.g., Maxwell et al. 2012; Ketchum et al. 2014). For example, tuna and oceanic sharks are known to congregate around seamounts, where productivity and biodiversity are both high. Protecting key portions of habitat for migratory species has been useful on land. Tens of thousands of migrating shorebirds benefit from the multiple national wildlife areas in the Bay of Fundy as well as Delaware Bay National Wildlife Refuge, where they eat the eggs of horseshoe crabs (Limulidae) and gain key resources for their successful migration to South America.[3] Thus, spatial protection can greatly benefit even large migratory fishes as they are increasingly targeted and damaged by fisheries (Johnston and Santillo 2004; De Forest and Drazen 2009; Clark and Dunn 2012; Bouchet et al. 2015).

Progress toward Protection

After decades of little attention to the ocean, with only minimal progress in establishing marine reserves, leaders from a number of nations have recently jump-started efforts to create large, fully or strongly protected areas (table 2.1, fig. 2.2) (Leenhardt et al. 2013; Toonen et al. 2013; Lubchenco and Grorud-Colvert 2015), bringing the total global protection at that level

---

3. See www.migratoryconnectivityproject.org (accessed 16 March 2016).

Table 2.1 Countries with the largest strongly or fully protected marine reserve

Country	EEZ size (km²)	Portion of EEZ in large, strongly or fully protected reserves (%)	EEZ size global rank	Large strongly or fully protected reserve	Commitment/designation year	Reserve area (km²)	Portion of EEZ (%)
Palau	603,978	82.8	42	Palau National Marine Sanctuary	2015	500,000	82.78
Seychelles	1,336,559	29.9	24	Seychelles Exclusive Economic Zone Marine Spatial Plan	2015	400,000	29.9
Chile	3,681,989	25.3	10	Motu Motiro Hiva Marine Park	2010	150,000	4.07
				Nazca-Desventuradas Marine Park	2015	300,000	8.15
				Easter Island Marine Park	2015	481,368	13.07
United Kingdom	6,805,160	21.9	5	Chagos Marine Reserve	2010	638,568	9.38
				South Georgia and South Sandwich Islands MPA	2012	20,431	0.30
				Pitcairn Islands Marine Reserve Proposal	2015	834,334	12.26
United States	11,351,000	15.5	1	Papahanaumokuakea Marine National Monument	2006	362,073	3.19
				Marianas Trench Marine National Monument	2009	42,000	0.37
				Pacific Remote Islands Marine National Monument	2009, 2014	1,321,466	11.64
				Rose Atoll Marine National Monument	2009	34,838	0.31
New Zealand	4,083,744	15.2	9	Kermadec Ocean Sanctuary	2015	620,000	15.18
Kiribati	3,500,533	11.9	12	Phoenix Islands Protected Area	2006	407,112	11.63
Australia	8,974,857	1.9	3	Great Barrier Reef Marine Park	1975, 2004	113,652	1.27
				Macquarie Island Commonwealth Reserve	1972, 1999	58,000	0.65

Note: Data from Lubchenco and Grorud-Colvert (2015) and www.MPAtlas.org.

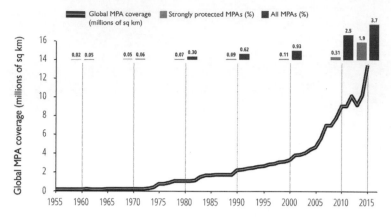

2.2. Increases in global MPA coverage over time. Line graph shows increasing MPA area. Bar graphs (decadal from 1960 to 2010, plus 2015) show progress toward ocean protection, with the percentage of the ocean in strongly or fully protected MPAs (*light gray*) and the total percentage of the ocean in MPAs across all levels of protection (*dark gray*). As of 2015, only 3.7% of the ocean receives any level of protection in MPAs, and only 1.9% is strongly protected.

from 0.08% a decade ago (Wood, Laughren, and Pauly 2008) to 1.9% today (Lubchenco and Grorud-Colvert 2015).[4] After early designations of protected areas in Australia, strong protection was finalized for both the Macquarie Island Commonwealth Marine Reserve in 1999 and the Great Barrier Reef Marine Park in 2004. In 2006 and 2009, President George W. Bush established four large, strongly protected marine national monuments around Pacific islands. In 2014, President Barack Obama expanded the Pacific Remote Islands Marine National Monument, creating the largest set of strongly protected marine reserves in the world. Meanwhile, Chile designated the Motu Motira Hiva Marine Park in 2010. But 2015 was a banner year for protected areas (fig. 2.2) (Lubchenco and Grorud-Colvert 2015). In January 2015, Kiribati strongly protected its previously declared Phoenix Island Protected Area, forbidding any commercial fishing. In March 2015, the United Kingdom celebrated the creation of the Pitcairn Islands Marine Reserve, which protects 332,138 square miles (834,334 km²) of the United Kingdom's overseas EEZ territory, making it the largest fully protected marine area in the world. In late September, New Zealand, home of one of the first marine reserves in the world (Ballantine 2014), designated the Kermadec Ocean Sanctuary, increasing its fully protected area to 15.2% of its EEZ. At the Our Ocean 2015 conference in early October, Chile

4. See also www.MPAtlas.org (accessed 16 March 2016).

announced its intention to increase the fraction of its EEZ that is fully protected from 4% to 25.3% with the designation of the Nazca-Desventuradas Marine Park and the initiation of the process to establish the Easter Island Marine Park. Chile led the pack in protecting the largest percentage of its EEZ until late October 2015, when Palau designated 82.8% of its maritime territory as a fully protected marine reserve with no extractive activities. In December 2015, the Seychelles ended the year with a commitment to set aside 30% of its EEZ as strongly protected MPAs, part of a marine spatial plan for its entire EEZ.[5]

The United States has the largest EEZ of any country in the world, an ocean area 1.5 times the size of the contiguous United States and an impressive 55% of the US total land and sea area (fig. 2.3). Just over 15.5% of the US EEZ is strongly or fully protected (see table 2.1). The United States ranks fifth in terms of the fraction of its EEZ strongly or fully protected, behind Palau, Seychelles, Chile, and the United Kingdom.

Other designations are proposed, but not yet finalized, in areas such as Chile, Panama, South Africa, and Australia, among others. For example, Australia's Coral Sea Commonwealth Marine Reserve was approved in 2012 (by proclamation under the Australian government's Environment Protection and Biodiversity Conservation Act 1999), but the reserve was later revoked in 2013, and is currently tabled in the Australian Parliament's House of Representatives while under review by the Commonwealth Marine Reserves Review Panel. Overall, such impressive progress within a decade demonstrates that more and more nations are realizing the importance of blue parks. However, with only 1.9% of the ocean fully or strongly protected, there is ample room for more action.

One reason why only a relatively small area of the ocean is protected is because currently the only legal tools for creating protected areas lie with individual countries, within their own EEZs. Jurisdiction over the oceans is more complicated than over Earth's land areas, which, with the exception of Antarctica, all lie within the sovereign control of individual nations. The UN Convention on the Law of the Sea allows each coastal nation to exercise control over ocean resources up to 200 miles from its shores within its EEZ. Ocean areas beyond any nation's EEZ are known as the high seas and traditionally have been open to exploitation by any nation. These high seas areas beyond national jurisdiction represent 64% of the surface of the ocean. In 2015, the UN General Assembly agreed to begin drafting a binding treaty that may provide options for increasing protection of the high

5. See www.MPAtlas.org (accessed 16 March 2016).

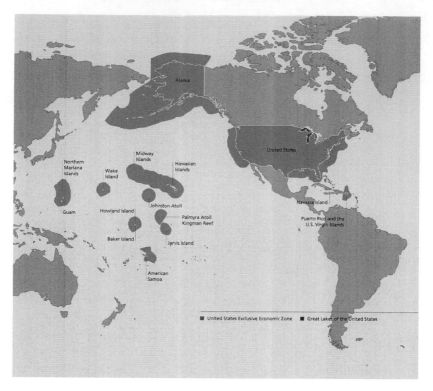

2.3. The US exclusive economic zone (EEZ), extending 200 nautical miles offshore.

seas.[6] To date, however, all existing fully protected areas are within countries' EEZs.

How much protected area is needed? Scientists have suggested that the answer depends on the goals in mind. Many analyses have suggested at least 20%–30% of an area should be fully protected to achieve conservation goals, but if it is desirable to achieve both conservation and fishery goals, as much as 50% may be needed, for example, to minimize the risk of fisheries collapse and maximize long-term sustainable catches. Moreover, and importantly, protected areas should be well distributed across different biogeographic regions to fully capture the variation in habitat and species diversity across different oceanographic basins (National Research Council 2001; Gaines, Carr, and Palumbi 2010). Despite the US EEZ spanning the central and eastern Pacific, Arctic, and Atlantic Oceans, and the Gulf of Mexico and the Caribbean Sea (see fig. 2.3), almost all of the fully and

6. UN General Assembly Resolution No. A/69/L.65.

strongly protected total area of the United States is in the central Pacific because of the presence of large protected areas there. The United States does have smaller fully or strongly protected areas, but most of these coastal sites are tiny in comparison. Small- and medium-sized, strategically placed protected areas can be very important for providing local species and habitat protection, but they cannot protect as much biodiversity or ecosystem functioning in their biogeographic regions because of their size. Moreover, an abundance of habitat types remain unprotected within the US EEZ. Overall, ample opportunities remain to designate large protected areas across the US EEZ in biogeographic regions other than the Pacific.

The United States has hundreds of small MPAs managed by federal or state agencies (table 2.2). Of the more than 1,700 federal and state MPAs in the United States, 86% of the total number is only partially protected (National MPA Center 2012). At the federal level, strongly protected MPAs make up only 15% of the total number of MPAs. The total area (vs. number of areas) protected is more difficult to determine for these smaller MPAs because of inconsistent terminology and levels of reporting. Florida has more than 340 state-managed MPAs in its state waters but only three of these are fully protected. However, some federally managed MPAs that straddle the state-federal boundary, such as the Florida Keys National Marine Sanctuary, include some fully protected marine reserves. For US national parks alone, terrestrial coverage is 34 times greater than the marine area that is protected (C. McCreedy, pers. comm.), although other marine areas are protected through alternate jurisdictions (e.g., national marine sanctuaries and other federal and state MPA authorities).

In the United States, some of the best-documented examples of the benefits of marine reserve protection come from national parks located in California, Alaska, and Florida. The natural sciences provide compelling evidence for the key role of parks in protecting and restoring the ocean. For example, Anacapa Island Marine Reserve, first established in California by the National Park Service in 1978 and currently part of the Channel Islands National Marine Sanctuary, has demonstrated strongly positive effects of protection on commercially targeted species such as the California lobster (*Panulirus interruptus*) (fig. 2.4) (Behrens and Lafferty 2004; Lubchenco et al. 2007; Babcock et al. 2010). Removal of lobsters outside the reserve as a result of heavy fishing pressure led to a boom in the population of sea urchins (Echinoidea). Urchins, which were held in check by lobster predation within the reserve, were over 13 times more abundant outside the reserve boundary. Outside the reserve, superabundant carpets of urchins overgrazed their primary food, kelp, virtually eliminating kelp forests and

Table 2.2 Number of permanent, year-round US federal and state MPAs by protection level

	Designated protection level (degree of protection)						
	No impact (fully protected)	No take (fully protected)	No access (strongly protected)	Uniform multiple use (partially protected)	Zoned multiple use (partially protected)	Zoned w/ no-take areas (partially + fully protected)	Total
Number of state MPAs							
Alaska	—	—	—	35	—	1	36
Florida	—	3	11	313	15	4	346
California	—	61	10	107	1	—	179
Louisiana	—	—	—	7	—	—	7
Hawai'i	1	3	2	27	9	4	46
Washington	—	5	7	28	—	1	41
Massachusetts	—	—	—	59	—	—	59
Maine	—	1	—	20	—	—	21
Virginia	—	—	—	18	6	—	24
Maryland	—	—	—	82	—	—	82
Number of federal MPAs	1	13	21	243	48	17	343

Note: States are the top ten by total area of territorial waters (http://www.census.gov/geo/reference/state-area.html). Data from National Marine Protected Areas Center (http://marineprotectedareas.noaa.gov/dataanalysis/mpainventory/).

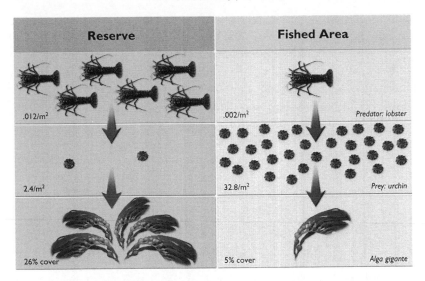

Reserve	Fished Area
.012/m²	.002/m² Predator: lobster
2.4/m²	32.8/m² Prey: urchin
26% cover	5% cover Alga gigante

2.4. In the Anacapa Island Marine Reserve in California, abundant lobsters keep their urchin prey in check, allowing kelp forests to flourish. Data from Behrens and Lafferty (2004); figure used with permission from Lubchenco et al. (2007).

leaving only barren seafloor (Behrens and Lafferty 2004). In stark contrast, protection of lobsters from fishing inside the reserve resulted in increased lobster abundance, which kept urchins in check and maintained healthy kelp forests that were resilient to stressful warming during El Niño events (Lafferty and Behrens 2005). Similar protection for multitrophic interactions occurred in Glacier Bay National Park. Sea otters (*Enhydra lutris*)— also predators on urchins—thrived inside the reserve, keeping urchins in check and protecting kelp forests, while decimation of otter populations by exploitation outside the reserve resulted in overabundance of urchins and elimination of kelp forests, together with the plethora of species that depend on kelp for habitat (Esslinger and Bodkin 2009). In the Dry Tortugas National Park in Florida, a fully protected marine reserve led to significant changes in density and abundance of both exploited and nontarget species, while decreases in density of exploited species were observed at a nearby area open to fishing (Ault et al. 2013).

Efforts are now underway to link the ecological effects of MPAs in the Dry Tortugas to economic indicators (Johns et al. 2014). As ecological data help build a clearer picture of the benefits of MPAs, data on the social and economic outcomes of protection help us understand how these benefits fit into the larger context of human well-being (e.g., Sala et al. 2013; Reithe, Armstrong, and Flaaten 2014; Daw et al. 2015). And as we expand our focus to include more blue parks on the map, this growing body of data can offer guidelines for developing effective protection and restoration of ocean systems.

Protected areas on land and in the ocean can provide insight into how places are changing through time and the impacts of extractive activities outside the protected areas. For example, historical and recent data from Biscayne National Park, Dry Tortugas National Park, and the Channel Islands National Marine Sanctuary—formerly a national park—have shown shifts in fish density and biomass over time (Hamilton et al. 2010; Kellison et al. 2012; Ault et al. 2013). Research on lionfish (*Pterois* spp.) in the national parks at Biscayne Bay and the Dry Tortugas is beginning to show the ecological impacts of a marine invasion (Ruttenberg et al. 2012). Because the abundance of fished predators increases inside protected areas, those MPAs offer an excellent natural laboratory to track the impact of predator return on community structure, as was the case with lobsters in the Channel Islands National Marine Sanctuary (Kay, Lenihan, Guenther, et al. 2012; Kay, Lenihan, Kotchen, and Miller 2012) and the return of sea otters to Glacier Bay National Park (Esslinger and Bodkin 2009). Looking into the future, we can predict that protected areas will become even more important as benchmarks and comparison areas to evaluate the potentially different responses of protected and unprotected (e.g., fished) areas to impacts of climate change like ocean acidification and coral bleaching, as well as other increasing anthropogenic pressures like pollution and invasive species. Research is already demonstrating the importance of protecting places that provide a natural refuge from coral bleaching (e.g., McClanahan et al. 2014). In addition, research indicates that more-intact ecosystems tend to have more trophic levels, healthier communities, and greater resistance to bleaching (such was the case with the unpopulated atolls of the northern Line Islands) (Sandin et al. 2008). Hence, fully protecting large areas appears to be a good strategy if the goal is to maintain or restore the resilience of an ecosystem in the face of environmental changes.

Pathways to Protection

MPAs within US federal and state waters have been created through a variety of authorities, both at the state and federal level, and with different stakeholder involvement, using both bottom-up and top-down approaches. To implement MPAs, states have used common law, constitutional authority, and statutory provisions. Federal pathways to designate MPAs include national marine sanctuaries under the National Marine

Sanctuaries Act, marine national monuments under the Antiquities Act, or other federally protected areas under such provisions as the Magnuson-Stevens Fishery Conservation and Management Act, the Wilderness Act, the National Wildlife Refuge System Administration and Refuge Improvement Act, and the Outer Continental Shelf Lands Act, among others (Baur et al. 2013).

Most recently, strongly protected areas in US federal waters (see table 2.1) have been designated through the Antiquities Act. Theodore Roosevelt signed the Antiquities Act into law in 1906, giving the president authority to designate national monuments for protection of "historic landmarks, historic and prehistoric structures, and other objects of historic or scientific interest."[7] Presidents George W. Bush and Barack Obama used this authority to establish or expand the very large, strongly protected areas in the central Pacific.

Both on land and in the water, numerous places have been initially designated using one authority and then later transitioned to a different authority. For example, Congress has redesignated 32 national monuments as national parks, including the iconic Grand Canyon National Park. Many parks that include MPAs—such as the Channel Islands National Park, Dry Tortugas National Park, and Glacier Bay National Park—also began as national monuments. Transitioning from a marine national monument to a national marine sanctuary is also feasible. For example, the Rose Atoll Marine National Monument was established in 2009 under the primary management authority of the National Oceanic and Atmospheric Administration. NOAA has been tasked with initiating the process to add the Rose Atoll Monument's marine areas to the Fagatele Bay National Marine Sanctuary in cooperation with the government of American Samoa.

This and other approaches using top-down authorities have led to many of our most beloved blue parks. In many of these cases, the top-down authority (Congress or the president) was responding to bottom-up requests or support for protection. A top-down approach alone can lead to controversy when the local communities are insufficiently involved (Christie 2004). For example, in the United Kingdom, top-down processes for the implementation of MPAs have caused concern among community groups, who worry that their interests will not be considered (Jones 2012).

Open ocean MPAs tend to be large and established with less stakeholder input, in many cases because these remote areas have fewer stakeholders to involve, although presidential proclamations such as those with the recent

7. 16 USC 431–433.

marine national monuments frequently involve stakeholder comment periods. Overall, issues of social justice still apply when designating large MPAs in more remote areas (De Santo 2013). Continued progress in this direction will require clear objectives and clear management authorities, resources, education, opportunities and funding for research, and outreach to engage and connect with stakeholders on an ongoing basis. Creative solutions to compliance and enforcement are also needed, including monitoring that involves local resource users and technology that makes it easier to enforce MPA rules (Maxwell et al. 2014).

In contrast, most existing coastal MPAs are small and affect diverse stakeholders. In coastal locales, it can be challenging to establish and implement MPAs when planning approaches are either solely stakeholder-based or solely science-based. For example, the first and second attempts to establish fully protected marine reserves and other MPAs via the Marine Life Protection Act in California failed because of a lack of significant public consultation (Weible 2008; Gleason et al. 2010). The subsequent and successful approach in California demonstrates the power of a combined top-down and bottom-up process, with state officials, stakeholders, and scientists each playing key roles. The state mandated the creation of a network of MPAs and articulated goals, scientists provided size and spacing guidelines, and stakeholders proposed specific places for protection that met the size and spacing criteria. In addition to the goal of involving stakeholders as true participants in the process, the use of science was mandated, requiring that scientists be engaged and transparent participants (Kirlin et al. 2013; Saarman et al. 2013). Recent research in Wales, United Kingdom, has also shown that stakeholders working with scientific guidelines can choose areas that protect a diversity of habitats to meet conservation goals, demonstrating how an integrated, science-and-stakeholder approach can succeed (Ruiz-Frau et al. 2015). Bringing both scientists and stakeholders to the table is key for successfully establishing blue parks in US waters.

MPAs are most likely to prove durably successful when their planning involves people who are passionate about the area and committed to protecting it. Protected areas include highly complex connections between ecological health and human well-being. Considering all species in an area, including humans, is critical (Lubchenco and Grorud-Colvert 2015). Those who currently use the resources in and around proposed MPAs should be engaged in planning, both to gain the benefit of their experiential knowledge and to get their buy-in. For example, when local community members see and experience the depletion of fisheries, they are more likely to seek ways to restore and protect the ecosystem they depend on, often leading to bottom-up management from the community members themselves (Cudney-Bueno et al. 2009; López-Angarita et al. 2014; Barner et al. 2015). Community-led MPA processes can evolve organically as local users see the potential benefits of closed areas to fishing, tourism, and other services provided by functioning marine systems. In fact, some fisheries cooperatives operating under rights-based fishery management programs in Fiji, Brazil, and Mexico have established their own fully protected marine reserves as a way to benefit directly from spillover from a no-take area into their fished areas (Afflerbach et al. 2014; Barner et al. 2015). Yet top-down, government-enforced rules are also likely to be necessary to make such bottom-up, community-implemented reserves effective. For example, in Mexico, local resource users and scientists partnered to identify and establish marine reserves that supported local mollusk fisheries near Puerto Peñasco in the Gulf of California. After a rapid increase in abundance of mollusks within these reserves, nonlocal fishermen poached the area and rapidly fished it out because there was no regional or federal governance in place to enforce the locally established rules (Cudney-Bueno and Basurto 2009). Without top-down regulatory endorsement, local stakeholders may not be able to capture the benefits of their voluntary self-restraint. Moving forward, small-and large-scale, top-down and bottom-up strategies are needed to effectively protect more of our marine heritage.

Scientific Lessons from the Past for the Future

We now have the opportunity to build on the wealth of scientific knowledge from MPAs in the United States and around the globe and from the lessons learned during the 100-year history of our National Park System. These lessons are equally applicable to terrestrial and ocean parks.

1. *No protected area is an island.* Decades of research show that it's not enough to just protect special places—we must also take care of the areas that surround them (DeFries, this volume, ch. 11). We need to consider parks in the context of their respective local and regional landscapes and seascapes if the goal is also to sustain long-term ecosystem health and services. For ocean parks, it is critical that activities in surrounding areas be sustainable and also compatible with park goals. Sustainable fisheries are an obvious example, but equally important are practices that minimize runoff of sediment, chemicals, nutrients, and plastics.

2. *We can't freeze areas in time.* No natural system, and no protected area, is static. We must expect and prepare for changes, not only as part of natural ecological cycles but also as part of changes in climate, ocean acidi-

fication, and human access and use. Since MPAs can promote ecosystem functioning by restoring processes like the trophic cascades in the Channel Islands and Glacier Bay, these biological communities may be more likely to resist or recover from human-caused disturbances. For example, during catastrophic flooding in eastern Australia, reefs in MPAs resisted the impacts of flooding while fished reefs were heavily degraded (Olds et al. 2014). Recovery may in fact be a key benefit of MPAs. For example, populations of fished species that are protected in these areas can seed fished areas outside after disturbances (Micheli et al. 2012). Scientists are beginning to synthesize information and prioritize sites that could act as resilience hot spots in the face of climate change (McClanahan et al. 2012). Although marine reserves may enhance resilience, they are not a cure-all for climate change impacts, especially in the case of such broad-reaching effects as ocean acidification and wide-scale temperature changes (e.g., Selig, Casey, and Bruno 2012).

3. Connectivity among protected areas is important. Parks are not isolated places—they are fundamentally connected to their surrounding areas. With more and more data from marine reserves, the importance of marine reserve networks has surfaced as a tool to protect multiple habitats by accounting for replication and connectivity (Gaines et al. 2010). In the fluid ocean, species, habitats, and ecosystems are connected via the movement of larvae, juveniles, or adults. Marine reserve networks can protect those connections, leading to even greater benefits than the individual reserves would provide (Grorud-Colvert et al. 2014). This is especially important when evaluating whether marine reserves can truly meet their conservation goals, such as with the Mediterranean Sea, for example, where regional reserves do not appear to be effectively connected through movements of larval or adult organisms (Andrello et al. 2013).

4. Humans use parks and will continue to use them. The future of parks depends in large part on the support of people. When visitors experience natural wonders and observe wildlife, their understanding, appreciation, and support for protection is enhanced. However, too many visitors, and certain kinds of use, can threaten the functioning of parks on land and in the ocean. Visitors may need instruction about how to enjoy the parks without harming them. For example, after hordes of uneducated swimmers and snorkelers caused significant damage to the coral reefs of Hanauma Bay State Park on Oʻahu, Hawaiʻi, the need for education programs became obvious. Adoption of mandatory user education, in the form of a film that must be watched before entering, and controls on the number of people in the water have resulted in significant improvement.[8] How much can we visit these places, be inspired and motivated by their beauty, yet still keep them healthy? When choosing between recreation and preservation, for example, the National Park Service prioritizes preservation "based on our mandate, policies, and good science" (Barna 2015). Yet with an increasing push to introduce even more people to protected areas, how can we balance the two? The business model approach is one creative way to establish MPAs that balance recreation and preservation. As healthy areas have more species and more diverse habitats, more tourists will want to visit these beautiful underwater areas, in turn bolstering the local economy and paying for park upkeep and monitoring (Sala et al. 2013).

5. *Managing protected areas is an active process.* It's not enough to simply set an area aside and assume everything will be fine, especially in light of climate change and ocean acidification. Active and adaptive management is required to deal with the challenges of shifting uses and shifting pressures. Data from MPAs have provided insight into what works and what doesn't work for successful protected areas, but even the best-laid plans can lead to unexpected results, as we've seen with the unanticipated overuse of some national parks on land or with the need to actively remove invasive species. Clear goals for an MPA, as well as monitoring data from before and after establishment, are critical for assessing whether an MPA is working and whether the management strategy needs to be modified (Ban et al. 2012). Management agencies must have the capacity and ongoing monetary support to use sound data and keep assessing MPA success. As human uses increase along with the many anticipated impacts of climate change, we need to learn from our past to protect our special underwater places in the future.

6. *We need good science across disciplines to guide decisions, but we also need art to inspire and ethics to guide us.* Scientific data and knowledge can show us how and why things change, but an emotional connection to a place or species is often essential for inspiring people to protect it (Bernbaum, this volume, ch. 14). The public won't be interested in protecting the nation's underwater treasures unless they're aware of them. Not only is it challenging to bring the ocean to those who don't live along the coast, but even in our coastal communities the vast majority of people will never spend much time underwater. All citizens should have the opportunity to understand and connect with healthy ocean habitats and species. Technology

8. See www.hanaumabaystatepark.com (accessed 16 March 2016).

is beginning to provide a window under the sea via outreach tools like Google Ocean Street View, scientists' real-time blogs from expeditions such as National Geographic's Pristine Seas, and deep-sea video from the community of scientists studying mid-ocean ridges (Goehring, Robigou, and Ellins 2012).

While scientific and outreach efforts highlight the need to protect these iconic places, many areas in the ocean are not pristine, and many of these degraded areas also need protection. Citizen science–based monitoring organizations such as REEF, Reef Check, and Coast Watch provide a way for a relatively small group of committed volunteers to witness these changing habitats first hand.

Art provides a powerful way for people to explore and connect with nature. Traveling installations such as *Reefs on the Edge* in Australia bring focus to the effects of ocean acidification and climate change. The grave problem of ocean pollution and climate change is strikingly described via sculpture made with ocean trash, including installations from the Plastic Ocean Project's *What Goes Around, Comes Around* art initiative, the Washed Ashore Project, artist Courtney Mattison (fig. 2.5), photographer Brian Skerry, and Alejandro Durán's *Washed Up* series, that inspire wonder, awe,

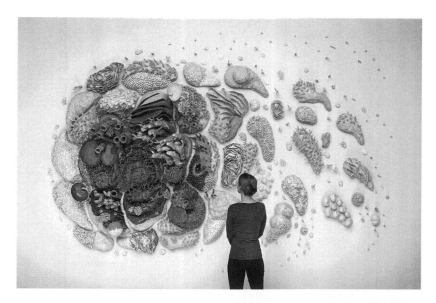

2.5. *Our Changing Seas III* by Courtney Mattison, March 2014, glazed stoneware and porcelain, 10 × 14 × 2 ft. Debut venue: Francis Young Tang Teaching Museum & Art Gallery, Saratoga Springs, New York. Upcoming venue: Virginia Museum of Contemporary Art, Virginia Beach, Virginia (winter 2016). Used with permission from the artist.

and the urgent need for protection of underwater seascapes and life. Moving forward, we must link images and issues to our underwater parks. A concrete first step is to establish artist-in-residence programs at existing parks that protect underwater areas—not only the terrestrial ones—and to engage with artists already focusing on other US underwater places that are priorities for protection.

Continuing discussions about the ethics of protected areas—for nature, for people, for both—are vital for keeping this process grounded in values and reality (Lubchenco et al. 2011; Kareiva and Marvier 2012; Soulé 2013, 2014; Kareiva 2014; Tallis and Lubchenco 2014). It is time for a new ocean ethic (Safina 1999; Kellert 2005; Earle and McKibben 2010; Lubchenco et al. 2011). All voices should be considered and heard in the conversation about what protection truly means.

A Blue Vision for American Parks

Given the state of our knowledge—the myriad data on benefits of fully protected marine reserves and other MPAs, the serious and escalating threats to life in the ocean, and the lessons learned from 100 years of protecting our national parks—now is the time to "seas the day" and expand our focus to include green plus blue. Scientific information and practical experience can guide us in establishing better, and more strategically designed, protection for special ocean places and ecosystem functioning.

Early visionaries such as Thomas Moran and John Muir called for protection of natural treasures before they were drastically altered by human pressures. The idea for a national system of protected areas on land was a bold vision, which continues to provide strong benefits today. The time is ripe for a similar call to action for the ocean. We have a golden opportunity on the anniversary of the National Park Service to expand the focus to our salty and wet treasures, regardless of the specific management authority overseeing an area. Now is the time to increase the coverage and representation of ocean spaces as protected places. In parallel with efforts to achieve sustainable fisheries, reduce pollution, and reduce greenhouse gas emissions, we call for expanded, and more effective, protection of the ocean through more blue parks. Seas the day!

Acknowledgments

For providing specific MPA information, we thank Sue Haig, Cliff Mc-Creedy, Russell Moffitt, Beth Pike, Dan Roby, Enric Sala, Greg Stone, Teuea

Toatu, Annie Turek, Matt Rand, and staff at the US National MPA Center, particularly Jordan Gass. Karen Garrison and Margaret Spring provided information on polling reports. Thanks also to Laura Cantral, Cyndi Cliff, Rebecca Jablonski-Diehl, Courtney Mattison, Susanne Duffner, and Monica Pessino for providing graphics and images. Holly Doremus provided insightful comments on an earlier draft of the manuscript. We thank Steve Beissinger and all the University of California, Berkeley, students, faculty, and staff for organizing the "Science for Parks, Parks for Science" symposium. And lastly, we thank all National Park Service, National Marine Sanctuaries, and US MPA personnel for their stewardship of our ocean.

Literature Cited

Afflerbach, J. C., S. E. Lester, D. T. Dougherty, and S. E. Poon. 2014. A global survey of "TURF-reserves," Territorial Use Rights for Fisheries coupled with marine reserves. Global Ecology and Conservation 2:97–106.

Andrello, M., D. Mouillot, J. Beuvier, C. Albouy, W. Thuiller, and S. Manel. 2013. Low connectivity between Mediterranean marine protected areas: a biophysical modeling approach for the dusky grouper *Epinephelus marginatus*. PLoS ONE 8:e68564.

Ault, J. S., S. G. Smith, J. A. Bohnsack, J. Luo, N. Zurcher, D. B. McClellan, T. A. Ziegler, et al. 2013. Assessing coral reef fish population and community changes in response to marine reserves in the Dry Tortugas, Florida, USA. Fisheries Research 144:28–37.

Babcock, R. C., A. C. Alcala, K. D. Lafferty, T. McClanahan, G. R. Russ, N. T. Shears, N. S. Barrett, and G. J. Edgar. 2010. Conservation or restoration: decadal trends in marine reserves. Proceedings of the National Academy of Sciences USA 107:18256–18261.

Ballantine, B. 2014. Fifty years on: lessons from marine reserves in New Zealand and principles for a worldwide network. Biological Conservation 176:297–307.

Ban, N. C., J. E. Cinner, V. M. Adams, M. Mills, G. R. Almany, S. S. Ban, L. J. McCook, and A. White. 2012. Recasting shortfalls of marine protected areas as opportunities through adaptive management. Aquatic Conservation: Marine and Freshwater Ecosystems 22:262–271.

Barna, D. 2015. Top 10 issues facing national parks. National Geographic: Travel. Accessed 22 May 2015. http://travel.nationalgeographic.com/travel/top-10/national-parks-issues/.

Barner, A. K., J. Lubchenco, C. Costello, S. D. Gaines, A. Leland, B. Jenks, S. A. Murawski, E. Scwabb, and M. Spring. 2015. Solutions for recovering and sustaining the bounty of the ocean: reforming fisheries and pairing rights-based fisheries management with marine reserves. Oceanography 28:252–263.

Baur, D., T. Lindley, A. Murphy, P. Hampton, P. Smyth, S. Higgs, A. Bromer, E. Merolli, and M. Hupp. 2013. Area-based management of marine resources: a comparative analysis of the National Marine Sanctuaries Act and other federal and state legal authorities. Perkins Coie LLP, Washington, DC.

Behrens, M. D., and K. D. Lafferty. 2004. Effects of marine reserves and urchin disease on southern Californian rocky reef communities. Marine Ecology Progress Series 279:129–139.

Benson, J. 1996. Wallace Stegner: his life and work. Viking, New York, New York.

Bouchet, P. J., J. J. Meeuwig, C. P. Salgado Kent, T. B. Letessier, and C. K. Jenner. 2015. Topographic determinants of mobile vertebrate predator hotspots: current knowledge and future directions. Biological Reviews 90:699–728.

Christie, P. 2004. Marine protected areas as biological successes and social failures in southeast Asia. American Fisheries Society Symposium 42:155–164.

Clark, M. R., and M. R. Dunn. 2012. Spatial management of deep-sea seamount fisheries: balancing sustainable exploitation and habitat conservation. Environmental Conservation 39:204–214.

Claudet, J., C. W. Osenberg, P. Domenici, F. Badalamenti, M. Milazzo, J. M. Falcón, I. Bertocci, et al. 2010. Marine reserves: fish life history and ecological traits matter. Ecological Applications 20:830–839.

Conservation Law Foundation, the Ocean Conservancy, Environmental Defense, World Wildlife Fund Canada, and the Canadian Parks and Wilderness Society. 2002. Northwest Atlantic Ocean needs more protection. WWF: News & Reports, 16 February. Accessed 22 May 2015. http://responsive.wwfca.panda.org/newsroom/?1101.

Cudney-Bueno, R., and X. Basurto. 2009. Lack of cross-scale linkages reduces robustness of community-based fisheries management. PLoS ONE 4:e6253.

Cudney-Bueno, R., L. Bourillon, A. Saenz-Arroyo, J. Torre-Cosio, P. Turk-Boyer, and W. W. Shaw. 2009. Governance and effects of marine reserves in the Gulf of California, Mexico. Ocean & Coastal Management 52:207–218

Daw, T. M., S. Coulthard, W. W. L. Cheung, K. Brown, C. Abunge, D. Galafassi, G. D. Peterson, T. R. McClanahan, J. O. Omukoto, and L. Munyi. 2015. Evaluating taboo trade-offs in ecosystems services and human well-being. Proceedings of the National Academy of Sciences USA 112:6949–6954.

Day, J., N. Dudley, M. T. Hockings, G. Holmes, D. Laffoley, S. Stolton, and S. Wells. 2012. Guidelines for applying the IUCN protected area management categories to marine protected areas. IUCN, Gland, Switzerland.

De Forest, L., and J. Drazen. 2009. The influence of a Hawaiian seamount on mesopelagic micronekton. Deep-Sea Research Part I: Oceanographic Research Papers 56:232–250.

Deguignet, M., D. Juffe-Bignoli, J. Harrison, B. MacSharry, N. Burgess, and N. Kingston. 2014. 2014 United Nations list of protected areas. UNEP-WCMC, Cambridge, United Kingdom.

De Santo, E. M. 2013. Missing marine protected area (MPA) targets: how the push for quantity over quality undermines sustainability and social justice. Journal of Environmental Management 124:137–146.

Duncan, D., and K. Burns. 2009. The national parks : America's best idea; an illustrated history. Alfred A. Knopf, New York, New York.

Earle, S., and B. McKibben. 2010. The world is blue: how our fate and the ocean's are one. Reprint. National Geographic, Washington, DC.

Edgar, G. J., R. D. Stuart-Smith, T. J. Willis, S. Kininmonth, S. C. Baker, S. Banks, N. S. Barrett, et al. 2014. Global conservation outcomes depend on marine protected areas with five key features. Nature 506:216–220.

Edge Research. 2002. Public opinion research and analysis. SeaWeb: Resources. Accessed 22 May 2015. http://www.seaweb.org/resources/puboresearch.php.

Esslinger, G. G., and J. L. Bodkin. 2009. Status and trends of sea otter populations in southeast Alaska, 1969–2003. US Geological Survey Scientific Investigations Report 2009-5045. US Geological Survey, Reston, Virginia.

Gaines, S. D., C. White, M. H. Carr, and S. Palumbi. 2010. Designing marine reserve networks for both conservation and fisheries management. Proceedings of the National Academy of Sciences USA **107**:18286–18293.

Gleason, M., E. Fox, S. Ashcraft, J. Vasques, E. Whiteman, P. Serpa, E. Saarman, et al. 2013. Designing a network of marine protected areas in California: achievements, costs, lessons learned, and challenges ahead. Ocean & Coastal Management **74**:90–101.

Gleason, M., S. McCreary, M. Miller-Henson, J. Ugoretz, E. Fox, M. Merrifield, W. McClintock, P. Serpa, and K. Hoffman. 2010. Science-based and stakeholder-driven marine protected area network planning: a successful case study from north central California. Ocean & Coastal Management **53**:52–68.

Goehring, L., V. Robigou, and K. Ellins. 2012. Bringing Mid-Ocean Ridge discoveries to audiences far and wide: emerging trends for the next generation. Oceanography **25**: 286–298.

Grorud-Colvert, K., J. Claudet, B. N. Tissot, J. E. Caselle, M. H. Carr, J. C. Day, A. M. Friedlander, et al. 2014. Marine protected area networks: assessing whether the whole is greater than the sum of its parts. PLoS ONE **9**:e102298.

Hamilton, S. L., J. E. Caselle, D. P. Malone, and M. H. Carr. 2010. Incorporating biogeography into evaluations of the Channel Islands marine reserve network. Proceedings of the National Academy of Sciences USA **107**:18272–18277.

Harmelin-Vivien, M., L. Le Direach, J. Bayle-Sempere, E. Charbonnel, J. A. Garcia-Charton, D. Ody, A. Perez-Ruzafa, et al. 2008. Gradients of abundance and biomass across reserve boundaries in six Mediterranean marine protected areas: evidence of fish spillover? Biological Conservation **141**:1829–1839.

Johns, G., D. J. Lee, V. Leeworthy, J. Boyer, and W. Nuttle. 2014. Developing economic indices to assess the human dimensions of the South Florida coastal marine ecosystem services. Ecological Indicators **44**:69–80.

Johnson, D. R., N. A. Funicelli, and J. A. Bohnsack. 1999. Effectiveness of an existing estuarine no-take fish sanctuary within the Kennedy Space Center, Florida. North American Journal of Fisheries Management **19**:436–453.

Johnston, P. A., and D. Santillo. 2004. Conservation of seamount ecosystems: application of a marine protected areas concept. Archive of Fishery and Marine Research **51**: 305–319.

Jones, P. J. S. 2012. Marine protected areas in the UK: challenges in combining top-down and bottom-up approaches to governance. Environmental Conservation **39**:248–258.

Kareiva, P. 2014. New conservation: setting the record straight and finding common ground. Conservation Biology **28**:634–636.

Kareiva, P., and M. Marvier. 2012. What is conservation science? BioScience **62**:962–969.

Kaunda-Arara, B., and G. A. Rose. 2004. Out-migration of tagged fishes from marine reef National Parks to fisheries in coastal Kenya. Environmental Biology of Fishes **70**:363–372.

Kay, M. C., H. S. Lenihan, C. M. Guenther, J. R. Wilson, C. J. Miller, and S. W. Shrout. 2012. Collaborative assessment of California spiny lobster population and fishery responses to a marine reserve network. Ecological Applications **22**:322–335.

Kay, M. C., H. S. Lenihan, M. J. Kotchen, and C. J. Miller. 2012. Effects of marine reserves on California spiny lobster are robust and modified by fine-scale habitat features and distance from reserve borders. Marine Ecology Progress Series **451**:137–150.

Kellert, S. R. 2005. Perspectives on an ethic toward the sea. American Fisheries Society Symposium **41**:703–711.

Kellison, G. T., V. McDonough, D. E. Harper, and J. T. Tilmant. 2012. Coral reef fish assemblage shifts and decline in Biscayne National Park, Florida, USA. Bulletin of Marine Science 88:147–182.

Ketchum, J. T., A. Hearn, A. P. Klimley, E. Espinoza, C. Penaherrera, and J. L. Largier. 2014. Seasonal changes in movements and habitat preferences of the scalloped hammerhead shark (*Sphyrna lewini*) while refuging near an oceanic island. Marine Biology 161:755–767.

Kirlin, J., M. Caldwell, M. Gleason, M. Weber, J. Ugoretz, E. Fox, and M. Miller-Henson. 2013. California's Marine Life Protection Act Initiative: supporting implementation of legislation establishing a statewide network of marine protected areas. Ocean & Coastal Management 74:3–13.

Lafferty, K. D., and M. D. Behrens. 2005. Temporal variation in the state of rocky reefs: does fishing increase the vulnerability of kelp forests to disturbance? Pages 511–520 *in* D. K. Garcelon and C. A. Schwemm, eds. Sixth California Islands Symposium. Institute for Wildlife Studies, Ventura, California.

Leenhardt, P., B. Cazalet, B. Salvat, J. Claudet, and F. Feral. 2013. The rise of large-scale marine protected areas: conservation or geopolitics? Ocean & Coastal Management 85:112–118.

Lester, S. E., and B. S. Halpern. 2008. Biological responses in marine no-take reserves versus partially protected areas. Marine Ecology Progress Series 367:49–56.

Lester, S. E., B. S. Halpern, K. Grorud-Colvert, J. Lubchenco, B. I. Ruttenberg, S. D. Gaines, S. Airame, and R. R. Warner. 2009. Biological effects within no-take marine reserves: a global synthesis. Marine Ecology Progress Series 384:33–46.

López-Angarita, J., R. Moreno-Sánchez, J. H. Maldonado, and J. A. Sánchez. 2014. Evaluating linked social-ecological systems in marine protected areas. Conservation Letters 7:241–252.

Love, M. S., M. Yoklavich, and L. Thorsteinson. 2002. The rockfishes of the Northeast Pacific. University of California Press, Oakland, California.

Lubchenco, J., R. Born, B. Simler, and K. Grorud-Colvert. 2011. Lessons from the land for protection in the sea: the need for a new ocean ethic. Pages 197–213 *in* P. Harrison, ed. Open spaces. University of Washington Press, Seattle, Washington.

Lubchenco, J., S. Gaines, K. Grorud-Colvert, S. Airame, S. R. Palumbi, R. R. Warner, and B. Simler Smith. 2007. The science of marine reserves. Partnership for Interdisciplinary Studies of Coastal Oceans (PISCO), Oregon State University, Corvallis, Oregon.

Lubchenco, J., and K. Grorud-Colvert. 2015. Making waves: the science and politics of ocean protection. Science 350:382–383 and supplemental materials.

Maxwell, S. M., N. C. Ban, and L. E. Morgan. 2014. Pragmatic approaches for effective management of pelagic marine protected areas. Endangered Species Research 26: 59–74.

Maxwell, S. M., J. J. Frank, G. A. Breed, P. W. Robinson, S. E. Simmons, D. E. Crocker, J. Pablo Gallo-Reynoso, and D. P. Costa. 2012. Benthic foraging on seamounts: a specialized foraging behavior in a deep-diving pinniped. Marine Mammal Science 28: E333–E344.

McClanahan, T. R., M. Ateweberhan, E. S. Darling, N. A. J. Graham, and N. A. Muthiga. 2014. Biogeography and change among regional coral communities across the Western Indian Ocean. PLoS ONE 9:e93385.

McClanahan, T. R., S. D. Donner, J. A. Maynard, M. A. MacNeil, N. A. J. Graham, J. Maina, A. C. Baker, et al. 2012. Prioritizing key resilience indicators to support coral reef management in a changing climate. PLoS ONE 7:e42884.

Meyer, D., A. Isakower, and B. Mott. 2015. An ocean of opportunities. The Ocean Project, Providence, Rhode Island.

Micheli, F., A. Saenz-Arroyo, A. Greenley, L. Vazquez, J. A. Espinoza Montes, M. Rossetto, and G. A. De Leo. 2012. Evidence that marine reserves enhance resilience to climatic impacts. PLoS ONE 7:e40832.

National MPA Center. 2012. Analysis of United States MPAs. National MPA Center, Silver Spring, Maryland.

National Research Council. 2001. Marine protected areas: tools for sustaining ocean ecosystems. National Academies Press, Washington, DC.

Olds, A. D., K. A. Pitt, P. S. Maxwell, R. C. Babcock, D. Rissik, and R. M. Connolly. 2014. Marine reserves help coastal ecosystems cope with extreme weather. Global Change Biology 20:3050–3058.

Pew Research Center. 2010. Global attitudes project. Pew Research Center, Washington, DC.

Public Broadcasting System. 2012. River of no return: national parks, national forests, and US wildernesses. PBS: Nature, 18 April. Accessed 12 July 2015. http://www.pbs.org/wnet/nature/river-of-no-return-national-parks-national-forests-and-u-s-wildernesses/7667/.

Quammen, D. 2006. The future of parks: an endangered idea. National Geographic Magazine, October, 62–67.

Reithe, S., C. W. Armstrong, and O. Flaaten. 2014. Marine protected areas in a welfare-based perspective. Marine Policy 49:29–36.

Roberts, C. M. 2007. The unnatural history of the sea. Island Press, Washington, DC.

Ruiz-Frau, A., H. P. Possingham, G. Edwards-Jones, C. J. Klein, D. Segan, and M. J. Kaiser. 2015. A multidisciplinary approach in the design of marine protected areas: integration of science and stakeholder based methods. Ocean & Coastal Management 103:86–93.

Ruttenberg, B. I., P. J. Schofield, J. L. Akins, A. Acosta, M. W. Feeley, J. Blondeau, S. G. Smith, and J. S. Ault. 2012. Rapid invasion of Indo-Pacific lionfishes (*Pterois volitans* and *Pterois miles*) in the Florida Keys, USA: evidence from multiple pre- and post-invasion data sets. Bulletin of Marine Science 88:1051–1059.

Saarman, E., M. Gleason, J. Ugoretz, S. Airamé, M. Carr, E. Fox, A. Frimodig, T. Mason, and J. Vasques. 2013. The role of science in supporting marine protected area network planning and design in California. Ocean & Coastal Management 74:45–56.

Safina, C. 1999. Song for the blue ocean: encounters along the world's coasts and beneath the seas. Holt Paperbacks, New York, New York.

Sala, E., C. Costello, D. Dougherty, G. Heal, K. Kelleher, J. H. Murray, A. A. Rosenberg, and R. Sumaila. 2013. A general business model for marine reserves. PLoS ONE 8:e58799.

Sandin, S. A., J. E. Smith, E. E. DeMartini, E. A. Dinsdale, S. D. Donner, A. M. Friedlander, T. Konotchick, et al. 2008. Baselines and degradation of coral reefs in the Northern Line Islands. PLoS ONE 3:e1548.

Sciberras, M., S. R. Jenkins, R. Mant, M. J. Kaiser, S. J. Hawkins, and A. S. Pullin. 2015. Evaluating the relative conservation value of fully and partially protected marine areas. Fish and Fisheries 16:58–77.

Secretariat of the Convention on Biological Diversity. 2014. Global biodiversity outlook 4. A mid-term assessment of progress towards the implementation of the strategic plan for biodiversity 2011–2020. Secretariat of the Convention on Biological Diversity, Montréal, Canada.

Selig, E. R., K. S. Casey, and J. F. Bruno. 2012. Temperature-driven coral decline: the role of marine protected areas. Global Change Biology 18:1561–1570.

Soulé, M. 2013. The "New Conservation." Conservation Biology 27:895–897.

———. 2014. Also seeking common ground in conservation. Conservation Biology 28: 637–638.

Starr, R. M., V. O'Connell, and S. Ralston. 2004. Movements of lingcod (*Ophiodon elongatus*) in southeast Alaska: potential for increased conservation and yield from marine reserves. Canadian Journal of Fisheries and Aquatic Sciences 61:1083–1094.

Steel, B. S., C. Smith, L. Opsommer, S. Curiel, and R. Warner-Steel. 2005. Public ocean literacy in the United States. Ocean & Coastal Management 48:97–114.

Tallis, H., and J. Lubchenco. 2014. Working together: a call for inclusive conservation. Nature 515:27–28.

Toonen, R. J., T. 'Aulani Wilhelm, S. M. Maxwell, D. Wagner, B. W. Bowen, C. R. C. Sheppard, S. M. Taei, et al. 2013. One size does not fit all: the emerging frontier in large-scale marine conservation. Marine Pollution Bulletin 77:7–10.

Vandeperre, F., R. M. Higgins, J. Sanchez-Meca, F. Maynou, R. Goni, P. Martin-Sosa, A. Perez-Ruzafa, et al. 2011. Effects of no-take area size and age of marine protected areas on fisheries yields: a meta-analytical approach. Fish and Fisheries 12:412–426.

Weible, C. M. 2008. Caught in a maelstrom: implementing California marine protected areas. Coastal Management 36:350–373.

Wood, L. J., L. Fish, J. Laughren, and D. Pauly. 2008. Assessing progress towards global marine protection targets: shortfalls in information and action. Oryx 42:340–351.

THREE

A Global Perspective on Parks and Protected Areas

ERNESTO C. ENKERLIN-HOEFLICH
AND STEVEN R. BEISSINGER

Introduction

Protected areas can be considered an expression of human values, choices, and decisions (Enkerlin-Hoeflich et al. 2015; Rozzi et al. 2015). The International Union for the Conservation of Nature (IUCN) defines a protected area as a "clearly defined geographical space, recognized, dedicated and managed, through legal or other effective means, to achieve the long term conservation of nature with associated ecosystem services and cultural values" (Dudley 2008). Yet, as human population growth, land conversion, and resource extraction accelerate, Earth is rapidly losing many of the last locations suitable for protected areas, and their establishment is literally a race against time (Edwards et al. 2014; Laurance et al. 2015). In many regions of the world, only fragments of nature remain in highly human-modified landscapes (Balmford et al. 2001), while other regions retain the basic functionality of ecosystems but with reduced biodiversity (Joppa and Pfaff 2009; Craigie, Pressey, and Barnes 2014). Pressure is mounting on protected areas, and many instances of degazetting, downsizing, and encroachment have occurred (Bernard, Penna, and Araújo 2014; Geldmann, Joppa, and Burgess 2014; Mascia et al. 2014). This race has greatly accelerated, but is not new.

In this chapter, we provide a perspective on protected areas around the world. We first examine how the values that protected areas serve have evolved over the past century. Then we consider global targets for protected area coverage, examine the state of protected areas globally, and review the accomplishments of the World Parks Congress in Sydney, Australia, in 2014. We conclude with some thoughts on the future directions for conservation of protected areas.

The Evolving Values of Protected Areas

Protecting special places is widely accepted in all cultures. Recent estimates suggest approximately 8 billion visits per year are made by people to the world's protected areas and that these visits generate approximately $600 billion per year in direct in-country expenditure (Balmford et al. 2015). The first protected areas were established long before the founding of the US National Park Service. Areas were set aside specifically for protection of natural resources over 2,000 years ago in India, and over 1,000 years ago in Europe hunting grounds for the wealthy were established (Holdgate 1999; Eagles, McCool, and Haynes 2002). Sacred groves and mountains, and *tapu*, or holy areas, have a long tradition in Africa, the Americas, Asia, and the Pacific (Bernbaum, this volume, ch. 14). Often these areas allowed very restricted or no public access. Starting with the designation of the early national parks in the United States, like Yellowstone and Yosemite, for public use and tourism, parks began to grow internationally over the 20th century (fig. 3.1), accelerating in area protected after 1960 and broadening their purposes (Watson et al. 2014).

Parks today are a mixture of "take" (i.e., use) and "no-take" (i.e., no use) protected areas, as illustrated by the different categories of protected areas recognized by the IUCN (table 3.1). These range from strict nature reserves and wilderness protection (Categories I, Ia, and Ib) to protected areas that are managed specifically for sustainable use of resources (Categories V and VI). Protected areas that allow for consumptive use of some goods and services, such as harvest of plant or animal populations, are sometimes hypothesized as being of lesser value or desirability than those where use is indirect or nonconsumptive, such as ecotourism. While this perspective views protection as part of a zero-sum game (i.e., either sites are totally protected from use or not), it is not particularly useful. When compared with lands receiving no protection, all categories of protected areas have made important contributions to conservation—even "paper parks" that exist legally but where on-the-ground conservation measures are not undertaken, and parks in "benign neglect" that are remote with minimal protection or a low potential to be converted to alternate land uses (Joppa and Pfaff 2009). There is now a wealth of rapidly increasing documentation that protected areas are effective and that society could not do without them in terms of their contributions to biodiversity conservation and sustainability (Bhagwat et al. 2005; Brandon and Wells 2009; Watson et al. 2014). Thus, for a century, parks and protected areas have

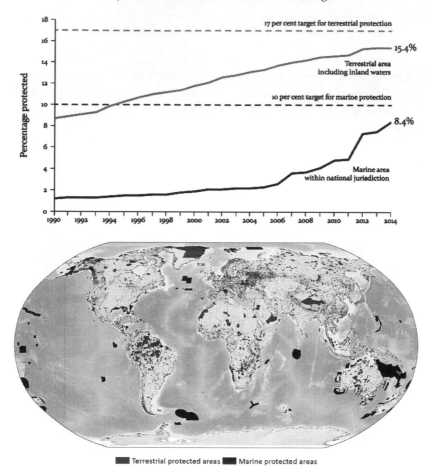

3.1. Recent trends in protected areas in the World Database on Protected Areas as of November 2014. *Top*, percentage of terrestrial area (including inland waters) and marine areas under national jurisdiction (0–200 nautical miles) covered by protected areas. *Bottom*, distribution of terrestrial and marine protected areas. Source: Juffe-Bignoli et al. (2014).

largely been established for the beauty, cultural significance, or biodiversity they protected, or for the resources they produced.

Recently, both biophysical and social changes are making protected area conservation an opportunity for contributing to societal priorities in addition to conservation. In the social change arena, parks and protected areas can contribute to poverty alleviation and economic gain (Stolton and Dudley 2010; Joppa and Pfaff 2011; Larsen, Turner, and Brooks 2012; Turner et al. 2012; Secretariat of the Convention on Biological Diversity 2008). A very large and rapidly growing set of studies, particularly over the past 20

Table 3.1 **Extent and distribution of protected areas reported to the World Database on Protected Areas according to the IUCN categories**

IUCN category	Description	Area protected (km²)	Proportion of total area protected (%)
Ia	Strict nature reserve	1,237,133	3.19
Ib	Wilderness area	1,187,003	3.06
II	National park	6,239,886	16.09
III	Natural monument or feature	310,482	0.80
IV	Habitat/species management area	3,479,212	8.97
V	Protected landscape/ seascape	3,094,296	7.98
VI	Protected area with sustainable use of resources	9,121,176	23.52
Not reported		11,283,175	29.09
Not applicable		2,831,002	7.30

Note: The total protected area coverage is not a global total because it includes overlap among protected areas of different categories.

years, has championed the notion of sustainability and ecosystem services that can be derived from protected areas (Millenium Ecosystem Assessment 2005; Durán et al. 2013; Dickson et al. 2014). Parks and protected areas are being viewed as generators of income for local communities, whether they are gateway communities of national parks in the United States or Africa, or indigenous communities living within biosphere reserves or their own traditionally managed territories. Parks and protected areas can be effective at serving societal needs and can be socially resilient if they are managed in an inclusive manner (Archabald and Naughton-Treves 2001).

In the case of biophysical change, climate change has already had a profound effect on parks and protected areas—not just in its potential to affect park natural and cultural resources, but in providing a new value for parks. Concern about the impact of climate change on societies and their economies had the immediate effect of devoting large amounts of resources to climate change adaptation and mitigation. Protected areas are now viewed as potential instruments for mitigating climate change by securing carbon-rich habitats in new or enhanced protected areas, and by facilitating adaptation through the provision of ecosystem services and cultural benefits that enable society to cope with the consequences of climate change (Jantz, Goetz, and Laporte 2014; Watson et al. 2014). There is an urgent need for understanding the critical role that protected area systems can play in climate change response strategies.

New audiences are also beginning to discover the value of parks and protected areas. The current crisis over biodiversity decline, climate change, and the vulnerability of the poor has created new energy between different social movements and the professionals engaged in conservation of landscapes and species. As this chapter was being prepared, Pope Francis, head of the Roman Catholic Church with its 1.2 billion followers worldwide, released a major science-informed policy statement on the need for a global dialogue to protect the environment and to stabilize the climate through switching to renewable energy sources. In the encyclical, the pope highlights the role of protected areas or sanctuaries to help conserve nature.[1] This overture could have a major impact on societal attitudes toward parks and protected areas. All major world religions contain specific scriptural obligations for followers to value, respect, and protect nature (Weeramantry 2009). Other faiths have also become increasingly involved with the need for protected areas. As Crawhall (2015) notes, "The societal value placed on nature conservation, as expressed in religion, national identity, political leadership, the media and so forth, will invariably determine where nature conservation fits within national priorities."

A second area of recently championed values of parks has been the link between parks and healthy people. Previously, health benefits of parks and protected areas tended to focus on ecosystem services, such as providing medicines and fresh water (Dudley et al. 2011). The recent advent of the "Healthy Parks Healthy People" approach, which is a collaboration between the medical profession and conservation, has established a broader understanding of the diverse health benefits of nature. Evidence for links between urban green space and physical and mental health and well-being are beginning to emerge (Maller et al. 2006; Lee and Maheswaran 2011; Romagosa, Eagles, and Lemieux 2015).

In summary, the numerous values and benefits that protected areas deliver for people and nature need to be more widely recognized alongside recreation and conservation. In addition to income derived directly from visitors (Balmford et al. 2015), parks are a source of cost-effective natural solutions for addressing many global threats because they can help contribute to water security, food security, climate change mitigation, and disaster risk reduction; combat desertification; regulate disease; mitigate climate events such as floods; and provide natural pollination services

---

1. Pope Francis, Encyclical letter Laudato si' of the Holy Father Francis on care for our common home, para. 37, 151, and 184, The Holy See, 24 May 2015, accessed 10 August 2015, http://w2.vatican.va/content/francesco/en/encyclicals/documents/papa-francesco_20150524 _enciclica-laudato-si.html.

(Vörösmarty et al. 2010). Parks also provide biocultural benefits of nature for physical, mental, and spiritual health, through provisions that respect cultural heritage and diversity, support livelihoods, and foster social well-being to sustain life.

Global Targets for Protected Area Coverage

One of the greatest scientific challenges is establishing thresholds for how much of the world needs to be conserved in a relatively undisturbed state so that nature does not irrevocably decline, and with it cause the collapse of societies (Diamond 2005). The question of "how much is enough" has vexed conservation biology since its inception with difficult problems such as estimating the amount of protected area coverage and designating the size of a viable population (Beissinger et al. 1996; Beissinger and Westphal 1998; Tear et al. 2005). Answers depend in part on willingness to accept risks and trade-offs among conflicting objectives (Svancara et al. 2005).

The protected areas community made its first concrete attempt to set a goal for the level of protection to be achieved worldwide in 1992 at the IVth World Parks Congress in Caracas, Venezuela. A "goal" of effectively protecting 10% of the world in healthy ecosystems as an "insurance policy" for biodiversity conservation was launched. From the beginning, the 10% goal was seen as grossly insufficient for conservation, but it was attractive and potentially attainable. Thus, it became effective policy guidance at the national level, a level at which most conservation policy decisions occur (Sarukhán et al. 2015). Considered a compromise, the 10% goal helped jump-start the first big wave of newly established protected areas, which we continue to ride today (Juffe-Bignoli et al. 2014).

A higher quantitative and qualitative target was established in 2010 as Target 11 of the 20 Aichi Biodiversity Targets for 2020, which were initiated by the Convention on Biological Diversity (CBD) and its Programme of Work on Protected Areas (Secretariat of the Convention on Biological Diversity 2004): "By 2020, at least 17 per cent of terrestrial and inland water areas and 10 per cent of coastal and marine areas, especially areas of particular importance for biodiversity and ecosystem services, are conserved through effectively and equitably managed, ecologically representative and well-connected systems of protected areas and other effective area-based conservation measures, and integrated into the wider landscape and seascape." The World Conservation Monitoring Centre, together with the IUCN and the IUCN World Commission on Protected Areas, was tasked to track progress on these goals (see next section).

Some scientists and conservationists have recently called for more-ambitious protection goals, moving toward "nature needs half" or similar concepts that call for one-half of the planet to bet set aside in healthy, functioning, and mostly undisturbed ecosystems (Locke 2014; Wilson, this volume, ch. 1). "Half the world for humanity, half for the rest of life, to make a planet both self-sustaining and pleasant," E. O. Wilson (2002) first stated in his book *The Future of Life*. This is an aspirational goal that needs to be seen through the lens of policy, in this case global policy.

Protected areas have experienced great growth over the past 20 years fueled by international agreements (see fig. 3.1), and are expected to continue to grow vigorously over the next decade. One of the drivers of the growth and improved delivery on conservation outcomes of protected areas has been the CBD's Programme of Work on Protected Areas (Secretariat of the Convention on Biological Diversity 2004). In 2010, the 192 state parties to the CBD adopted a strategic plan to halt biodiversity loss and to ensure the sustainable and equitable use of natural resources. The plan includes the 20 Aichi Biodiversity Targets, most of which are to be achieved by 2020. The United States is one of only a handful of countries that have not ratified the CBD. Yet, in its observer status, the United States has been a committed and relevant participant in supporting and leading the work of this convention in general and particularly for protected areas.

Two very different scenarios exist for the establishment of future protected areas at the global scale. One scenario is for western Europe, parts of Asia, and the Middle East, where ecosystems were mostly to entirely converted from pristine states to other land uses many centuries ago and now suffer from the effects of habitat loss and fragmentation and invasive species (Crooks et al. 2011; Foxcroft et al. 2013). The other scenario is for the Americas, Africa, Oceania, and parts of Asia and eastern Europe, where there are still large areas of naturally functioning ecosystems, though many are rapidly being lost and fragmented. When global conservation targets are set, keep in mind that these represent averages and that achieving them must consider local conditions and land-use history.

In the oceans, paradoxically, "pristine, unspoiled, and healthy" ecosystems are even less common than on land (Edgar et al. 2014; Thomas et al. 2014). So while the prospect of securing one-half of the oceans for conservation might seem like a feasible goal across most marine ecosystems, few areas would conform to the concept of "pristine, unspoiled, and healthy" ecosystems.

Most protected areas recognized by the IUCN are small, and there is a great need to protect large areas (Craigie, Pressey, and Barnes 2014). Large protected areas deliver a set of different values that complement smaller protected areas located in mosaics of working landscapes and seascapes. At sea, no-take zones are frequently a fisheries management tool that allow some degree of use in other zones such that societal benefits and values are maximized (Lester et al. 2009; Grorud-Colvert, Lubchenco, and Barner, this volume, ch. 2).

State of Protected Areas Globally

The United Nations established a list of parks and protected areas that has evolved into the World Database on Protected Areas (Deguignet et al. 2014), a joint initiative of the United Nations Environment Programme, through its World Conservation Monitoring Centre, and the IUCN, including the World Commission on Protected Areas. The *Protected Planet Report* is powered by the World Database on Protected Areas and has become extremely useful in aiding the evaluation of progress toward globally agreed targets in protected areas. Particularly in the last decade, the World Database on Protected Areas has improved dramatically in the number of protected areas and actual polygons are constantly being updated regularly, allowing observers to closely track the rapid increase in protected area coverage (Bertsky et al. 2012). The *Protected Planet Report 2014* follows the recommendation of the *Protected Planet Report 2012* to provide a comprehensive overview for each of the elements of Aichi Target 11 (Lopoukhine and de Souza Dias 2012). It summarizes current knowledge and progress toward achieving each element of the target, and provides further guidance for implementation based on data from the World Database on Protected Areas (Deguignet et al. 2014), a review of published literature, and expert review.

Global protected area coverage (see fig. 3.1) is currently at about 209,000 protected areas covering 15.4% of the planet's terrestrial areas and inland water areas and 3.4% of the oceans (Tittensor et al. 2014; Juffe-Bignoli et al. 2014 and June 2015 update by the World Conservation Monitoring Centre). In the ocean, 8.4% of all marine areas within national jurisdiction (200 nautical miles offshore) are covered by protected areas, while only 0.25% of areas beyond national jurisdiction are protected. In total, another 2.2 million km^2 of land and inland water areas and 2.2 million km^2 of marine area within national jurisdiction (Thomas et al. 2014) will need to be designated as protected areas to cover 17% of the land and 10% of the marine and coastal areas.

Unfortunately, protected areas do not sufficiently cover areas of importance for biodiversity. Only 22%–23% of recognized "key biodiversity areas" are completely covered by protected areas (Butchart et al. 2012), and many terrestrial and marine ecoregions are considered to be still poorly represented (Butchart et al. 2015). The same problem plagues protected lands and biodiversity in the United States (Jenkins et al. 2015). Targeted expansion of protected area networks is needed to include these key areas on land, and especially at sea. Thus, it seems likely that more than 17% of the land and 10% of the sea, the percentages called for in the Aichi Biodiversity Targets, will need to be protected to adequately conserve biodiversity.

There is strong and increasing evidence that well-supported protected areas conserve biodiversity and habitats, both on land and at sea. By 2013, 29% of the area of nationally designated protected areas had been assessed for the standard of protected area management effectiveness (Coad et al. 2013). Most protected areas in the United States have not been formally evaluated for effectiveness, even though most have internal methodologies and planning to achieve effectiveness. Furthermore, few studies have specifically assessed biodiversity outcomes linked to conservation actions in protected areas, and results on how management inputs relate to conservation outcomes are still equivocal.

Linking protected areas through corridors has been a major emphasis in global conservation. Available evidence for the outcomes of corridors indicates they generally have a positive conservation benefit (Jongman and Pungetti 2004). Despite a growing number of large projects promoting connectivity for conservation around the world in recent years (Crooks et al. 2011; Opermanis et al. 2012), there is no agreed-on standardized method to measure connectivity at a global level, and we have little knowledge of the level of connectivity between conservation areas across the wider landscapes and seascapes (Wegmann et al. 2014).

Protected areas are unlikely to be effective if they are managed as isolated elements or islands in human-dominated landscapes (Baron et al., this volume, ch. 7; DeFries, this volume, ch. 11). They need to be integrated into all aspects of landscape planning, especially into development planning. In 2014, 92% of the parties to the CBD had developed national biodiversity strategies and action plans. Nevertheless, the level of integration of protected areas into national planning has not yet been assessed globally.

There is no global indicator for measuring social equity in protected areas. In protected areas management, equity refers to the distribution among groups of people of (1) costs, benefits, and risk; (2) involvement in decision making; and (3) access to decision-making procedures (Juffe-Bignoli et al. 2014). Governance types provide limited information on enabling conditions for equity, and the *Protected Planet Report 2014* considers four classes: governance by government, shared governance, private governance, and governance by indigenous peoples. In 2014, 85% of the area of protected areas for which a governance type was reported were governed by governments, with the remainder governed by other arrangements. There are few published assessments of governance quality (Juffe-Bignoli et al. 2014).

The Promise of Sydney: A Protected Areas Charter for the Next Generation

The World Parks Congress (WPC), organized by the IUCN, occurs every 10 years and has been crucial in establishing the protected areas agenda. WPCs represent points of departure, and perhaps rupture, in a continually evolving science and practice of conservation. The WPC, by design, aims to bring conservation science and practice together with conservation policy. The resulting mix is well described in one of the opening statements given by Achim Steiner, executive director of the United Nations Environment Programme, at the most recent WPC in Sydney, Australia, in November 2014:

> Commit to bold, transformative actions and effective implementation at site, national and international levels. Let us learn from the past, but also recognize that it is today's youth that will inherit our protected area legacy, and the responsibility for managing the protected areas of the future. They will also bear the cost of our decisions today. Such decisions must ensure that the protected areas of the future will not be fenced off last frontiers that ward off humans to keep in what is left of our natural heritage. But rather that the Parks of the future will be a place where multiple values interact: ecological, biological, cultural, societal, economic and aesthetic—brought together by sound management and sustainable financing, as a basis for biodiversity conservation and sustainable development. Now this is a promise worth keeping.[2]

Each WPC also has created a groundswell of change by introducing new ideas (Phillips 2003; Dudley et al. 2014), launching new commitments,

---

2. Statement by Achim Steiner at the Opening Plenary of the IUCN World Parks Congress, 13 November 2014, accessed 11 March 2015, http://www.unep.org/newscentre/Default.aspx?DocumentID=2813&ArticleID=11069&l=en.

and signaling important developments in policy. These congresses stand out as a series of milestones in the development of the world's protected area systems (see table 3.1) (Phillips 2003; Dudley, Higgins-Zogib, and Mansourian 2005).

The WPC in Sydney represented as much an exercise in continuity on the tradition of protected areas conservation as it represented a point of departure with the inward-looking nature of previous congresses (Dudley et al. 2014). Organized around eight streams and four crosscutting themes (table 3.2), the Sydney WPC attracted a very diverse variety of participants numbering more than 6,000 from 170 countries. The congress design purposely limited the valuable but "business as usual" components of biodiversity conservation to one of the eight streams in order to induce a more comprehensive integration of new and increasingly relevant protected area themes and players. The main outcome to influence the protected areas agenda is the Promise of Sydney. It can more rightly be called "evolutionary" rather than "revolutionary" in what it promotes; many of the aspirations and innovations contribute to augmenting delivery in scale and influence rather than provide "new" ways of solving the challenges (Sandwith et al. 2014; Enkerlin-Hoeflich et al. 2015). The slogan for the congress— "Parks, People, Planet: Inspiring Solutions"—was meant to provide inspiration, and to give a sense of balance between biocentric and anthropocentric views of nature conservation.

The Promise of Sydney consists of four distinct elements that function together to advance protected area conservation and position protected areas as strategic assets at new levels of decision making. There is a core vision, which contains a series of aspirational statements that capture the moment and energy of the event in a broad and inclusive way. It recognizes threats to protected areas, but mostly concentrates on what needs to be done to accomplish the protected area goals discussed above. The second component consists of innovative approaches to transformative change that were drafted by the participants of the 12 streams and crosscutting themes (table 3.2). It includes close to 150 recommendations on approaches that will lead to the transformations in the decision making, practice, policy, capacity, and financing needed to demonstrate the full value of protected areas. The third component of the Promise of Sydney concentrates on developing a platform for sharing and exchanging inspiring solutions. A web-based "panorama" of solutions is available that uses a peer-to-peer tool for interchange.[3] This provides opportunities to learn

---

3. See http://www.panorama.solutions/ (accessed 17 March 2016).

Table 3.2   Streams and crosscutting themes of the VIth World Parks Congress

Streams	Crosscutting themes
Reaching conservation goals	Marine
Responding to climate change	World heritage
Improving health and well-being	Capacity development
Supporting human life	New social compact
Reconciling development challenges	
Enhancing diversity and quality of governance	
Respecting indigenous and traditional knowledge and culture	
Inspiring a new generation	

from others and to contribute from real-life examples of solutions to park and protected area problems and challenges.

Finally, the fourth element of the Promise of Sydney comprises the "promises." It contains an annex of commitments and pledges made by countries, funders, organizations, and other partners that are contributions to support accelerated success and implementation of protected area growth. In essence, the fourth element is where science and policy meet, and results in measurable deliverables whose progress can be tracked over time. For instance, Brazil committed to increasing protection of its marine territory from 1.5% to 5%, Palau committed to restricting commercial fisheries in its entire exclusive economic zone of 600,000 km^2, and Russia committed over the next decade to expanding its protected area network by establishing at least 27 federal protected areas and expanding 12 others, which would increase the total federal protected areas by 22% (or 13 million ha).

How the Promise of Sydney affects protected areas conservation over the next decade remains to be seen. Nevertheless, participants left energized, and new networks and collaborations were created among those involved in protected areas.

Future Directions for Protected Area Conservation

Ultimately, we should be speaking about science and parks to ensure the sustainability of our planet (Ostrom 2009). Yet, beyond our own community, conservation is still viewed in many nations and sectors as an optional and philanthropic endeavor, rather than an investment to ensure local or national competitiveness, to maintain human well-being, and to ensure future options. Protected areas have always been about conservation, and for a very long time protection was enough to guarantee conservation. Today, as threats to protected areas and their biodiversity mount, there is an increased need for active management that goes beyond simply designating borders of protected areas. Moreover, in the next century, restoration will likely become the most prevalent activity in the continual cycle of modern conservation: protection, management, and restoration.

What this means in practice is that achieving a protected area coverage goal—be it Aichi Target 11 or "nature needs half"—will require more than simply setting aside that amount of area. A large proportion of these areas will be conserved by including some form of direct or indirect use, whether to maintain functionality or to justify the investments. A large proportion of these areas will also likely be under control of their indigenous or local community owners or rights holders. Moreover, it is likely that many countries could eventually reach a goal of setting aside at least 50% of terrestrial areas as healthy functioning ecosystems if we included the entire gamut of natural resource management areas in this tally. Less than half of that area will likely consist of national parks or equivalent conservation regimes. Instead we will move to more flexible and adaptive systems of protected and conserved areas, in which the outcomes can be maximized at the landscape and seascape level. Nevertheless, many countries will not be able to reach a goal of protecting 50% of terrestrial areas, even with massive restoration commitments and investments, without large social upheaval (e.g., most of Europe and large parts of Asia).

We must accelerate the protection of many sites, but especially those few that still maintain a wilderness character. We must use the precautionary principle and swiftly go for protecting as much as possible of the global oceans in both take and no-take areas, and develop an international regime for their conservation and conservation financing. While a targeted expansion is highly desirable, we should not ignore a protected area proposal just because the site is not large enough, diverse enough, connected enough, pristine enough, or a top priority. Aspire for the best and accept the most we can get.

We must swiftly move the science and the policy from individual protected areas to systems of protected and conserved areas that function as networks embedded in landscapes and seascapes beyond ecological, geopolitical, administrative, institutional, cultural, and ideological boundaries. The concept of transboundary conservation, while a tradition in protected area management, must be redefined, as nearly all conservation today is "transboundary," especially in relation to climate change. Agencies within countries and between countries must work together. The new vision must replace the business-as-usual approach that "does conservation in our areas and cooperates internationally as needed" with systems that are cooperatively designed and administered. The NATURA 2000 network of protected sites in Europe is a good example of this kind of forward-looking international cooperation among countries (Opermanis et al. 2012).

New forms of cooperation should extend not just to countries but to new organizational partners. As the climate and biodiversity crises deepen, religious organizations are beginning to support conservation science by calling for a major reorganization of our relationship with nature. From an ethical and philosophical perspective, natural law may regain its early meaning, perhaps with a result that realigns human law with ecosystem capacity. In 2012, the IUCN World Conservation Congress adopted Resolution 009 on cooperation in climate advocacy and nature conservation with faith-based and religious organizations and networks (Crawhall 2015). Perhaps new bedfellows can produce important breakthroughs for protected areas that exceed those accomplished by traditional alliances.

Acknowledgments

Many thanks to David Ackerly and Gary Machlis for improving our chapter. The VIth World Parks Congress convened under the auspices of the IUCN was organized by the IUCN Global Protected Areas Program directed by Trevor Sandwith with support from the World Commission of Protected Areas network of volunteer experts around the world, particularly the World Commission of Protected Areas leadership led by the program chair, Kathy MacKinnon, who engaged a large number of partner organizations and their leaders.

Literature Cited

Archabald, K., and L. Naughton-Treves. 2001. Tourism revenue-sharing around national parks in Western Uganda: early efforts to identify and reward local communities. Environmental Conservation 28:135–149.

Balmford, A., J. M. H. Green, M. Anderson, J. Beresford, C. Huang, R. Naidoo, M. Walpole, and A. Manica. 2015. Walk on the wild side: estimating the global magnitude of visits to protected areas. PLoS Biology 13:e1002074.

Balmford, A., L. J. Moore, T. Brooks, N. Burgess, L. A. Hansen, P. Williams, and C. Rahbek. 2001. Conservation conflicts across Africa. Science 291:2616–2619.

Beissinger, S. R., E. C. Steadman, T. Wohlgenant, G. Blate, and S. Zack. 1996. Null models for assessing ecosystem conservation priorities: threatened birds as titers of threatened ecosystems. Conservation Biology 10:1343–1352.

Beissinger, S. R., and M. I. Westphal. 1998. On the use of demographic models of population viability analysis in endangered species management. Journal of Wildlife Management 62:821–841.

Bernard, E., L. A. O. Penna, and E. Araújo. 2014. Downgrading, downsizing, degazettement, and reclassification of protected areas in Brazil. Conservation Biology 28:
939–950.

Bertsky, B., C. Corrigan, J. Kemsey, S. Kenney, C. Ravilious, C. Besançon, and N. Burgess. 2012. Protected planet report 2012: tracking progress towards global targets
for protected areas. IUCN, Gland, Switzerland, and United Nations Environment
Programme World Conservation Monitoring Centre (UNEP-WCMC), Cambridge,
United Kingdom.

Bhagwat, S. A., C. G. Kushalappa, P. H. Williams, and N. D. Brown. 2005. The role of
informal protected areas in maintaining biodiversity in the Western Ghats of India.
Ecology and Society 10:8.

Brandon, K., and M. Wells. 2009. Lessons for REDD+ from protected areas and integrated conservation and development projects. Pages 225–235 in A. Angelsen and
M. Brockhaus, eds. Realising REDD+: national strategy and policy options. Center
for International Forestry Research (CIFOR), Bogor, Indonesia.

Butchart, S. H. M., M. Clarke, J. S. Robert, R. E. Sykes, J. P. W Scharlemann, M. Harfoot,
G. M. Buchanan, et al. 2015. Shortfalls and solutions for meeting national and global
conservation area targets. Conservation Letters 8:329–337.

Butchart, S. H. M., J. P. W. Scharlemann, M. I. Evans, S. Quader, S. Aricò, M. B. Arinaitwe,
L. A. Bennun, et al. 2012. Protecting important sites for biodiversity contributes to
meeting global conservation targets. PLoS ONE 7:1–8.

Coad, L., F. Leverington, N. D. Burgess, I. C. Cuadros, J. Geldmann, T. R. Marthews,
J. Mee, et al. 2013. Progress towards the CBD protected area management effectiveness targets. Parks 19:13–24.

Craigie, I. D., R. L. Pressey, and M. Barnes. 2014. Remote regions—the last places where
conservation efforts should be intensified. A reply to McCauley et al. (2013). Biological Conservation 172:221–222.

Crawhall, N. 2015. Social and economic influences shaping protected areas. Pages 117–
144 in G. L. Worboys, M. Lockwood, A. Kothari, S. Feary, and I. Pulsford, eds. Protected area governance and management. ANU Press, Canberra, Australia.

Crooks, K. R., C. L. Burdett, D. M. Theobald, C. Rondinini, and L. Boitani. 2011. Global
patterns of fragmentation and connectivity of mammalian carnivore habitat. Philosophical Transactions of the Royal Society B 366:2642–2651.

Deguignet, M., D. Juffe-Bignoli, J. Harrison, B. MacSharry, N. D. Burgess, and N. Kings-
ton. 2014. 2014 United Nations List of Protected Areas. United Nations Environment
Programme World Conservation Monitoring Centre (UNEP-WCMC), Cambridge,
United Kingdom.

Diamond, J. 2005. Collapse: how societies choose to fail or succeed. Penguin, New York,
New York.

Dickson, B., R. Blaney, L. Miles, E. Regan, A. van Soesbergen, E. Väänänen, S. Blyth, et al.
2014. Towards a global map of natural capital: key ecosystem assets. United Nations
Environment Programme (UNEP), Nairobi, Kenya.

Dudley, N., ed. 2008. Guidelines for applying protected area management categories.
IUCN, Gland, Switzerland.

Dudley, N., C. Groves, K. H. Redford, and S. Stolton. 2014. Where now for protected areas? Setting the stage for the 2014 World Parks Congress. Oryx 48:496–503.

Dudley, N., L. Higgins-Zogib, M. Hockings, K. MacKinnon, T. Sandwith, and S. Stolton.
2011. National parks with benefits: how protecting the planet's biodiversity also provides ecosystem services. Solutions 2:87–95.

Dudley N., L. Higgins-Zogib, and S. Mansourian, eds. 2005. Beyond belief: linking faiths and protected areas to support biodiversity conservation. WWF International, Gland, Switzerland.

Durán, A. P., S. Casalegno, P. A. Marquet, and K. J. Gaston. 2013. Representation of ecosystem services by terrestrial protected areas: Chile as a case study. PLoS ONE 8:e82643.

Eagles, P. F. J., S. F. McCool, and C. D. A. Haynes. 2002. Sustainable tourism in protected areas: guidelines for planning and management. World Commission on Protected Areas (WCPA) Best Practice Protected Area Guideline Series No. 8. IUCN, Gland, Switzerland.

Edgar, G. J., R. D. Stuart-Smith, T. J. Willis, S. Kininmonth, S. C. Baker, S. Banks, N. S. Barret, et al. 2014. Global conservation outcomes depend on marine protected area with five key features. Nature 506:216–220.

Edwards, D. P., S. Sloan, L. Weng, P. Dirks, J. Sayer, and W. F. Laurance. 2014. Mining and the African environment. Conservation Letters 7:302–311.

Enkerlin-Hoeflich, E. C., T. Sandwith, K. MacKinnon, D. Diana, A. Andrade, T. Badman, P. Bueno, et al. 2015. IUCN/WCPA protected areas program: making space for people and biodiversity in the Anthropocene. Pages 339–350 in R. Rozzi, F. S. Chapin III, J. B. Callicott, S. T. A. Pickett, M. E. Power, J. J. Armesto, and R. H. May Jr., eds. Earth stewardship. Linking ecology and ethics in theory and practice. Springer International Publishing, Cham, Switzerland.

Foxcroft, L. C., P. Pyšek, D. M. Richardson, and P. Genovesi, eds. 2013. Plant invasions in protected areas: patterns, problems and challenges. Springer Netherlands, Dordrecht, Netherlands.

Geldmann, J., L. N. Joppa, and N. D. Burgess. 2014. Mapping change in human pressure globally on land and within protected areas. Conservation Biology 28:1604–1616.

Holdgate, M. 1999. The green web: a union for world conservation. Earthscan, London, United Kingdom.

Jantz, P., S. Goetz, and N. Laporte. 2014. Carbon stock corridors to mitigate climate change and promote biodiversity in the tropics. Nature Climate Change 4:138–142.

Jenkins, C. N., K. S. Van Houtan, S. L. Pimm, and J. O. Sexton. 2015. US protected lands mismatch biodiversity priorities. Proceedings of the National Academy of Sciences USA 112:5081–5086.

Jongman, R. H. G., and G. Pungetti, eds. 2004. Ecological networks and greenways: concept, design, implementation. Cambridge University Press, Cambridge, United Kingdom.

Joppa, L. N., and A. Pfaff. 2009. High and far: biases in the location of protected areas. PLoS ONE 4:e8273.

———. 2011. Global protected area impacts. Proceedings of the Royal Society B 278: 1633–1638.

Juffe-Bignoli, D., N. D. Burgess, H. Bingham, E. M. S. Belle, M. G. de Lima, M. Deguignet, B. Bertzky, et al. 2014. Protected planet report 2014. United Nations Environment Programme World Conservation Monitoring Centre (UNEP-WCMC), Cambridge, United Kingdom.

Larsen, F. W., W. R. Turner, and T. M. Brooks. 2012. Conserving critical sites for biodiversity provides disproportionate benefits to people. PLoS ONE 7:e36971.

Laurance, W. F., A. Peletier-Jellema, B. Geenen, H. Koster, P. Verweij, P. Van Dijck, T. E. Lovejoy, J. Schleicher, and M. Van Kuijk, M. 2015. Reducing the global environmental impacts of rapid infrastructure expansion. Current Biology 25:R259–R262.

Lee, A. C. K., and R. Maheswaran. 2011. The health benefits of urban green spaces: a review of the evidence. Journal of Public Health 33:212–222.

Lester, S., B. Halpern, K. Grourd-Colvert, J. Lubchenco, B. Ruttenberg, S. Gaines, S. Airamé, and R. Warner. 2009. Biological effects within no-take marine reserves: a global synthesis. Marine Ecology Progress Series 384:33–46.

Locke, H. 2014. Nature needs half: a necessary and hopeful new agenda for protected areas in North America and around the world. The George Wright Forum 31:359–371.

Lopoukhine, N., and B. F. de Souza Dias. 2012. Editorial: what does target 11 really mean? Parks 18:5–8.

Maller, C., M. Townsend, A. Pryor, P. Brown, and L. St Leger. 2006. Healthy nature healthy people: 'contact with nature' as an upstream health promotion intervention for populations. Health Promotion International 21:45–54.

Mascia, M. B., S. Pailler, R. Krithivasan, V. Roschchanka, D. Burns, M. J. Mlotha, D. R. Murray, and N. Peng. 2014. Protected area downgrading, downsizing, and degazettement (PADDD) in Africa, Asia, and Latin America and the Caribbean, 1900–2010. Biological Conservation 169:355–361.

Millenium Ecosystem Assessment. 2005. Ecosystems and human well-being: biodiversity synthesis. World Resources Institute, Washington, DC.

Opermanis, O., B. MacSharry, A. Aunins, and Z. Sipkova. 2012. Connectedness and connectivity of the Natura 2000 network of protected areas across country borders in the European Union. Biological Conservation 153:227–238.

Ostrom, E. 2009. A general framework for analyzing sustainability of social-ecological systems. Science 325:419–422.

Phillips, A. 2003. Turning ideas on their heads: a new paradigm for protected areas. The George Wright Forum 20:8–32.

Romagosa, F., P. F. J. Eagles, and C. J. Lemieux. 2015. From the inside out to the outside in: exploring the role of parks and protected areas as providers of human health and well-being. Journal of Outdoor Recreation and Tourism. doi:10.1016/j.jort.2015.06.009.

Rozzi, R., F. S. Chapin III, J. B. Callicott, S. T. A. Pickett, M. E. Power, J. J. Armesto, and R. H. May Jr. 2015. Earth stewardship. Linking ecology and ethics in theory and practice. Springer International Publishing, Cham, Switzerland.

Sandwith, T., E. Enkerlin, K. MacKinnon, D. Allen, A. Andrade, T. Badman, T. Brooks, et al. 2014. The promise of Sydney: an editorial essay. Parks 20:7–18.

Sarukhán, J., T. Urquiza-Haas, P. Koleff, J. Carabias, R. Dirzo, E. Ezcurra, S. Cerdeira-Estrada, and J. Soberón. 2015. Strategic actions to value, conserve, and restore the natural capital of megadiversity countries: the case of Mexico. BioScience 65:164–173.

Secretariat of the Convention on Biological Diversity. 2004. Programme of work on protected areas (CBD programmes of work). Secretariat of the Convention on Biological Diversity, Montreal, Canada.

———. 2008. Protected areas in today's world: their values and benefits for the welfare of the planet. CBD Technical Series No. 36. Secretariat of the Convention on Biological Diversity, Montreal, Canada.

Stolton, S., and N. Dudley, eds. 2010. Arguments for protected areas: multiple benefits for conservation and use. Earthscan, London, United Kingdom.

Svancara, L. K., R. Brannon, J. M. Scott, C. R. Groves, R. F. Noss, and R. L. Pressey. 2005. Policy-driven versus evidence-based conservation: a review of political targets and biological needs. BioScience 55:989–995.

Tear, T. H., P. Kareiva, P. L. Angermeier, P. Comer, B. Czech, R. Kautz, L. Landon, et al. 2005. How much is enough? The recurrent problem of setting measurable objectives in conservation. BioScience 55:835–849.

Thomas, H. L., B. MacSharry, L. Morgan, N. Kingston, R. Moffitt, D. Stanwell-Smith, and L. Wood. 2014. Evaluating official marine protected area coverage for Aichi Target 11: appraising the data and methods that define our progress. Aquatic Conservation: Marine and Freshwater Ecosystems 24(S2):8–23.

Tittensor, D. P., M. Walpole, S. L. L. Hill, D. G. Boyce, G. L. Britten, N. D. Burgess, S. H. M. Butchart, et al. 2014. A mid-term analysis of progress toward international biodiversity targets. Science 346:241–244.

Turner, W. R., K. Brandon, T. M. Brooks, C. Gascon, H. K. Gibbs, K. S. Lawrence, R. A. Mittermeier, and E. R. Selig. 2012. Global biodiversity conservation and the alleviaton of poverty. BioScience 62:85–92.

Vörösmarty, C. J., P. B. McIntyre, M. O. Gessner, D. Dudgeon, A. Prusevich, P. Green, S. Glidden, et al. 2010. Global threats to human water security and river biodiversity. Nature 467:555–561.

Watson, J. E. M., N. Dudley, D. B. Segan, and M. Hockings. 2014. The performance and potential of protected areas. Nature 515:67–73.

Wegmann, M., L. Santini, B. Leutner, K. Safi, D. Rocchini, M. Bevanda, H. Latifi, S. Dech, and C. Rondinini. 2014. Role of African protected areas in maintaining connectivity for large mammals. Philosophical Transactions of the Royal Society B 369:20130193.

Weeramantry, C. G. 2009. Tread lightly on the Earth: religion, the environment and the human future. Stamford Lake, Pannipitiya, Sri Lanka.

Wilson, E. O. 2002. The future of life. Random House, New York, New York.

FOUR

Strategic Conversation: Mission and Relevance of National Parks

EDITED BY KELLY A. KULHANEK, LAUREN C. PONISIO, ADAM C. SCHNEIDER, AND RACHEL E. WALSH

On 25 August 1916, the National Park Service Organic Act was signed into law by President Woodrow Wilson, thereby establishing the agency and its mission in a mere 731 words. The key mission, still in force today, is "to conserve the scenery and the natural and historic objects and the wild life therein and to provide for the enjoyment of the same in such manner and by such means as will leave them unimpaired for the enjoyment of future generations."

Since then, the demographic, social, political, environmental, and economic landscape of the United States has dramatically changed. This strategic discussion, which transpired at the Berkeley summit "Science for Parks, Parks for Science" on 26 March 2015, focuses on the legacy of the National Park Service mission, as well as its relevance in the 21st century. The discussion panel consisted of three members: Denis Galvin, who retired from the National Park Service after 38 years of working for the agency in many capacities, including serving as deputy director from 1985 to 1989 and 1997 to 2002; George Miller, who recently retired from the US Congress after 40 years of service, which included 30 years on the Natural Resources Committee; and Frances Roberts-Gregory, a PhD student studying science communication, greenspace accessibility, and environmental racism in the Department of Environmental Science, Policy, and Management at the University of California, Berkeley. This conversation was moderated by Holly Doremus, a professor of environmental law at the University of California, Berkeley.

HOLLY DOREMUS: *In your view, what is the most important mission of the National Park System, how has that changed over the last 100 years, and how do you think it might change in the next 100?*

GEORGE MILLER: I think the mission has changed dramatically in magnitude if not in purpose. To me, it's a continued challenge of presenting this incredibly complex platform that we call the National Park System that is managed by the National Park Service. How do you recognize the complexity of the resources, and the complexity of the agency with that system of parks and with the American public that holds those parks in very high esteem? The public looks at the National Park Service as authoritative, determining how public lands should be managed. The National Park Service transfers a huge amount of culture across generations. When we come up with plans to revise the operations of national parks, we bump into generational and cultural habits, and into traditions in families that probably dominate an incredible chunk of that debate. So to continue to protect, to preserve, to open greater access—that's the challenge. This is an incredibly complex platform. You may not have designed it this way in the very beginning, but that's what it is today, growing in complexity within our society.

FRANCES ROBERTS-GREGORY: The National Park System mission is very complex. I think that, in order to engage a more diverse public in the future, we have to talk about the diverse histories, stories, and narratives that have existed in the past, and also exist in the present. We just have to uncover what's already there, and bring voice to understandings of the world and identifications with the National Park Service that previously have been underdocumented.

DENIS GALVIN: Often, the mission of the National Park Service is simply the restatement of the Organic Act of 25 August 1916. Actually, there are dozens of pieces of legislation that give the National Park Service other missions, such as the Endangered Species Act, the Land and Water Conservation Fund, and the National Environmental Protection Act that require parks to manage wildlife, pollution, et cetera. But there is officially a second sentence to the mission in the Organic Act. It says that the National Park Service cooperates with partners to extend the benefits of natural and cultural resource conservation and outdoor recreation throughout this country and the world. So in thinking about those two sentences—the first to conserve the scenery and the second to cooperate—and in thinking about time and changes over time, over the next 100 years the second sentence may be more important than the first. What's happened over the last 100 or 150 years to a place like Yellowstone National Park that's out there all alone? The forces that acted on Yellowstone 100 or 150 years ago are enormously different than the forces that act on it today. Many of the forces, perhaps the most important forces on Yellowstone, Gettysburg, Cape Cod, or Cape Hatteras National Parks, come from outside the boundaries of the parks. As several of today's speakers mentioned, these parks are islands in a very complex matrix. The National Park Service has to cooperate with people living outside the parks. It becomes important not only for people to build conservation in their own communities, which is very important, but it's also important to protect the park system itself.

DOREMUS: Let me follow up on that then. As you've noted, national parks are necessarily embedded in a larger landscape. That can bring both spillover benefits, which are important economic engines for the communities around them, and spillover costs, such as bison that move out of Yellowstone National Park and are perceived to be a source of disease for livestock. *How should the neighborhoods of the parks influence their management? Or I might put that question a little bit differently and ask, whom are our national parks for? Should local communities have a special voice in their management, and if so, what should that look like? Should there be special efforts to connect local communities with their parks, and how can that be done?*

ROBERTS-GREGORY: I definitely think it's very important. The local communities surrounding a park are influenced by the management strategies pursued by park rangers and park officials, and vice versa. Too often in the past there's been a particular idea of who should enjoy a park, and what enjoyment of a particular site should look like. We have to take into account that different people have different ways of enjoying even a so-called wilderness and the areas that we want to protect. I think that it is really important to take into account what local communities want and not just rely on a unidirectional model of communication. You must actually have conversations with these individuals. Learning to be bilingual and trilingual is really important—not just in terms of what we think of as languages, but in terms of different ways of viewing the world or viewing what is science.

GALVIN: You had to bring up bison! In the mid-1990s, Secretary of the Interior Bruce Babbitt called me up to his office about six o'clock at night and said, "I want you to go out to Yellowstone and stop the slaughter." We were on network television every night. One of the biologists said, "Shooting bison is like shooting a sofa." Also, it was very easy to televise. The bison management plan, with which I was deeply involved, goes directly to the issue of whom the parks are for. In that instance, we probably had 65,000 comments from around the world on the bison situation. You know, 64,900 of them were for the bison. About 100 of them, mostly from a few ranchers around Wyoming and the governor of Montana, were much more influential in the management of bison than the 64,900 people who said, "Yes, save the bison at all costs." The science was pretty clear that brucellosis didn't really threaten cattle outside the parks. Well, not entirely clear, but probably 95%

clear. For one thing, there were no cattle around the park when the bison went out. So then the question became, how long does the *Brucella* last? The science influenced our decision, but as earlier speakers indicated, it's a policy decision and you've got to make trade-offs. The solution developed is a very imperfect solution. We're still killing bison outside the park, but it's getting better. One of the things we said was that we're going to start intensive management when there are 2,900 bison. Well now there are 4,500 bison, so at least there are more bison. The state has become more open to accepting bison outside the park that are working their way up to a place called Yankee Jim Canyon. There are two things about the bison example. One is that local people tend to have much more influence on park decisions than a national or international constituency. The second is that science influences policy, but it doesn't set policy.

MILLER: If we start to think about the parks in this day and age as islands, we're doomed. Certainly the parks are doomed. I think there are rings of intensity and there are rings of ownership, to some extent, of those parks. Obviously, the communities and the activities around the parks have much more concern about the operation of those parks and the planning and development of those parks. That's the progression. We didn't need the buffers when we created the parks, and now we look at the impacts of population growth and the rest of it.

I think also you have to understand that those parks have to run. When you look at the state of California, knowledge and awareness of the parks has to run all the way to South Central Los Angeles and back to Kings Canyon Park or to Sequoia or to Yosemite. The fact is that the National Park System has to think of local parks, like the East Bay Regional Parks,[1] as a "farm club" for how people conduct themselves outdoors, how they interact together. If you walk on the great trails of the East Bay Regional Parks, it's the bikes versus the horses versus the dogs versus the people versus the runners, and it happens every day. You have the same kind of complexity inside national parks. The stakes may be somewhat higher, and the national parks are somewhat more overwhelmed for three months of the year, but you have to think about operating them in that system.

I would say that the role of science in that process is to completely and continuously revise the park operational plan, if you will, so we can do the least amount of harm and hopefully provide for the positive recovery of those parks. It's very hard when you look at Yosemite Valley in July and think,

---

1. The East Bay Regional Parks are a system of 65 parks covering 119,000 acres across two counties on the east side of San Francisco Bay.

how are you going to do this? But the fact of the matter is, scientists should involve park service personnel in the design and implementation of the science. Then, with the results of the science in hand, National Park Service personnel can start to think about how they can refigure the parks. We have institutions within those parks that are cherished and historical, but they are also threatening to the parks. The question is, can you keep the tradition? Can you diminish the adverse impact on cultural memories and at the same time allow for a lighter footprint of many of those activities in these big iconic parks in the system. That, to me, is the challenge.

DOREMUS: *What's distinctive about the National Park System, and what should be distinctive as opposed to the many other kinds of protected or partially protected lands that we have, ranging from the local level, including some private lands, to state lands and other sorts of federal lands?*

GALVIN: A couple of thoughts. One is the mandate going back to the Organic Act. The national parks are to preserve everything. I'm always correcting texts when they get to the word "wildlife" in the Organic Act, because in the original act it's two words: "wild life." So "wildlife," one word, is often considered to be elk, bison, and other game animals. The National Park Service has always interpreted "wild life" as every living thing in the park. That's one distinction. Not like the US Fish and Wildlife Service, which has the mission to protect particular species. Not like the US Forest Service, which does multiple use. The other thing is that the interpretive and education programs of the National Park Service are extensive. That's not to say the other agencies don't have them, but there is certainly a more extensive and a longer history in the National Park Service. These programs are real resources for science and conservation. So, to sum up, what is distinctive about our national parks is interpretation, education, and the mission to preserve everything.

MILLER: I think that's all part of it. For the last 30 years, I've walked back and forth across the tops of Kings Canyon, Sequoia, and Yosemite National Parks. Over those years, I've picked up a lot of cowboys, wranglers, sawyers, trail crews, and convicts—all engaged in the park—and professional park personnel to sit around the campfire and discuss the complexity of the park and the challenges of the mission. In some cases, where they would admit, the park professionals may be flying blind because they really don't know the impact of the changes they might have to make. That's harder to do today with all the regulatory requirements.

National parks are different, and I think Americans probably want them treated somewhat differently. Yet we have to constantly decide whether national parks will survive. Is there a design on some of the assets that are in those parks? Think of California in this historic drought, and then think of wild and scenic rivers. People are saying, "Maybe we ought to go back to dam building." But the many dams already in place didn't do much to avoid this drought.

The point of this conference is that science has a huge contribution to make. Some people think it is bad to support science for science's sake. But with all due respect, science for science's sake took us to the next generation of a lot of things. Also, the application of science in the administering and the enjoyment of the parks is absolutely critical. Often we've managed by the seat of our pants. We've made huge mistakes in the administration of the parks that we've come back to try to repair. But I would also like to continue providing a quilt of protective environments around those parks, so that the parks have more flexibility and the bison have more flexibility for survival.

ROBERTS-GREGORY: Throughout my life, I've visited local parks, state parks, and national parks. I've even traveled to parks internationally. I find that there are unique facets that I enjoy about all these different types of sites. Obviously, there is something we consider very sacred about our national parks, but I would push back against that. There is something to be gained by visiting and supporting parks that perhaps don't have as many resources and perhaps are most valued by a certain demographic of people—people that unfortunately might not be seen as valuable to folks who view national parks as sacrosanct. For example, I don't think urban parks get as much attention as they probably deserve, and few of them are classified as national parks. Yet urban parks are extremely important for the myriad social, public health, economic, and environmental benefits they provide, in addition to the opportunities they provide as sites for community engagement in scientific research.

DOREMUS: *Should we be adding to the National Park System, either in terms of new units or new lands?* We heard from E. O. Wilson that we need to protect much more of the globe and from Jane Lubchenco that we need to protect much more of the seas, but then we heard from Hugh Possingham some skepticism about whether what we need to do is to protect more or to protect better. *What do you think about adding parks to the National Park System?*

MILLER: I voted for them all. In some areas, there are a lot of different attitudes about public ownership of the lands, the interface of public ownership and private ownership, and the interface culturally with the idea that this land is going to be a federally administered area in some fashion. But very often what you find is that the creation of the park is like an icebreaker in the spring! It opens the path for a lot of other good things to happen in that geographic area. Parks are very important for the long-term vision of saving our assets, whether from the perspective of conservation that Professor Wilson discussed or for the immediate idea that we can spend the weekend there. Those are kind of competing views.

Some of the areas waiting for park designation should be given absolutely serious consideration. The first guys got the best sites, right? There's El Capitan, and then there's Half Dome, and then there's the Grand Canyon of the Yellowstone, and then there's the Grand Canyon. You're not going to find another Grand Canyon, but you can find a lot of important areas that should be protected. If we're only going to have the elite schools, then a lot of people are going to get left out of an education. So there are areas waiting for park designation that rise to the same mission, the same purpose, and the same protection, and I think that we should strive to include them.

GALVIN: Let me just say that a lot of people are going to get left out. I've been on two groups looking at the future of the National Park System in the last five years, and yes, there's opportunity for robust growth. Just thinking about people, there are 100 acres of national park for every person in Alaska. There are one million people in Illinois for every acre of national park in Illinois. There are going to be different kinds of parks. There are going to be parks like the Upper Mississippi where you have a 72-mile river corridor, and the National Park Service only owns 75 acres, but it coordinates the planning in that corridor. So we're not going to be building Yellowstones, but there's ample opportunity for growth in both the natural and cultural spheres.

DOREMUS: *Finally, what's the single biggest management challenge for national parks in the next 100 years?*

ROBERTS-GREGORY: I think the biggest management challenges are really engaging the public (a lot of people have talked about it) and also resources (making sure there are enough resources for the parks in the future). I think we talk about engagement but we don't actually implement it. We rely on some of the same methods to talk about what is science and why people should be involved, but we are not being effective. I think that if you want to reach changing demographics in America, you have to also change yourself. If you want an individual who has never thought about parks to get involved, you yourself might have to get involved in, let's say, human rights issues or social justice. And these might be issues you've never thought about previously in regard to park management. We need to cut down these binaries between what is science and what is not, what is a citizen and what is a noncitizen, what is the public and what is the government or the private sector. *All* of us have to change ourselves—we can no longer just say that it is solely the public that needs to change.

GALVIN: Absolutely. I agree with that completely. In fact, I think one of the great opportunities for science is the opportunity for citizens to engage in science in national parks. Not to make them scientists, but to have them participate. You think about things like the theory of evolution. Many people in this country do not support the theory of evolution. You go to a place like Grand Canyon, and it hits you in the face. Let me just finish with a quote from a panel that Dr. Wilson was on (John Hope Franklin was the chair): "By caring for the parks and conveying the park ethic, we care for ourselves and act on behalf of the future. The larger purpose of this mission is to build a citizenry that is committed to conserving its heritage and its home on earth." That's the opportunity for parks. Parks can't do it alone, but they can move the citizenry to protect their community and the planet.

MILLER: I think the challenge is climate, and not only because of the direct impact on the park. If you go to the iconic parks on the Canadian-American border, we're building freeway overpasses for bears, elk, and other species. They're already starting to migrate and move, and if you go to look at the glacier in Glacier National Park, it's a long hike. It's not leaning over the edges anymore. I think it is climate. Climate is also going to have an impact outside the park. It's going to conceivably disturb populations. Parks have to be managed in that context; it's not just in the valley of the park, in the center of the park, the canyon of the park. I think climate is going to be a huge challenge to the general national park ecosystem, which is much larger, of course, than the park.

Stewardship of Parks in a Changing World

From the outset, with the passage of the Organic Act in 1916, the US National Park Service (NPS) has pursued two parallel, and at times conflicting, goals—visitor enjoyment and conservation of natural resources in the parks. The role of science in support of conservation, and the role of parks as natural laboratories for science, has waxed and waned over the years (see Beissinger and Ackerly, this volume, ch. 18). In the early years, resource management efforts focused on wildlife and fisheries and the oft-quoted goal articulated in the 1963 Leopold Report to maintain or re-create conditions that prevailed before the arrival of Europeans. National parks stood as islands of nature, for wildlife and people, in a sea of development and multiuse lands of other federal agencies.

Yet, by the time many parks were set aside, they had already suffered significant ecological deterioration, and populations of many wildlife species had been decimated across the continent. Conditions in even the largest parks could never be isolated from the surrounding landscape, or from social and political forces at the local and national levels. With the rise of outdoor recreation through the 20th century, many parks were victims of their own success, as direct impacts of visitation created challenges to the mandate of conservation. As we look ahead to the next century, the necessity to view parks in this larger environmental and geographic context is clear. Anthropogenic threats, from air pollution to climate change, do not respect political or administrative boundaries, nor do natural processes, such as wildfire and animal migration. The most successful conservation solutions also have to be pursued at larger scales, both in geographic scope and in the social network of stakeholders whose lives and livelihoods are linked to the parks.

In recent years, the NPS has increasingly turned its attention to the changing landscape of conservation and management in the 21st century. In its 2010 report entitled *Climate Change Response Strategy*, the NPS identified a four-pronged approach focused on science, adaptation, mitigation, and communication to respond to "the long-range and cascading effects of climate change [which] are just beginning to be understood." In the 2012 report entitled *Revisiting Leopold: Resource Stewardship in the National Parks*, an NPS advisory board reexamined the principles that have guided the agency for 50 years, proposing a revised vision: "The overarching goal of NPS resource management should be *to steward NPS resources for continuous change that is not yet fully understood, in order to preserve ecological integrity and cultural and historical authenticity, provide visitors with transformative experiences, and form the core of a national conservation land- and seascape*" (italics in original). Implementation of this vision represents a fundamental shift from a more retrospective view of parks as museums of the past to a forward-looking vision of parks as crucibles of change. The embrace of uncertainty about the rate and trajectory of these changes reflects a fundamental humility about both our understanding of nature and our ability to shape the future, a lesson that we may encounter often, yet is hard to fully assimilate. The vision of parks as the core of a national landscape-scale conservation plan also reflects a growing attention to the broader spatial, temporal, and societal context that informs NPS policies and priorities.

The first five chapters in this section, while selected to capture a wide range of issues related to resource management, share a common theme of parks coupled to their surroundings, including both the conservation challenges and the solutions that emerge in this larger context. While these chapters focus their attention on the US national parks, the lessons learned and challenges that lie ahead are shared more broadly by other parks and protected areas in the United States and in other countries.

Monica Turner and colleagues summarize 20+ years of research and lessons learned since the 1988 Yellowstone fires. It is now well established that infrequent, stand-replacing fires are typical of the historical disturbance regime in this system. Heterogeneity in landscape features and fire severity is a critical feature and contributes to recovery and resilience of the ecosystem. Their work illustrates a critical lesson from long-term research: management interventions are not always necessary in response to disturbance and environmental change, and we need a greatly expanded understanding of how other systems will respond to changing conditions.

In that context, Patrick Gonzalez provides a comprehensive review of documented and projected impacts of global climate change on the US

National Park System. Changes in physical climate have affected most of the area under NPS jurisdiction, and weather stations and other long-term monitoring at parks have contributed valuable data for detection of these changes. Biotic responses to climate change are also widespread, and many other changes observed in US national parks are consistent with, though not yet documented as attributed to, climate change. Yet these changes still pale in comparison with the magnitude of projected impacts in the 21st century, if we stay on our current trajectories of greenhouse gas emissions. While national parks provide critical opportunities to document effects of climate change, negative impacts to iconic landscape features, such as glaciers, or species, such as redwoods, have the potential to undermine the core mission of the park system. Eventually this may change our appreciation of and relationship to these exemplary areas.

Climate change is the most recent manifestation of changes in atmospheric conditions where the causes of change lie outside protected areas but the impacts traverse the boundaries. Jill Baron and colleagues document the remarkable history and role of the NPS in tackling air pollution problems in the latter half of the 20th century. Spurred by the discovery of acid rain in the eastern United States, and deteriorating visibility in western US parks, such as the Grand Canyon, the NPS embarked on long-term watershed studies that uncovered a hitherto unknown and widespread problem of nitrogen deposition. Backed by legislative authority to address pollution sources that affected the air over federal lands, the NPS was a key player in the development of regional policy initiatives that have led to improved air quality across much of the country. This story, which is still unfolding in the regulation of NO_x emissions in the western United States, is a remarkable example of the successful integration of science and policy toward the protection of natural resources.

Addressing a biotic threat that also crosses park borders with impunity, Daniel Simberloff reviews the long and troubled history of alien and invasive species in US national parks, and the contributions of NPS scientists to better understand their impacts and enhance eradication efforts. Approaches to management of invasives reflect changing mores and conservation goals, and for many years nonnative fish were stocked in parks for recreational fishing. Parks have also served as important study sites for pathbreaking research on the impacts of invasives on ecosystem function, and this work has provided the scientific basis to strengthen the case for nonnative removal and exclusion. Unfortunately, facilitated by increased human traffic and disturbance, new invaders continue to arrive in parks and other natural areas. Nonnative control and eradication can be one of the largest expenses in resource management budgets, and 21st-century climate change will present ever greater challenges in this regard.

In the penultimate chapter of this section, we revisit contemporary challenges in the conservation of large mammals, the iconic species that inspire scientists and citizens alike and have often provided the greatest impetus for conservation action and investment. Drawing on case studies of the Florida panther, bison, and muskoxen, Joel Berger addresses the problems of fragmentation and population isolation, long-distance migration, and climate change. In each of these cases, protection in a park, even very large parks such as those of the Alaskan Arctic, is insufficient to sustain viable populations of these large mammals. Research on isolation and genetic variation, migration pathways, and causes of mortality for animals inside and outside parks continues to provide essential information for effective conservation. The protection of long-distance migration pathways, especially the spectacular Path of the Pronghorn, has set new precedents for successful conservation strategies that transcend individual protected areas and draw together managers, land owners, and other stakeholders across regional landscapes. The coming century will offer the last, best chances for similar efforts to maintain connectivity across the world's remaining wilderness areas. National parks, in the United States and across the world, will continue to play a vital leadership role in the future of conservation.

The section concludes with a strategic conversation on stewardship challenges from diverse perspectives: Josh Donlan (Advanced Conservation Strategies), Laurel Larsen (UC Berkeley), Stephanie Carlson (UC Berkeley), and Raymond Sauvajot (National Park Service). The panelists discuss several of the challenges for parks in a larger landscape context, examining species introductions in a historical context and managed gene flow, as well as the challenges of integrating research and restoration at landscape scales. As Sauvajot concludes, there's a "window of opportunity" to unite the two park missions, sharing with the many visitors to the parks the dramatic changes underway, the management challenges ahead, and the unique role of science for parks and parks for science.

FIVE

Climate Change and Novel Disturbance Regimes in National Park Landscapes

MONICA G. TURNER, DANIEL C. DONATO,
WINSLOW D. HANSEN, BRIAN J. HARVEY, WILLIAM H.
ROMME, AND A. LEROY WESTERLING

Introduction

National parks preserve unique elements of the American landscape and are highly valued components of our national heritage. These protected areas provide reference conditions along the continuum of land use from pristine to rural to urban, and their ecological value grows as surrounding landscapes become increasingly developed, fragmented, or degraded (Hansen et al. 2014). Large national parks such as Yellowstone anchor many of our last intact landscapes, and their scientific value for understanding the structure and function of natural ecosystems is unparalleled because management interventions are minimal. As drivers of global change alter ecosystems worldwide, national parks offer irreplaceable opportunities for scientists and resource managers to understand ecological responses to environmental change. Of particular importance is the need to understand consequences of changing climate and disturbance regimes (Turner 2010).

Disturbance is a key process in ecological systems, affecting terrestrial, aquatic, and marine ecosystems over a wide range of scales. Disturbances alter ecosystem states and trajectories, and they can shape ecosystem dynamics long into the future. Scientific understanding of natural disturbances and appropriate management of disturbance-prone landscapes evolved considerably during the 20th century. Ecologists had long upheld balance-of-nature concepts and believed that ecosystems could be maintained in desired but static states over the long term. Natural disturbances were not considered integral or desirable in many ecosystems. Reflecting the science of the time and that widely held equilibrium worldview, the 1963 Leopold Report, *Wildlife Management in the National Parks*, stated: "A national park should present a vignette of primitive America" (Leopold et al. 1963). Understanding of how natural disturbances structure ecosystems increased in subsequent decades (Pickett and White 1985), and ecologists recognized that few ecosystems were ever at equilibrium (Turner et al. 1993; Wu and Loucks 1995). Conventional wisdom about steady-state conditions also was challenged by occurrences of large, severe natural disturbances that captured public attention (Turner, Dale, and Everham 1997). The 1992 Risser report, *Science and the National Parks*, recognized these advances in scientific understanding when it stated: "Ecological science now recognizes that change is central to the structure and functioning of all ecosystems, and it is now evident that the managers of the parks must understand the changes—both natural and anthropogenic—that occur. To conserve ecosystems unchanged is simply impossible" (National Research Council 1992). By the end of the 20th century, disturbance was recognized as ecologically important, and maintaining dynamic ecosystems within their historical range of variability was widely embraced as a management goal (Keane et al. 2009). However, baselines are once again shifting in science and management as global changes accelerate. The magnitude and rate of climate warming make it more difficult to project the future based on past knowledge, and effects on national parks and other protected areas are highly uncertain. What does this imply for national parks? How much will they change? Will future dynamics exceed historical ranges of variation? The 2012 report of the National Park System Advisory Board Science Committee, *Revisiting Leopold: Resource Stewardship in the National Parks*, now states: "National Park Service . . . should . . . steward resources for continuous change that is not yet fully understood" (Colwell et al. 2012).

Climate and disturbance regimes are both changing rapidly, and it is increasingly important for ecologists and park managers to understand the past and anticipate what lies ahead. The frequency, severity, and extent of natural disturbances are changing substantially as climate warms; effects on many ecosystems may be profound (Westerling et al. 2006; Seidl, Schelhaas, and Lexer 2011; Parks, Parisien, and Miller 2012; Weed, Ayres, and Hicke 2013; Moritz et al. 2014). In the Northern Rocky Mountains, a region with several national parks, fire and insect outbreaks are key drivers of landscape pattern and ecosystem function. Climate-driven changes in these disturbances will affect most western national parks; indeed, changes may already be underway. Long-term studies in Greater Yellowstone have documented tremendous ecological resilience to these natural disturbances over centuries to millennia, but projected climate change may lead to novel disturbance regimes and unforeseen ecological responses. Understanding the how, when, where, and why of these dynamics is urgent for park management and conservation.

Drawing primarily from our research in Greater Yellowstone and the Northern Rocky Mountains, we highlight the critical role of national parks as living laboratories for scientific research during these times of rapid change, as well as the importance of science for park management. We provide an overview of Greater Yellowstone and its dominant natural disturbances, summarize general lessons that emerged from long-term basic scientific studies, and then consider how future change in climate and disturbance dynamics may affect the landscape. We conclude by advocating for an even stronger commitment to the value of parks for science.

Natural Disturbances in Greater Yellowstone

The 80,000 km² Greater Yellowstone Ecosystem is centered on Yellowstone National Park and straddles portions of Wyoming, Montana, and Idaho (fig. 5.1). It includes Grand Teton National Park, seven national forests, the National Elk Refuge, and parts of the Wind River Indian Reservation. Greater Yellowstone is unique in some respects—notably the extensive geothermal features and abundant wildlife for which the region is famous—but it is also representative of temperate mountain ecosystems throughout western North America. Therefore, lessons from Yellowstone are relevant for other regions that are less well studied. Yellowstone National Park encompasses ~9,000 km², most of which lies on a high-elevation (~2,100–2,700 m) volcanic plateau with relatively gentle topography. Surrounding the plateau are higher, rugged mountains of various crystalline, sedimentary, and volcanic substrates, as well as broad river valleys and basins characterized by a semiarid climate. Approximately 80% of Yellowstone National Park is dominated by lodgepole pine (*Pinus contorta* var. *latifolia*) forest, although subalpine fir (*Abies lasiocarpa*), Engelmann spruce (*Picea engelmannii*), and whitebark pine (*Pinus albicaulis*) can be locally abundant at high elevations. At lower elevations, Douglas-fir (*Pseudotsuga menziesii*) and aspen (*Populus tremuloides*) forests grade into sagebrush (*Artemisia* spp.) steppe and grasslands. The climate is characterized by cold, snowy winters and dry, mild summers. Some ungulate populations were controlled in the past, and wolves were extirpated and subsequently reintroduced. Nonetheless, in contrast to much of the Rocky Mountain region, the pre-Columbian flora and fauna of Greater Yellowstone remain largely intact, in part because it is one of the largest tracts of wild, undeveloped land in the continental United States (Gude et al. 2006). This largely

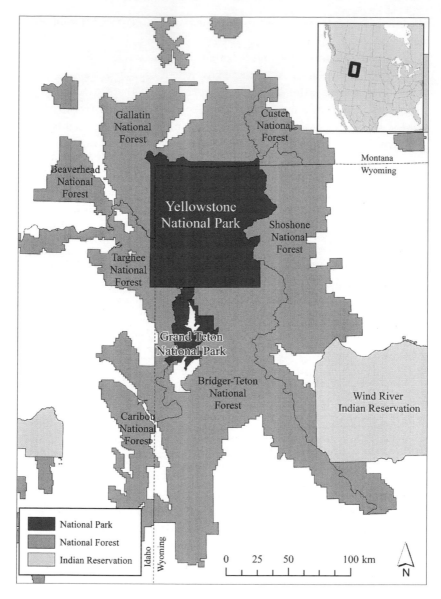

5.1. Map of Greater Yellowstone

pristine condition makes Yellowstone invaluable for research into natural patterns and processes at multiple spatial and temporal scales.

Fire

The role of fire has been recognized in Yellowstone for a long time. The early explorers of the Yellowstone region even mentioned it—in his diary of the 1870 Washburn Expedition, Nathaniel Pitt Langford (who later became the first superintendent of Yellowstone National Park) wrote: "Tuesday, September 20—We broke camp at half past 9 o'clock, traveling along the rocky edge of the [Firehole] river bank by the rapids, passing thence through a beautiful pine wood and over a long stretch of fallen timber, blackened by fire, for about four miles" (reprinted by Miller 2009). Based on their route that day, the expedition likely traversed a large fire that occurred circa 1862, the date of origin for lodgepole pine forests along the east side of the Firehole River. In addition, numerous entries in the Washburn Expedition diary report exceedingly slow and difficult travel through areas with abundant downfall—much of which was likely legacy wood from past fires. For example, Langford described pine forests they navigated along the eastern shores of Yellowstone Lake a couple of weeks earlier:

> Tuesday, September 8—Our journey for the entire day has been most trying. . . . The difficulty of . . . making choice of routes, extricating the horses when wedged between the trees, and readjusting the packs so that they would not project beyond the sides of the horses, required constant patience and untiring toil.

> Wednesday, September 9— . . . through fallen timber almost impassable in the estimation of pilgrims. . . . Frequently, we were obliged to rearrange the packs and narrow them, so as to admit of their passage between the standing trees. (reprinted by Miller 2009)

Based again on their route, the expedition was probably slogging through dense lodgepole pine regeneration and fallen, fire-killed trees where the forest had burned circa 1840. (The even-aged pines were about 160 years old in 1999, and Langford's description well matches our recent attempts to traverse impenetrably dense 25-year-old postfire lodgepole pine forests.)

Fire-history studies based on extensive tree-ring analyses found that large stand-replacing fires had burned in Yellowstone during the 18th and 19th centuries (Romme 1982; Romme and Knight 1982; Romme and

Despain 1989). This work also revealed a dynamic landscape mosaic of stand ages in response to infrequent, high-severity fire. Romme's research had been designed to address fundamental questions in ecology about disturbances and equilibrium, and such studies could only be addressed in large wildland landscapes like Yellowstone. His results were of great interest to forest landscape ecologists because his studies quantified spatial-temporal dynamics over a large landscape and documented a non-steady-state system. However, this basic scientific understanding also proved essential for park managers when the hot, dry summer of 1988 produced large wildfires throughout Greater Yellowstone. The science was crucial for placing those fires in context and recognizing that they were consistent with the historical disturbance regime.

The 1988 Yellowstone fires were among the first in what has proven to be an upsurge in large severe fires in the western United States during the past 20 years. The fires burned under extreme drought and high winds, and ultimately they affected ~600,000 ha in Greater Yellowstone. Compared with previous 20th-century fires, their size and severity were a surprise to scientists and managers, and ecological effects of the fires were highly uncertain. Little was known at that time about the impacts of such a large severe disturbance because scientists had had few previous opportunities to study such an event. Soon after the fires, ecologists generated testable predictions regarding short- and long-term effects on vegetation, wildlife, aquatic ecosystems, biogeochemistry, and primary productivity based on scientific understanding of the time (see the November 1989 special issue of *BioScience*). Many studies were initiated to evaluate these ideas, and results of this body of research were synthesized at postfire milestones of 10 years (Turner, Romme, and Tinker 2003; Wallace 2004) and 20 years (Schoennagel, Smithwick, and Turner 2008; Turner 2010; Romme et al. 2011). The new understanding gained from those studies has proven extremely valuable and relevant to fire policy throughout the western United States (Weeks 2012; Stephens et al. 2013). The 1988 fires created novel opportunities to study postfire succession and ecosystem processes in a wilderness setting. In particular, they offered a natural landscape-level experiment in which ecological effects of spatial patterns could be tested (fig. 5.2a). Results established benchmarks for early postfire dynamics in western conifer forests, and Turner's and Romme's studies provided compelling examples of the ecological role of landscape pattern (e.g., Turner et al. 1997). After more than 25 years, ongoing studies of the young postfire forests continue to add new knowledge and insights. Young forests are increasing in extent throughout the western United States in response to

5.2. Disturbance-created heterogeneity in Greater Yellowstone. *A*, the 1988 fires created a mosaic of patches that vary in size, shape, and severity across the landscape. Photo by M. G. Turner, October 1988. *B*, bark beetle outbreaks create a fine-grained mosaic of tree mortality, as shown here for spruce beetle outbreaks in Engelmann spruce. Photo by M. G. Turner, June 2006.

greater fire activity, and understanding their dynamics is essential for good stewardship of these rapidly changing landscapes.

Bark Beetle Outbreaks

Outbreaks of native species of bark beetle (Dendroctonae) have also been part of Greater Yellowstone for a long time. Native bark beetles of the genus *Dendroctonus* undergo episodic population outbreaks that result in widespread mortality of host trees through pheromone-mediated mass attacks (Wallin and Raffa 2004; Raffa et al. 2008). From about 2003 to 2012, Greater Yellowstone experienced widespread outbreaks of bark beetles, including the mountain pine beetle (*Dendroctonus ponderosae*) in lodgepole and whitebark pine, spruce beetle (*Dendroctonus rufipennis*) in Engelmann spruce, and Douglas-fir beetle (*Dendroctonus pseudotsugae*) in Douglas-fir. The recent outbreak was mostly in the eastern and northern parts of Greater Yellowstone and involved multiple tree and beetle species (Simard et al. 2012), whereas an earlier outbreak in the 1970s and 1980s affected the western and southern portions of Greater Yellowstone and involved mostly lodgepole pine and the mountain pine beetle (Furniss and Renkin 2003; Lynch et al. 2006). Across the western United States, recent outbreaks appear to be more extensive, more homogeneous, and more severe in their effects on stand and landscape structure compared with previous outbreaks (Raffa et al. 2008; Meddens, Hicke, and A. Ferguson 2012). It was widely believed that tree mortality resulting from beetle outbreaks would increase the likelihood of severe fires, and likewise that trees injured by fire would be more susceptible to beetle attack. Empirical evidence for this conventional wisdom was lacking, and testing it required extensive intact forests in which both disturbances occurred in the absence of intensive forest management.

Greater Yellowstone again provided an opportunity for basic landscape-level research on these potential disturbance interactions (fig. 5.2b). Empirical studies documented changes in stand structure and ecosystem process rates and revealed substantial capacity of the forests to withstand beetle outbreaks (Simard et al. 2011; Griffin, Turner, and Simard 2011; Donato, Harvey, et al. 2013). Modeling studies suggested that the likelihood of severe fire might not be worsened by beetle outbreaks (Simard et al. 2011), and subsequent empirical (Harvey et al. 2013, 2014) and modeling studies (Donato, Simard, et al. 2013) in Greater Yellowstone supported this notion. Research in national park and wilderness areas also provided a baseline for evaluating effects of postdisturbance management (e.g., Griffin, Simard, and Turner 2013; Donato, Simard, et al. 2013). Like the fire studies, these studies in Greater Yellowstone are providing valuable insights about disturbance in western forests (Harvey, Donato, and Turner 2014) and informing regional land management (Wells 2012; Carswell 2014).

Lessons from Yellowstone about Natural Disturbances

Given the wealth of disturbance studies in Greater Yellowstone, what general lessons have been learned that apply to other places, to other national parks, and to the expansive forests of the western United States? Here, we summarize six general scientific lessons that have emerged from our long-term studies in Yellowstone.

1. *Large, infrequent, severe fires are "business as usual" in subalpine forest landscapes.* Although the 1988 fires were large and severe, we have learned that such fires are not unusual in Greater Yellowstone. There is *no evidence* that the size or severity of the 1988 fires resulted from human activities, such as fire suppression. Large, stand-replacing fires have occurred during warm, dry periods in the historical past (Romme and Despain 1989) and during past millennia (Meyer and Pierce 2003; Millspaugh, Whitlock, and Bartlein 2004; Whitlock et al. 2008; Higuera, Whitlock, and Gage 2011), and the biota are well adapted to these events. Fire return interval varies with elevation, averaging about 170 years at sites less than 2,300 m above sea level and about 290 years at sites more than 2,300 m (Schoennagel, Turner, and Romme 2003). Many subalpine and boreal forests have similar infrequent, high-severity fire regimes (Turner and Romme 1994). Thus, it is not so surprising after all that the region's forests have regenerated rapidly following recent large fires.

2. *Natural disturbances are important sources of landscape heterogeneity.* In contrast to claims made by some observers of the 1988 fires and recent beetle outbreaks, even large, high-severity disturbances are spatially heterogeneous. The 1988 fires created a complex (and, to many observers, even beautiful) mosaic of burned and unburned patches across the landscape (Turner et al. 1994), and patterns created by natural fires differed markedly from patterns of forest harvesting in Greater Yellowstone (Tinker, Romme, and Despain 2003). New vistas were revealed, wildflowers bloomed prolifically, and openings in the forest offered new resource patches to be used for wildlife. The bark beetle outbreaks created a very fine-grained mosaic, because tree mortality is not complete within stands. For example, outbreak severity (percentage of basal area killed by beetles) ranged from 36% to 82% in lodgepole pine stands sampled in 1981 and 2007, during each of

the two most recent outbreaks in Greater Yellowstone (Simard et al. 2012), and from 38% to 83% in Douglas-fir stands attacked between 1980 and 2010 (Donato, Harvey, et al. 2013). Disturbance-created heterogeneity (see fig. 5.2) is functionally important, establishing patterns of stand and landscape structure that sustain ecosystem processes for decades to centuries.

3. *Beetle outbreaks kill trees but do not destroy forests.* Bark beetles attack large trees, and conspicuous red crowns of beetle-killed trees can make it appear as if the entire forest is dying. However, this is not the case. Even in very severe outbreaks (e.g., when more than 90% of tree basal area is killed by beetles), postoutbreak forests usually contain many more live than dead trees. Trees underneath the canopy are often too small to be killed by beetles, and these trees experience accelerated growth rates postoutbreak. In addition, mature nonhost trees often escape an outbreak unscathed (Simard et al. 2011; Donato, Harvey, et al. 2013). Rapid growth of surviving trees, coupled with slow decay of beetle-killed trees, results in recovery of preoutbreak biomass carbon storage within a few decades postoutbreak (Donato, Simard, et al. 2013). Wildflowers and grasses also respond rapidly when mature trees die, taking advantage of newly available resources (e.g., nutrients, water, space) and effectively conserving nutrients in disturbed stands (Griffin, Turner, and Simard 2011). These outbreak-induced changes may also benefit forest wildlife. High-quality forage provided by nutrient-rich herbaceous plants, coupled with increased habitat structure complexity from snags and falling beetle-killed trees, attracts elk, deer, moose, and birds across many guilds (Saab et al. 2014). In short, the death and decadence following beetle outbreaks is counteracted by rapid stimulation of life and growth.

4. *Climate is an important driver of fire and bark beetle outbreaks.* Studies continue to demonstrate that climate—particularly warm, dry conditions—is the key driver of large, stand-replacing fires as well as bark beetle outbreaks (Westerling et al. 2006; Raffa et al. 2008; Bentz et al. 2010; Krause and Whitlock 2013). In western conifer forests, it is the extremely warm, dry, and windy summers that are responsible for most of the area burned (Westerling et al. 2006). Historically, most summers were too moist and cool to support large fires, even though fuels were abundant, and fires were not enormous in moderately dry years (Turner and Romme 1994). Warm, dry conditions also foster bark beetle outbreaks because drought-stressed trees are more vulnerable to beetle attack (Raffa et al. 2008). Of course, climate also interacts with other variables, such as topography, past disturbance history, and antecedent forest structure, to determine the size and severity of a given disturbance event. However, climate is often the strongest

influence among candidate variables, especially in mid- to high-elevation western conifers (Westerling et al. 2006; Harvey et al. 2014). Because of its importance, changes in climate are likely to alter fire and bark beetle outbreak dynamics in western landscapes.

5. *Beetle outbreaks do not cause or worsen fire impacts, and fires do not cause or worsen beetle outbreaks.* Bark beetle outbreaks and fires both occur under warm, dry conditions that stress trees, weakening their defenses against insects, and increase flammability, raising the likelihood of fire occurrence. Warm temperatures during winter also increase overwinter survival of bark beetles and can sustain an outbreak from one year to the next. However, both disturbances are responding to a similar driver, rather than directly affecting one another. Beetle outbreaks do alter the fuel structure of forests at the stand scale (Simard et al. 2011; Donato, Harvey, et al. 2013) and may affect the way fire behaves (Jenkins et al. 2012). However, contrary to expectations, when wildfires burn through beetle-affected stands, most measures of fire severity (effects on the ecosystem) are unrelated to outbreak severity and are largely similar to those in unaffected stands (Harvey et al. 2013, 2014; Harvey, Donato, and Turner 2014). Instead, fire severity in beetle affected landscapes is driven by two of the main factors affecting any wildfire: weather and topography. Further, postfire tree regeneration is generally robust in previously beetle-affected landscapes as long as seed sources remain (i.e., surviving trees or viable cones)—as in any wildfire. Postfire regeneration was robust in beetle-killed lodgepole pine forests because serotinous cones were still present (Harvey et al. 2014), but poor in beetle-killed Douglas-fir forests because seed sources were absent (Harvey et al. 2013). Regarding the converse interaction in which fires are expected to cause beetle outbreaks in surrounding forests, recent research in Greater Yellowstone demonstrates that fire-injured trees provide local refugia for beetle populations but generally do not generate extensive outbreaks in healthy trees because reproductive success is low (Powell, Townsend, and Raffa 2012).

6. *Forests of Greater Yellowstone have been remarkably resilient to natural disturbances.* Collectively, our long-term studies of natural disturbances in Greater Yellowstone have documented tremendous ecological resilience. Paleoecological studies have also demonstrated long-term resilience in disturbance and vegetation dynamics over the past 10,000 years (Whitlock and Bartlein 1993; Whitlock, Shafer, and Marlon 2003; Millspaugh, Whitlock, and Bartlein 2004; Higuera, Whitlock, and Gage 2011). Natural disturbance has not been an ecological catastrophe. Disturbances structure this landscape; between 1984 and 2010, most of Yellowstone has been influenced by disturbance (fig. 5.3). These disturbances produce a dynamic

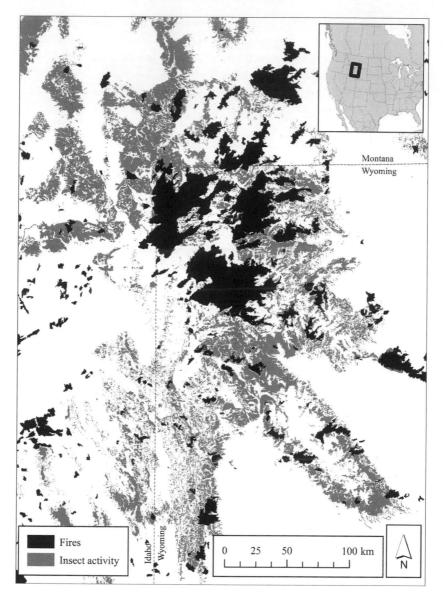

5.3. Greater Yellowstone is strongly influenced by natural disturbances, as shown by areas affected by fire and insect outbreaks between 1984 and 2010. Map generated by B. J. Harvey from USDA aerial detection survey data (http://www.fs.fed.us/foresthealth/) and Monitoring Trends in Burn Severity (MTBS) data (Eidenshink et al. 2007; www.mtbs.gov).

mosaic in which forest ages and tree densities vary substantially across the landscape and through time. Ecosystem recovery from natural disturbances has not required any management intervention (Romme and Turner 2004). However, accelerating rates of environmental change pose new challenges.

Future Climate Change and Disturbance in Yellowstone

Earth's climate is warming, and this warming can only be explained by accounting for human-caused emissions of greenhouse gases, especially carbon dioxide. Warming will continue throughout the 21st century, even if greenhouse gas emissions are reduced. The rate and magnitude of projected climate change heighten the urgency for scientists to anticipate and managers to prepare for changes in national parks. In Yellowstone, forests have been resilient to past changes in climate and disturbance regimes, as forests have regenerated well following past disturbances. Will resilience be guaranteed if the magnitude of future changes exceeds variability during the past 10,000 years? Projected climate changes could lead to novel disturbance regimes and unforeseen ecological responses. Answers to questions of resilience will depend on the variables used to assess change in the system and the scales at which resilience is measured. An environmental change that leads to a state transition, such as from forest to nonforest, would indicate a lack of *forest* resilience at particular locations. However, the *ecosystem* might be considered resilient if other native species expand in place of trees, and ecosystem functions are maintained (e.g., carbon sequestration, nutrient cycling, and provision of wildlife habitat). Furthermore, habitats (such as forests) could retreat from some places but expand at others while maintaining their extent at a regional scale. Thus, resilience is a multifaceted concept.

As climate warms, park managers will likely consider whether to let changes occur as they will or to intervene to try to redirect or slow rates of change (Marris 2011). *We assert that parks and protected areas are not the place for management to redirect or alter ecosystem responses to climate change.* Such activities can be implemented in many other landscapes and may be desirable in more intensively used areas. However, national parks and protected areas provide critical reference conditions for understanding how ecosystems respond to rapid change, and knowledge gained will ultimately inform what is done in other places. To maintain the capacity for ecosystems to adapt to environmental change, park managers could focus on minimizing other threats that would limit the ability of native species to respond.

For example, managers might intensify efforts to control aggressive non-native invaders such as cheatgrass (*Bromus tectorum*), which we have observed at low-elevation, dry topographic positions following recent fires in Greater Yellowstone. Affording native species the opportunities to disperse and shift ranges will be critical for ecosystems to adapt to climate change. Providing for connectivity of natural areas over large landscapes is essential, and securing regional connectivity will require cooperation among multiple land managers.

As for future climate and disturbance, what is expected for Yellowstone? Temperatures in the Northern Rocky Mountains have warmed over the past few decades, especially at middle elevations (Westerling et al. 2006; Shuman 2012). This warming is associated with earlier timing of spring snowmelt (Pederson et al. 2011), warmer summer conditions, and a longer growing season and fire season. Climate models predict continued warming, with average spring and summer temperatures increasing 4°C–6°C by the end of the 21st century (Westerling et al. 2011). The pace of current warming is much faster than the warming at the end of the Pleistocene and happening in a world affected by other human impacts, such as habitat fragmentation. Future precipitation remains uncertain, but recent trends in observed climate indicate an overriding effect of temperature that exacerbates drought during the growing (and fire) season. A warmer, drier future for Greater Yellowstone appears most likely for the coming decades. Summers as hot and dry as 1988 are likely to occur with increasing frequency throughout the 21st century, and to become the norm by the latter part of the century (Westerling et al. 2011).

Implications of climate warming for natural disturbance regimes are substantial. The frequency, extent, and severity of fires in the western United States have already increased with warming (Westerling et al. 2006; Weed, Ayres, and Hicke 2013), and landscapes are changing rapidly as mature conifer forests are increasingly reset by severe fire to early successional stages (Johnstone, Chapin, et al. 2010; Johnstone, Hollingsworth, et al. 2010; O'Connor et al. 2014). In the Northern Rocky Mountains, novel fire regimes that are well outside even paleoecological ranges of variability are predicted during the 21st century (Westerling et al. 2011; Liu, Goodrick, and Stanturf 2013). Peterson and Littell (2014) projected a more than 600% increase in median area burned in Greater Yellowstone and the Southern Rocky Mountain region with only a 1°C rise in temperature. Westerling et al. (2011) projected an even greater increase in burning. Summers conducive to widespread burning, like 1988, would become common, and years without any large fires, which were frequent historically, would become rare. Consequences of such changes for forest landscapes may be profound.

Greater Yellowstone continues to offer an unparalleled opportunity to understand how intact ecosystems and landscapes respond to changing climate and disturbance regimes. In such large heterogeneous landscapes, scientists can measure responses of the biota to changing conditions, evaluate mechanisms of resilience that may apply broadly across ecosystems, and potentially identify early indicators of ecosystem change. The need for creative, long-term measurement programs that are sensitive to anticipated changes in climate and disturbance regimes is more important now than ever before. We suggest two priorities.

First, the importance of long-term study in Greater Yellowstone and other national parks cannot be overemphasized. Long-term study of the ecological consequences of the 1988 Yellowstone fires has already produced a tremendous amount of new knowledge (Turner 2010; Romme et al. 2011), and these data now provide the benchmarks against which the consequences of future fires can be compared. The 1988 fires and ecological responses to those fires represent the historical fire regime that characterized this region throughout most of the Holocene. The fires burned mostly in mature and old-growth forests, also typical of previous large fires in Yellowstone. Postfire trajectories after mid-21st-century fires may differ significantly from those measured following the 1988 fires, and it will be important to document these future postfire dynamics, as well as to continue following long-term development of the post-1988 forests. Future fires will likely burn in younger stands, and postfire recovery will occur under substantially warmer and possibly drier conditions. Comparing future fires and fire effects with what we saw after 1988 will allow the magnitude of departure from the historical fire regime to be measured.

Second, there is a critical need to understand mechanisms and identify early warning signs of major qualitative changes in the landscape. For instance, forests could be converted to shrublands or grasslands after fire if fire intervals become so short that trees cannot reach reproductive age before the next fire occurs, or the climate becomes unsuitable for survival of postfire tree seedlings. What conditions lead to loss of forest resilience, and the nature and rates of species responses to changing tree distributions, are not known; indeed, long-term studies in protected areas may provide the basis for new understanding of what constitutes ecosystem resilience. Large national parks and protected areas are ideal places for studying such patterns because they capture a wide range of disturbances, environmental conditions, and genetic diversity, and landscape management interventions are minimal. The value of national parks for such studies is exemplified by a recent study of tree regeneration following recent fires in Yellowstone, Grand Teton, and Glacier National Parks (Harvey, Donato, and Turner 2016). Sampling was conducted in 184 plots that burned as stand-replacing fire, and data shows that subsequent years of drought substantially reduced postfire tree establishment. Detecting such signals of gradual environmental change cannot be readily done in managed ecosystems. In another example, direct effects of climate on postfire tree establishment are being addressed experimentally in Yellowstone (W. D. Hansen et al., unpublished data). Seed-germination experiments that compare current and projected midcentury climate will identify temperature and moisture conditions that allow tree seedlings to establish in recently burned forest soils. Initial results suggest that warmer climatic conditions at lower treeline may be dangerously close to conditions that preclude successful lodgepole pine establishment. Observational and experimental field studies are also providing the basis for modeling the longer-term implications of alternative mechanisms and rates of change across larger landscapes. Detecting change is but a first step; understanding how ecosystems respond, and which ecological patterns and processes are resilient to future perturbations, is critical, and national parks offer irreplaceable opportunities for such study.

As climate and disturbance regimes change, Yellowstone will become increasingly valuable for its critical role in allowing processes and changes to play out with minimal intervention, providing a benchmark for understanding how natural systems will change and adapt. Forests of Greater Yellowstone may be less resilient to future fires than they were to the massive fires of 1988. However, Yellowstone will continue to evolve as environmental conditions change, just as it did at the end of the Pleistocene and throughout the Holocene. It will not be "destroyed" in the future, only changed. Native plants and animals will still be present, even though relative abundances may change and some new species may arrive. Moreover, because so much of the western landscape has been altered by human land use, Greater Yellowstone, with its large area of contiguous and diverse natural habitats, will be crucial for sustaining a wide variety of species that cannot persist elsewhere. Yellowstone is a dynamic, vital, intact ecosystem that holds many secrets yet to be revealed.

Parks for Science, and Science for Parks

Climate warming and changing disturbance regimes are inevitable; changes are coming fast, and many are already underway. Ecological effects of climate change are likely to be much more substantial and far-reaching than we realized even just a decade ago. The past may not predict the future— we may well be heading beyond the range of climatic and ecological conditions that have characterized the last 10,000 years and moving quickly into uncharted territory. Scientists and managers must be alert to potential tipping points and thresholds beyond which major qualitative changes will take place. During these times of rapid change, the importance of national parks as living laboratories for scientific research only increases.

Parks for Science

We strongly advocate for a renewed and strengthened commitment to "parks for science." As we have shown for Greater Yellowstone, national parks represent some of the best places for research designed to understand causes and consequences of environmental change independent of management effects. Because they contain ecosystems shaped primarily by natural processes, national parks can be sensitive sentinels of change. For example, climate-driven changes in range distributions of species may be detectable sooner in national parks than in highly developed landscapes. Many large national parks include high-elevation and high-latitude systems that have already been identified as extremely vulnerable to effects of global climate change. Changes in the biota and in ecosystem processes and services must be understood in the absence of the myriad other factors that confound attribution of cause and effect in human-dominated landscapes. National parks serve as natural laboratories for studying effects of environmental change in areas not confounded by management or direct human impacts. In essence, national parks provide the reference conditions against which the effects of manipulating nature elsewhere can be assessed.

Research on how national parks sustain ecological processes, ecosystem services, and integrity of the larger landscape is also of high priority. National parks are often key to maintaining benefits from nature that are valued well beyond the park boundaries. With ongoing climate change, national parks will be increasingly important for sustaining the regional biota (e.g., migratory animal populations, vegetation communities, and genetic diversity). Parks are often of significance for delivering clean water to downstream aquatic systems and sustaining hydrologic ecosystem services valued by human communities. Parks may serve as refugia for aquatic populations and as source populations for degraded biota found in downstream aquatic ecosystems. The need to understand how changing landscape mosaics will influence future delivery of ecosystem services is now widely recognized (e.g., Turner, Donato, and Romme 2013). National parks provide an array of benefits and values to people—even to people who never visit—including benefits to local economies, land values, ecosystem services, and existence value. Understanding how these benefits of nature may change in the future is important for park management.

Resource interpretation programs implemented by the US National Park Service (NPS) could emphasize the importance of parks for science, taking advantage of the unique opportunities to educate visitors. Ecological literacy and scientific understanding is arguably at a low point in the United States (Mooney and Kirshenbaum 2010), and opportunities for the public to understand the role of science are desperately needed. Rather than having all evidence of scientific study hidden from visitors, research in the parks could be publicized with pride, with an emphasis on how much can be learned from these intact landscapes. Research-related interpretation could accomplish two goals. First, it would create opportunities for the public to be exposed to the process of science, for them to witness the human side, the creativity, trial and error, and innovation that go into research. By humanizing science, we may be able to foster greater understanding and appreciation for science among the general public. Second, showcasing science in the parks could engage the public in discussion of regional conservation and resource issues. Research could be a conversation starter that leads visitors to a deeper understanding of the park and its surroundings.

Greater opportunities for comparative study across national parks in the United States and worldwide could also be explored. The United States is recognized throughout the world for leadership in establishing and protecting national parks, and scientific studies that compare and contrast ecological responses to global change in different protected areas could add even greater value to research in any one park. For example, the European Union recently funded a study of 22 national parks and protected areas spanning a range of biogeographic regions in Europe and beyond. What about partnering with international protected areas to develop an even more comprehensive understanding of our changing planet? The National Park System includes 30 parks recognized as UNESCO Biosphere Reserves and 10 designated as World Heritage Sites (National Research Council 1992). Such networks offer opportunities yet to be explored.

Science for Parks

Of course, science also should continue to address issues related directly to park management, and NPS decision makers should seek and use the best available science. Managers must rely on science for guidance in understanding novel conditions and risks to parks, now and in the future (Colwell et al. 2012). For climate change and disturbance, this will require computer-based modeling to explore potential future scenarios along with observational studies that may detect early indicators of ecological change in particular national parks. Science should inform the stewardship and management of national parks. For example, management-relevant questions might include the following: Should parks be actively managed in response to changing climate, or should a hands-off policy be continued? How should nearby lands be managed so that the integrity and character of our national parks do not degrade as environmental conditions change?

Externally funded research conducted in national parks will help strengthen the foundation of park management and complement management-oriented research. For example, externally funded and peer-reviewed science on fire history in Yellowstone provided critical information needed by park managers in 1988. Externally funded and peer-reviewed research also provided crucial data about drivers and dynamics of Yellowstone's northern range when Congress mandated an independent review of ungulate management (National Research Council 2002). These and many other examples show that research helps rather than hinders park management. However, the management relevance of curiosity-driven science may not be immediately obvious, and national parks can sometimes seem unsupportive of science, considering it to conflict with the NPS mission. It is notable that even the Wilderness Act explicitly recognizes science as an appropriate purpose for and use of wilderness. National parks are often managed as wilderness, and it can be difficult to have research approved. We strongly support the imperative to respect the resource and mission of the NPS, but processes for conducting scientific research in national parks have become increasingly bureaucratic in recent decades—just when the urgency to understand causes and consequences of regional change is growing. Appropriate experimentation and research installations should be encouraged, rather than considered a nuisance. National parks will benefit by actively embracing scientific research, recognizing that today's basic science may well provide the foundation for tomorrow's policy decisions.

Conclusions

Accelerating rates of environmental change will affect our national parks during coming decades. As we have shown for Greater Yellowstone, biotic communities are often well adapted to particular disturbances occurring in a given climate space. As we look to the future, the potential for interacting, novel disturbance regimes and climates to fundamentally change ecosystem structure and function is real. Consequences of such changes are important but difficult to anticipate, and how climate-disturbance interactions will affect regional landscapes and our national parks is only beginning to be explored. National parks play a critical role as living laboratories for scientific research, and science is crucial for park management during these times of rapid change. We wholeheartedly endorse the following statement from the 2012 *Revisiting Leopold* report: "The need for science—to understand how park ecosystems function, monitor impacts of change (even from afar), inform decision makers and their decisions, and enrich public appreciation of park values—has never been greater. In addition, the National Park System is an extraordinary national asset for advancing science and scholarship—from new discoveries of valuable genetic resources to monitoring benchmarks for environmental change" (Colwell et al. 2012).

Our research on disturbance and changing climates in Greater Yellowstone has led to general lessons about natural disturbances and demonstrated the importance of science as climate and disturbance regimes change in the 21st century. Many national parks offer comparable opportunities to understand effects of changing climate and disturbance regimes in natural ecosystems. For example, other large western national parks, including Yosemite, Rocky Mountain, and Glacier, can yield additional insights into effects of fire, insect outbreaks, and climate. Coastal US national parks, such as Virgin Islands, Everglades, Cumberland Island, Cape Lookout, and Acadia, offer opportunities to study hurricanes and climate. In desert parks, such as Canyonlands and Joshua Tree, we can learn the limits of resilience to changing temperature and precipitation regimes in drought-tolerant organisms and ecosystems. National parks remain among the best places for scientists to understand ecological responses to environmental change in the absence of factors that confound attribution of cause and effect. Studying these majestic landscapes is an honor, a privilege, and a responsibility. We hope that our research and that of the many other scientists studying national parks will aid stewardship of these national treasures in the years ahead.

Acknowledgments

We thank Steve Beissinger and David Ackerly for comments on this manuscript. Support for research underpinning this manuscript is gratefully acknowledged from the following sources: the National Science Foundation (BSR-9016281, BSR-9018381, and DEB-9806440), the National Geographic Society, the Andrew W. Mellon Foundation, the Joint Fire Science Program (grant numbers 06-2-1-20, 09-1-06-3, 09-3-01-47, 11-1-1-7, and 12-3-01-3), a National Park Service George Melendez Wright Climate Change Fellowship to BJH, and a National Science Foundation Graduate Research Fellowship to WDH (DGE-1242789).

Literature Cited

Bentz, B. J., J. Régniére, C. J. Fettig, E. M. Hansen, J. L. Hayes, J. A. Hicke, R. G. Kelsey, J. F. Negron, and J. F. Seybold. 2010. Climate change and bark beetles of the western United States and Canada: direct and indirect effects. BioScience 60:602–613.

Carswell, C. 2014. Don't blame the beetles. Science 346:154–156.

Colwell, R., S. Avery, J. Berger, G. E. Davis, H. Hamilton, T. Lovejoy, S. Malcolm, et al. 2012. Revisiting Leopold: resource stewardship in the national parks. A report of the National Park System Advisory Board Science Committee. http://www.nps.gov/calltoaction/PDF/LeopoldReport_2012.pdf.

Donato, D. C., B. J. Harvey, M. Simard, W. H. Romme, and M. G. Turner. 2013. Bark beetle effects on fuel profiles across a range of stand structures in Douglas-fir forests of Greater Yellowstone, USA. Ecological Applications 23:3–20.

Donato, D. C., M. Simard, W. H. Romme, B. J. Harvey, and M. G. Turner. 2013. Evaluating post-outbreak management effects on future fuel profiles and stand structure in bark beetle-impacts forests of Greater Yellowstone. Forest Ecology and Management 303:160–174.

Eidenshink, J., B. Schwind, K. Brewer, Z. L. Zhu, B. Quayle, and S. Howard. 2007. A project for monitoring trends in burn severity. Fire Ecology 3:3–21.

Furniss, M. M., and R. Renkin. 2003. Forest entomology in Yellowstone National Park, 1923–1957: a time of discovery and learning to let live. American Entomologist 49:198–209.

Griffin, J. M., M. Simard, and M. G. Turner. 2013. Salvage harvest effects on advance tree regeneration, soil nitrogen, and fuels following mountain pine beetle outbreak in lodgepole pine. Forest Ecology and Management 291:228–239.

Griffin, J. M., M. G. Turner, and M. Simard. 2011. Nitrogen cycling following mountain pine beetle disturbance in lodgepole pine forests of Greater Yellowstone. Forest Ecology and Management 261:1077–1089.

Gude, P. H., A. J. Hansen, R. Rasker, and B. Maxwell. 2006. Rates and drivers of rural residential development in the Greater Yellowstone. Landscape and Urban Planning 77:131–151.

Hansen, A. J., N. Piekielek, C. Davis, J. Haas, D. M. Theobald, J. E. Gross, W. B. Monahan, T. Olliff, and S. W. Running. 2014. Exposure of US national parks to land use and climate change 1900–2100. Ecological Applications 24:484–502.

Harvey, B. J., D. C. Donato, W. H. Romme, and M. G. Turner. 2013. Influence of recent bark beetle outbreak on fire severity and postfire tree regeneration in montane Douglas-fir forests. Ecology 94:2475–2486.

———. 2014. Fire severity and tree regeneration following bark beetle outbreaks: the role of outbreak stage and burning conditions. Ecological Applications 24:1608–1625.

Harvey, B. J., D. C. Donato, and M. G. Turner. 2014. Recent mountain pine beetle out-breaks, wildfire severity, and postfire tree regeneration in the US Northern Rockies. Proceedings of the National Academy of Sciences USA 111:15120–15125.

———. 2016. High and dry: post-fire tree seedling establishment in subalpine forests de-creases with post-fire drought and large stand-replacing burn patches. Global Ecol. Biogogr. doi:10.1111/geb.12443.

Higuera, P. E., C. Whitlock, and J. A. Gage. 2011. Linking tree-ring and sediment-charcoal records to reconstruct fire occurrence and area burned in subalpine forests of Yellow-stone National Park, USA. The Holocene 21:327–341.

Jenkins, M. J., W. G. Page, E. G. Hebertson, and M. E. Alexander. 2012. Fuels and fire be-havior dynamics in bark beetle-attacked forests in western North America and impli-cations for fire management. Forest Ecology and Management 275:23–34.

Johnstone, J. F., F. S. Chapin, T. N. Hollingsworth, M. C. Mack, V. Romanovsky, and M. Turetsky. 2010a. Fire, climate change and forest resilience in interior Alaska. Cana-dian Journal of Forest Research 40:1302–1312.

Johnstone, J. F., T. N. Hollingsworth, F. S. Chapin III, and M. C. Mack. 2010b. Changes in fire regime break the legacy lock on successional trajectories in Alaskan boreal forest. Global Change Biology 16:1281–1295.

Keane, R. E., P. F. Hessburg, P. B. Landres, and F. J. Swanson. 2009. The use of historical range and variability (HRV) in landscape management. Forest Ecology and Manage-ment 258:1025–1037.

Krause, T. R., and C. Whitlock. 2013. Climate and vegetation change during the late-glacial/early Holocene transition inferred from multiple proxy records from Blacktail Pond, Yellowstone National Park, USA. Quaternary Research 79:391–402.

Leopold, A. S., S. A. Cain, C. M. Cottam, I. N. Gabrielson, and T. L. Kimball. 1963. Wild-life management in the national parks: the Leopold Report. Report delivered to the US Secretary of the Interior.

Liu, Y., S. L. Goodrick, and J. A. Stanturf. 2013. Future US wildfire potential trends pro-jected using a dynamically downscaled climate change scenario. Forest Ecology and Management 294:120–135.

Lynch, H. J., R. A. Renkin, R. L. Crabtree, and P. R. Moorcroft. 2006. The influence of previous mountain pine beetle (*Dendroctonus ponderosae*) activity on the 1988 Yellow-stone fires. Ecosystems 9:1318–1327.

Marris, E. 2011. The end of the wild. Nature 150:150–152.

Meddens, A. J. H., J. A. Hicke, and C. A. Ferguson. 2012. Spatiotemporal patterns of ob-served bark beetle-caused tree mortality in British Columbia and the western United States. Ecological Applications 22:1876–1891.

Meyer, G. A., and J. L. Pierce. 2003. Climatic controls on fire-induced sediment pulses in Yellowstone National Park and central Idaho: a long-term perspective. Forest Ecology and Management 178:89–104.

Miller, M. M. 2009. Adventures in Yellowstone: early travelers tell their tales. Globe Pe-quot Press, Guilford, Connecticut.

Millspaugh, S. H., C. Whitlock, and P. Bartlein. 2004. Postglacial fire, vegetation, and climate history of the Yellowstone-Lamar and Central Plateau provinces, Yellowstone National Park. Pages 10–28 in L. L. Wallace, ed. After the fires: the ecology of change in Yellowstone National Park. Yale University Press, New Haven, Connecticut.

Mooney, C., and S. Kirshenbaum. 2010. Unscientific America: how scientific illiteracy threatens our future. Basic Books, New York, New York.

Moritz, M. A., E. Batllori, R. A. Bradstock, A. M. Gill, J. Handmer, P. F. Hessburg, J. Leonard, et al. 2014. Learning to coexist with wildfire. Nature 515:10–18.

National Research Council. 1992. Science and the national parks. National Academies Press, Washington, DC.

——. 2002. Ecological dynamics on Yellowstone's northern range. National Academies Press, Washington, DC.

O'Connor, C. D., D. A. Falk, A. M. Lynch, and T. W. Swetnam. 2014. Fire severity, size, and climate associations diverge from historical precedent along an ecological gradient in the Pinaleno Mountains, Arizona. Forest Ecology and Management 329:264–278.

Parks, S. A., M. A. Parisien, and C. Miller. 2012. Spatial bottom-up controls on fire likelihood vary across western North America. Ecosphere 3:art12. doi:10.189/ES11-00298.1.

Pederson, G. T., S. T. Gray, C. A. Woodhouse, J. L. Betancourt, D. B. Fagre, J. S. Littell, E. Watson, B. H. Luckman, and L. J. Graumlich. 2011. The unusual nature of recent snowpack declines in the North American cordillera. Science 333:332–335.

Peterson, D. L., and J. S. Littell. 2014. Risk assessment for wildfire in the western United States. Pages 232–235 in D. L. Peterson, J. M. Vose, and T. Patel-Weynand, eds. Climate change and United States forests. Springer, New York, New York.

Pickett, S. T. A., and P. S. White, eds. 1985. The ecology of natural disturbance and patch dynamics. Academic Press, New York, New York.

Powell, E. N., P. A. Townsend, and K. F. Raffa. 2012. Wildfire provides refuge from local extinction but is an unlikely driver of outbreaks by mountain pine beetle. Ecological Monographs 82:69–84

Raffa, K. F., B. H. Aukema, B. J. Bentz, A. L. Carroll, J. A. Hicke, M. G. Turner, and W. H. Romme. 2008. Cross-scale drivers of natural disturbances prone to anthropogenic amplification: the dynamics of bark beetle eruptions. BioScience 58:501–517.

Romme, W. H. 1982. Fire and landscape diversity in subalpine forests of Yellowstone National Park. Ecological Monographs 52:199–221.

Romme, W. H., M. S. Boyce, R. E. Gresswell, E. H. Merrill, G. W. Minshall, C. Whitlock, and M. G. Turner. 2011. Twenty years after the 1988 Yellowstone fires: lessons about disturbance and ecosystems. Ecosystems 14:1196–1215.

Romme, W. H., and D. G. Despain. 1989. Historical perspective on the Yellowstone fires of 1988. BioScience 39:695–699.

Romme, W. H., and D. H. Knight. 1982. Landscape diversity: the concept applied to Yellowstone Park. BioScience 32:664–670.

Romme, W. H., and M. G. Turner. 2004. Ten years after the 1988 Yellowstone fires: is restoration needed? Pages 318–361 in L. L. Wallace, ed. After the fires: the ecology of change in Yellowstone National Park. Yale University Press, New Haven, Connecticut.

Saab, V. A., Q. S. Latif, M. M. Rowland, T. N. Johnson, A. D. Chalfoun, S. W. Buskirk, J. E. Heyward, and M. A. Dresser. 2014. Ecological consequences of mountain pine beetle outbreaks for wildlife in western North American forests. Forest Science 60:539–559.

Schoennagel, T., E. A. H. Smithwick, and M. G. Turner. 2008. Landscape heterogeneity following large fires: insights from Yellowstone National Park, USA. International Journal of Wildland Fire 17:742–753.

Schoennagel, T., M. G. Turner, and W. H. Romme. 2003. The influence of fire interval and serotiny on postfire lodgepole pine density in Yellowstone National Park. Ecology 84:2967–2978.

Seidl, R., M. J. Schelhaas, and M. J. Lexer. 2011. Unraveling the drivers of intensifying forest disturbance regimes in Europe. Global Change Biology 17:2842–2852.

Shuman, B. 2012. Recent Wyoming temperature trends, their drivers, and impacts in a 14,000-year context. Climatic Change 112:429–447.

Simard, M., E. N. Powell, K. F. Raffa, and M. G. Turner. 2012. What explains landscape patterns of bark beetle outbreaks in Greater Yellowstone? Global Ecology and Biogeography 21:556–567.

Simard, M., W. H. Romme. J. M. Griffin, and M. G. Turner. 2011. Do mountain pine beetle outbreaks change the probability of active crown fire in lodgepole pine forests? Ecological Monographs 81:3–24.

Stephens, S. L., J. K. Agee, P. Z. Fulé, M. P. North, W. H. Romme, T. W. Swetnam, and M. G. Turner. 2013. Managing forests and fire in changing climates. Science 342:41–42.

Tinker, D. B., W. H. Romme, and D. G. Despain. 2003. Historic range of variability in landscape structure in subalpine forests of the Greater Yellowstone Area, USA. Landscape Ecology 18:427–439.

Turner, M. G. 2010. Disturbance and landscape dynamics in a changing world. Ecology 91:2833–2849.

Turner, M. G., V. H. Dale, and E. E. Everham III. 1997. Fires, hurricanes and volcanoes: comparing large-scale disturbances. BioScience 47:758–768.

Turner, M. G., D. C. Donato, and W. H. Romme. 2013. Consequences of spatial heterogeneity for ecosystem services in changing forest landscapes: priorities for future research. Landscape Ecology 28:1081–1097.

Turner, M. G., W. H. Hargrove, R. H. Gardner, and W. H. Romme. 1994. Effects of fire on landscape heterogeneity in Yellowstone National Park, Wyoming. Journal of Vegetation Science 5:731–742.

Turner, M. G., and W. H. Romme. 1994. Landscape dynamics in crown fire ecosystems. Landscape Ecology 9:59–77.

Turner, M. G., W. H. Romme, R. H. Gardner, and W. W. Hargrove. 1997. Effects of fire size and pattern on early succession in Yellowstone National Park. Ecological Monographs 67:411–433.

Turner, M. G., W. H. Romme, R. H. Gardner, R. V. O'Neill, and T. K. Kratz. 1993. A revised concept of landscape equilibrium: disturbance and stability on scaled landscapes. Landscape Ecology 8:213–227.

Turner, M. G., W. H. Romme, and D. B. Tinker. 2003. Surprises and lessons from the 1988 Yellowstone fires. Frontiers in Ecology and the Environment 1:351–358.

Wallace, L. L., ed. 2004. After the fires: the ecology of change in Yellowstone National Park. Yale University Press, New Haven, Connecticut.

Wallin, K. F., and K. F. Raffa. 2004. Feedback between individual host selection behavior and population dynamics in an eruptive herbivore. Ecological Monographs 74:101–116.

Weed, A. S., M. P. Ayres, and J. A. Hicke. 2013. Consequences of climate change for biotic disturbances in North American forests. Ecological Monographs 83:441–470.

Weeks, J. 2012. Managing wildfires. CQ Researcher 22:941–964.

Wells, G. 2012. Bark beetles and fire: two forces of nature transforming western forests. Fire Science Digest, Issue 12. Joint Fire Science Program, Boise, Idaho.

Westerling, A. L., H. G. Hidalgo, D. R. Cayan, and T. W. Swetnam. 2006. Warming and earlier spring increase western US forest wildfire activity. Science 313:940–943.

Westerling, A. L., M. G. Turner, E. A. H. Smithwick, W. H. Romme, and M. G. Ryan. 2011. Continued warming could transform Greater Yellowstone fire regimes by mid-21st century. Proceedings of the National Academy of Sciences USA 108:13165–13170.

Whitlock, C., and P. J. Bartlein. 1993. Spatial variations of Holocene climatic change in the Yellowstone region. Quaternary Research 39:231–238.

Whitlock, C., J. Marlon, C. Briles, A. Brunelle, C. Long, and P. Bartlein. 2008. Long-term relations among fire, fuel, and climate in the northwestern US based on lake-sediment studies. International Journal of Wildland Fire 17:72–83.

Whitlock, C., S. L. Shafer, and J. Marlon. 2003. The role of climate and vegetation change in shaping past and future fire regimes in the northwestern US and the implications for ecosystem management. Forest Ecology and Management 178:5–21.

Wu, J., and O. L. Loucks. 1995. From balance of nature to hierarchical patch dynamics: a paradigm shift in ecology. Quarterly Review of Biology 70:439–466.

SIX

Climate Change Trends, Impacts, and Vulnerabilities in US National Parks

PATRICK GONZALEZ

Introduction

Field measurements have detected glaciers melting in Glacier National Park (Vaughan et al. 2013), sea level rising in Golden Gate National Recreation Area (Church and White 2011), trees dying in Sequoia National Park (van Mantgem et al. 2009), vegetation shifting upslope in Yosemite National Park (Millar et al. 2004) and poleward in Noatak National Preserve (Suarez 1999), wildfire changing in Yellowstone National Park (Littell et al. 2009), and corals bleaching in Virgin Islands National Park (Eakin et al. 2010). Published analyses of these and similar cases around the world have attributed the cause to human-induced climate change (Intergovernmental Panel on Climate Change [IPCC] 2013, 2014a). If we do not reduce greenhouse gas emissions from power plants, cars, and deforestation, continued climate change may fundamentally alter many of the globally unique ecosystems, endangered plant and animal species, and physical and cultural resources that national parks protect.

A growing collection of scientific research focuses specifically on climate change in US national parks. This chapter reviews climate change research published in peer-reviewed journals and IPCC reports that uses data from US national parks. The chapter covers climate trends, historical impacts, and projected vulnerabilities.

Published field research from national parks has contributed to the detection of 20th-century physical and ecological changes and to the attribution of the cause of those changes to human-induced climate change. The section on historical impacts first reviews research that has employed the research procedures of detection and attribution (IPCC 2001a). Detection is the finding of statistically significant changes over time. Attribution is the analysis of the relative weights of different causes and the determination of human-induced climate change as the primary cause. Attribution requires examination of causal factors and a time series of at least 30 years, the minimum statistically significant size for a time series and a period long enough to rule out short-term variations (von Storch and Zwiers 1999).

Detection answers the basic question of whether a species, ecosystem, or other resource is changing. Attribution guides resource management toward the predominant factor that is causing change. Whereas resource managers have developed many actions to address urbanization and other nonclimate factors, changes attributed to human-induced climate change may require new adaptation measures. A subsection in the historical impacts section reviews research that has found other changes that are consistent with, but not formally attributed to, human-induced climate change.

Analyses of climate and resources in US national parks project potential future vulnerabilities to climate change. The section on projected vulnerabilities reviews research that has specifically analyzed national parks. The potential future magnitude of climate change depends on human population size, the magnitude and efficiency of energy use and industrial activity, the extent of deforestation, and feedbacks among climate and biogeochemical cycles. The IPCC has defined greenhouse gas emissions scenarios—discrete sets of potential future conditions that provide standard situations for vulnerability analyses. The most recently updated emissions scenarios are the four representative concentration pathways (RCPs) (Moss et al. 2010; IPCC 2013), ranging from a low emissions scenario (RCP2.6) in an environmentally favorable society to a very high emissions scenario (RCP8.5) due to lack of improvements in practices and policies. General circulation models (GCMs) of the atmosphere provide projections of potential future climate. The two major uncertainties of future climate projections are the extent to which society changes its practices and policies to reduce greenhouse gas emissions and the varying skill among the GCMs to accurately portray spatial and temporal patterns of climate.

Vulnerability is "the propensity or predisposition to be adversely affected" (IPCC 2014a). Three components of vulnerability most relevant to national park resources are exposure, sensitivity, and adaptive capacity (IPCC 2007a). The most effective vulnerability analyses combine historical observations and future projections of climate and resources to identify locations of vulnerable areas and potential refugia and to quantify uncertainties (Gonzalez 2011). In addition to the uncertainties of societal changes to reduce emissions and the varying skill of GCMs, vulnerability analyses are also subject to uncertainty in the accuracy of models used to project responses of species, ecosystems, and other resources to climate change.

Effective vulnerability analyses provide spatial data for prioritizing the location of future adaptation measures.

Historical Impacts

Climate Trends

Analyses of weather station measurements have detected a statistically significant increase of global temperature and other climate changes since the beginning of the instrumental record in 1850, and the analyses of causal factors have attributed the cause to an increase of atmospheric concentrations of carbon dioxide (CO_2) and other greenhouse gases emitted from power plants, cars, deforestation, and other human sources (IPCC 2013). Analyses of spatial data interpolated from weather stations (Daly et al. 2008) show that the average annual temperature of the area of the US National Park System (this chapter refers to the 410 national parks existing in April 2016) increased at a statistically significant rate of 0.9°C ±0.2°C per century (mean ±SE) from 1895 to 2010, with 96% of system area experiencing increases and two-thirds experiencing statistically significant increases (F. Wang et al., unpublished data). In the 20th century, the National Park System experienced heating at a rate three times greater than the United States as a whole (fig. 6.1), mainly because 60% of National Park System area is in Alaska and temperature increases have been greater at higher latitudes (IPCC 2013).

For a sample of the largest national parks and 30 km buffer zones around those parks, average annual temperature increased at rates greater than 1°C per century from 1895 to 2009 (Hansen et al. 2014), and temperature from 1982 to 2012 was higher than other periods from 1901 to 2012 (Monahan and Fisichelli 2014). Some smaller national parks in the southeastern United States lie in an anomalous area where temperature has not increased because of local cooling effects of increased precipitation, the El Niño–Southern Oscillation, and other factors (Portmann, Solomon, and Hegerl 2009).

Because warmer air can hold more moisture, climate change has been increasing precipitation globally (IPCC 2013). In the United States as a whole, precipitation increased at a statistically significant rate of +4% ±2% per century from 1895 to 2010, with only one-fifth of the area experiencing decreases (F. Wang et al., unpublished data). In contrast, total annual precipitation of the area of the National Park System decreased at a rate of –2% ±2% per century from 1895 to 2010, with half of system area experiencing decreases and 16% experiencing statistically significant changes.

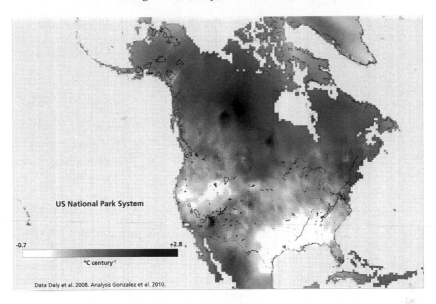

6.1. Trend in average annual temperature, from linear regres
sion, 1901–2002 (Daly et al. 2008; Gonzalez et al 2010).

Across the western United States, 53 national parks host National
Weather Service stations that have contributed data to the detection of
global climate change in the last half of the 20th century and to the attribution to human emissions. Changes in climate in the western United
States include increases in winter minimum temperatures at rates of 2.8°C
to 4.3°C per century (Barnett et al. 2008; Bonfils et al. 2008), decreases in
ratio of snow to rain at rates of –24% to –79% per century (Barnett et al.
2008; Bonfils et al. 200), and an advance of spring warmth of one week
from 1950 to 2005 (Ault et al. 2011).

Weather stations in national parks across the United States that are part
of the Global Historical Climatology Network have contributed to the
global detection of extreme temperature and precipitation events. In the
United States, the number of warm nights per year (minimum daily temperature greater than the 90th percentile) increased by up to 20 days from
1951 to 2010 (IPCC 2013). In the northeastern United States, total annual
precipitation falling in heavy storms (daily precipitation greater than the
95th percentile) increased by 50% from 1951 to 2010 (IPCC 2013).

National parks on the Atlantic and Pacific coasts lie in the path of
tropical cyclones (also called hurricanes). Although historical observations show an increase in the intensity of North Atlantic hurricanes after

1970, changing historical methods, incomplete understanding of physical mechanisms, and tropical cyclone variability prevent direct attribution to climate change (IPCC 2012, 2013).

Physical Impacts

Field data from numerous national parks have contributed to detection of physical changes and to attribution to human-induced climate change (table 6.1). Measurements from National Weather Service stations in 53 western national parks and from Natural Resources Conservation Service snow courses in many of those parks contributed to detection of decreased snowpack (Barnett et al. 2008; Pierce et al. 2008) and advances of spring stream flow of one week (Barnett et al. 2008). Analyses of snow measurements and tree rings from across the western United States, including sites in nine national parks, detected snowpack levels in the 20th century lower than any time since the 13th century and attributed the low snowpack to human-induced climate change (Pederson et al. 2011).

IPCC analyses of measurements of 168,000 glaciers around the world, including glaciers in Denali National Park and Preserve, Glacier National Park, Glacier Bay National Park and Preserve, Kenai Fjords National Park, Lake Clark National Park and Preserve, North Cascades National Park, and Wrangell–Saint Elias National Park and Preserve, have detected decreases in length, area, volume, and mass for almost all the glaciers since 1960 (Vaughan et al. 2013). The IPCC has shown that the cause is attributable to human-induced climate change more than natural variation or other non-human factors (Bindoff et al. 2013). Further analyses confirm that the loss of mass from Alaskan and western North American glaciers in the period 1960–2010 is attributable to human-induced climate change (Marzeion et al. 2014). In Glacier National Park, Agassiz Glacier receded 1.5 km from 1926 to 1979 (Pederson et al. 2004). In Glacier Bay National Park and Preserve, the greatest ice loss has occurred from Muir Glacier, which lost 640 m in its lower reaches from 1948 to 2000 (Larsen et al. 2007) (fig. 6.2).

Analyses of tidal gauge measurements around the world have detected a statistically significant rise in global sea level (Church and White 2011; Church et al. 2013), with IPCC analyses of potential causal factors attributing the rise to human-induced climate change (Bindoff et al. 2013). Golden Gate National Recreation Area in San Francisco, California, hosts the tidal gauge with the longest time series in the Western Hemisphere, operated by the National Oceanic and Atmospheric Administration (NOAA). Sea level there rose at a statistically significant rate of 14 cm ± 0.8 cm per

Table 6.1 Historical changes detected in US national parks and attributed to human-induced climate change

Resource	Impact	National park	Years	Reference
Physical				
Glaciers	Decreased length, area, volume, and mass of most glaciers	Global; Denali NP, Glacier NP, Glacier Bay NP Pres., Kenai Fjords NP, Lake Clark NP Pres., North Cascades NP, Wrangell–Saint Elias NP Pres	1960–2012	IPCC 2013; Marzeion et al. 2014
	Disappearance of 4 glaciers, thinning of 12 glaciers by 6 m	Lake Chelan NRA, North Cascades NP, Ross Lake NRA	1984–2004	Pelto 2006; IPCC 2013
	Recession of Agassiz Glacier by 1.5 km, recession of Jackson Glacier by 1 km	Glacier NP	1926–1979	Pederson et al. 2004; IPCC 2013
	Recession of Toklat River glaciers greater than 400 m, thinning 5–6 m year^{-1}	Denali NP	2000–2009	Crossman, Futter, Whitehead 2013; IPCC 2013
	Reduction in total area of 7%	Lake Chelan NRA, North Cascades NP, Ross Lake NRA	1958–1998	Granshaw and Fountain 2006; IPCC 2013
	Reductions in area among 30 glaciers up to 40%	Rocky Mountain NP	1888–2005	Hoffman, Fountain, and Achuff 2007; IPCC 2013
	Thinning up to 7.5 m year^{-1}	Wrangell–Saint Elias NP Pres.	2000–2007	Arendt et al. 2008; IPCC 2013
	Thinning up to 640 m	Glacier Bay NP Pres.	1948–2000	Larsen et al. 2007; IPCC 2013
Sea level	Rise 17 cm ±2 cm century^{-1}	Global; Golden Gate NRA	1901–2010	Church and White 2011; IPCC 2013
	Rise 1–37 cm century^{-1}	19 Atlantic and Pacific coast NPs	1854–1999	Pendleton, Thieler, and Williams 2010; IPCC 2013
Sea temperatures	Increase 1.1 °C ±0.2 °C century^{-1}	Global; Buck Island Reef NM, Channel Islands NP, Virgin Islands Coral Reef NM	1971–2010	IPCC 2013
	Increase ~0.8 °C century^{-1}	Biscayne NP	1878–2012	IPCC 2013; Kuffner et al. 2015

(continued)

Table 6.1 *(continued)*

Resource	Impact	National park	Years	Reference
Snowpack	Decrease to lowest extent in 8 centuries	Western United States; 9 NPs	1200–2000	Pederson et al. 2011
	Decrease up to 8% decade^{-1}	Western United States; 50+ NPs	1950–1999	Barnett et al. 2008; Pierce et al. 2008
Streams	Peak stream flow advance up to 1.7 days decade^{-1}	Western United States; 50+ NPs	1950–1999	Barnett et al. 2008
Ecological Biomes	Shift northward of boreal conifer forest into tundra 80–100 m	Noatak N. Pres.	1700–1990	Suarez et al. 1999
	Shift upslope of subalpine forest into alpine meadows	California; Yosemite NP	1880–2002	Millar et al. 2004
Birds	Shift northward of winter ranges of 254 species 0.5 km ±0.3 km year^{-1}	Contiguous United States; 50+ NPs	1975–2004	La Sorte and Thompson 2007
Corals	Bleaching of corals due to highest temperatures in the period 1855–2008	Caribbean Sea; Biscayne NP, Buck Island Reef NM, Salt River Bay NHP and Ecological Pres., Virgin Islands NP, Virgin Islands Coral Reef NM	2005	Eakin et al. 2010; IPCC 2014a
Insect pests	Bark beetles in the largest outbreak in time period	Western United States; Yellowstone NP, other NPs	1880–2005	Raffa et al. 2008; Logan, Macfarlane, and Willcox 2010; Macfarlane, Logan, and Kern 2013
Mammals	Shift upslope of ranges of half of 28 small mammal ranges ~500 m	Yosemite NP	1914–1920	Moritz et al. 2008
Trees	Mortality doubled in old conifer forests	Kings Canyon NP, Lassen Volcanic NP, Mount Rainier NP, Rocky Mountain NP, Sequoia NP, Yosemite NP	1955–2007	van Mantgem et al. 2009
Wildfire	Climate dominant factor controlling burned area	Western United States; NPs	1916–2003	Littell et al. 2009

Note: NHP = National Historical Park; NHS = National Historic Site; NL = National Lakeshore; NM = National Monument; NP = National Park; NP Pres. = National Park and Preserve; N. Pres. = National Preserve; NRA = National Recreation Area; NS = National Seashore.

6.2. Muir Glacier, Glacier Bay National Park and Preserve, Alaska: August 13, 1941 (photo by William O. Field, courtesy of the National Park Service, National Snow and Ice Data Center, and US Geological Survey), and August 31, 2004 (photo by Bruce F. Molnia, courtesy of the US Geological Survey). IPCC has analyzed a global database of 168,000 glaciers, including Muir Glacier, and attributed melting since the 1960s to human-induced climate change (Bindoff et al. 2013; Vaughan et al. 2013).

century from 1855 to 2014. At the NOAA tidal gauge in Washington, DC, not far from the Jefferson Memorial and numerous other national parks in the capital, sea level rose at a rate of 31 cm ±1 cm per century from 1931 to 2013. Sea level has been rising at rates of up to 37 cm per century in 19 national parks along the Atlantic and Pacific coasts (Pendleton, Thieler, and Williams 2010).

Global measurements of sea surface temperatures, including measurements in Buck Island Reef National Monument, Channel Islands National Park, and Virgin Islands Coral Reef National Monument, have detected an increase in the top 75 m of ocean water of 1.1°C ±0.2°C per century from 1971 to 2010 (Rhein et al. 2013), with IPCC analyses attributing the cause to human-induced climate change (Bindoff et al. 2013). In 1878, the US government built Fowey Rocks Lighthouse in the Florida Keys in what would later become Biscayne National Park. Comparison of sea surface temperatures taken by lighthouse keepers from 1879 to 1912 with measurements by electronic sensors from 1991 to 2012 showed a statistically significant warming of ~0.8°C per century (Kuffner et al. 2015). Summer sea surface temperatures from 1991 to 2012 exceeded 29°C, a threshold of stress for many coral species.

Increased atmospheric CO_2 concentrations from human activities have increased the acidity of ocean water around the world by 0.1 pH units since ~1750 (Rhein et al. 2013). Ocean acidification occurs when CO_2 dissolves in water and forms carbonic acid. High acidity can dissolve the shells of many marine species. Research on past acidification in national parks has not been published.

Ecological Impacts

Field data from numerous national parks have contributed to detection of ecological changes and attribution to human-induced climate change (see table 6.1). Vegetation at the level of the biome (10–15 major global vegetation types) has shifted upslope or toward the poles or the equator at sites around the world, and analyses of possible causes have attributed most of the shifts to human-induced climate change (Gonzalez et al. 2010; Settele et al. 2014). In Noatak National Preserve, Alaska, boreal conifer forest shifted northward 80–100 m into tundra between 1800 and 1990 (Suarez et al. 1999). In Yosemite National Park, subalpine forest shifted upslope into alpine meadows between 1880 and 2002 (Millar et al. 2004).

Multivariate analysis of wildfire across the western United States from 1916 to 2003, using data from numerous national parks and other areas, indicates that climate was the dominant factor controlling the extent of burned area, even during periods of active fire suppression (Littell et al. 2009). Reconstruction of fires of the past 400 to 3,000 years in the western United States (Trouet et al. 2010; Marlon et al. 2012) and in Sequoia and Yosemite National Parks (Swetnam 1993; Swetnam et al. 2009; Taylor and Scholl 2012) confirms that temperature and drought are the dominant factors explaining fire occurrence.

Field and remote sensing data from across western North America, including numerous national parks, have also documented how climate change has caused bark beetle outbreaks leading to the most extensive tree mortality across western North America in the last 125 years (Raffa et al. 2008). Tracking of trees in permanent old-growth conifer forest plots across the western United States, including plots in Kings Canyon, Lassen Volcanic, Mount Rainier, Rocky Mountain, Sequoia, and Yosemite National Parks, found a statistically significant doubling of tree mortality between 1955 and 2007 (van Mantgem et al. 2009). Analyses of fire, mortality of small trees, forest fragmentation, air pollution, and climate attributed the mortality to warming due to climate change.

Climate change has caused shifts in latitude or elevation of the ranges of numerous animal species around the world (Settele et al. 2014). In Yosemite National Park, small mammal resurveys in 2006 of the Grinnell surveys from 1914 to 1920 showed that the ranges of half of 28 small mammal species shifted upslope an average of ~500 m (Moritz et al. 2008). Because the national park had protected the survey transect, land-use change or other factors were not major factors. Therefore, the authors could attribute the shift to a 3°C increase in minimum temperature caused by climate change. Analyses of Audubon Christmas Bird Count data across the United States, including sites in 54 national parks, detected a northward shift of winter ranges of a set of 254 bird species at an average rate of 0.5 km ±0.3 km per year from 1975 to 2004, attributable to human-induced climate change (La Sorte and Thompson 2007). Examples include northward shifts of the Evening Grosbeak (*Coccothraustes vespertinus*) in Shenandoah National Park and the Canyon Wren (*Catherpes mexicanus*) in Santa Monica Mountains National Recreation Area.

High ocean temperatures due to climate change have bleached and killed coral around the world (Wong et al. 2014). In 2005, the hottest sea surface temperatures recorded in the Caribbean Sea in the period 1855–2008 caused coral bleaching and the death of up to 80% of coral area at sites in Biscayne National Park, Buck Island Reef National Monument, Salt River Bay National Historical Park and Ecological Preserve, Virgin Islands

National Park, and Virgin Islands Coral Reef National Monument (Eakin et al. 2010).

Other Changes Consistent with, but Not Attributed to, Climate Change

Researchers have observed other 20th-century changes to resources in national parks that have not been explicitly attributed to human-induced climate change but are consistent with responses to climate change (table 6.2). The most prominent physical change is melting of permafrost in Alaskan national parks (Riordan, Verbyla, and McGuire 2006; Jones et al. 2011; Necsoiu et al. 2013; Balser, Jones, and Gens 2014). Ecological changes include upslope shifts of forests into alpine meadows, vegetation dieback in areas of increased aridity, changes to amphibians, range shifts of birds, and declines of mammal species. Changes in phenology include advances of cherry tree (*Prunus × yedoensis*) blooming in Washington, DC (Abu-Asab et al. 2001), White-tailed Ptarmigan (*Lagopus leucurus*) hatching in Rocky Mountain National Park (Wang et al. 2002), and loggerhead sea turtle (*Caretta caretta*) nesting in Canaveral National Seashore (Pike, Antworth, and Stiner 2006). One change in cultural resources has occurred in Wrangell–Saint Elias and Lake Clark National Parks and Preserves, where melting glaciers are revealing archaeological artifacts, such as wooden arrow shafts and a birch bark basket fragment, dating from circa 500 to 1770 (Dixon, Manley, and Lee 2005; VanderHoek et al. 2012).

Future Vulnerabilities

Climate Projections

Spatial analyses of the output of the 33 GCMs used by the IPCC (2013) provide climate projections for the area of the National Park System. The ensemble of GCMs projects an average annual temperature increase (1971–2000 to 2071–2100) of 2.2°C ±0.9°C per century (mean ±SD) under RCP2.6, and 5.6°C ±1.3°C per century under RCP8.5 (F. Wang et al., unpublished data). This potential 21st-century heating would be two to six times the magnitude of historical 20th-century warming. Temperature projections are highest for the national parks in Alaska, with projected increases up to 10°C per century under RCP8.5.

The ensemble of GCMs projects a total annual precipitation increase (1971–2000 to 2071–2100) of 9% ±13% per century (mean ±SD) under RCP2.6, and 21% ±5% per century under RCP8.5 (F. Wang et al.,

Table 6.2 Changes in US national parks consistent with, but not formally attributed to, climate change

Resource	Change	National park	Years	Reference
Physical				
Permafrost	Contraction of ponds due to draining	Denali NP Pres., Wrangell–Saint Elias NP Pres.	1950–2002	Riordan et al. 2006
	Melting edges and contraction of water bodies due to draining	Kobuk Valley NP	1951–2005	Necsoiu et al. 2013
	Thaw and retrogressive thaw slump initiation	Noatak N. Pres.	1992–2011	Balser, Jones, and Gens 2014
	Thermokarst lakes increased in number and decreased in surface area due to draining	Bering Land Bridge N. Pres.	1950–2007	Jones et al. 2011
Streams	Beetle kill increased, forest cover decreased, groundwater contributions to streams increased	Rocky Mountain NP	1994–2012	Bearup et al. 2014
	Stream nitrate concentration increased as glacial melt exposed sediments	Rocky Mountain NP	1991–2006	Baron et al. 2009
Plants				
Biomes	Shift northward of mangrove forest	Biscayne NP, Canaveral NS	1984–2011	Cavanaugh et al. 2014
	Shift upslope of piñon-juniper woodland (*Pinus edulis*, *Juniperus monosperma*) into ponderosa pine forest (*Pinus ponderosa*)	Bandelier NM	1935–1975	Allen and Breshears 1998
	Shift upslope of subalpine forest	California; Yosemite NP	1929–2009	Dolanc, Thorne, and Safford 2013
		Glacier NP	1945–1991	Klasner and Fagre 2002
		Glacier NP	1925–2003	Roush, Munroe, and Fagre 2007
		Lassen Volcanic NP	1840–1990	Taylor 1995
	Shift upslope of subalpine forest into alpine meadows	Mount Rainier NP	1916–1969	Franklin et al. 1971
		Mount Rainier NP	1930–1990	Rochefort and Peterson 1996

(continued)

Table 6.2 (*continued*)

Resource	Change	National park	Years	Reference
Plants				
	Shift upslope of temperate conifer forest into subalpine meadows	Olympic NP	1905–1991	Woodward, Schreiner, and Silsbee. 1995
		Rocky Mountain NP	1930–1990	Hessl and Baker 1997
		Yellowstone NP	1860–1986	Jakubos and Romme 1993
Plants, nonwoody	Alpine plants declined at high elevation	Glacier NP	1988–2011	Lesica and McCune 2004; Lesica 2014
	Haleakala silversword (*Argyroxiphium sandwicense macrocephalum*) decreased as rainfall decreased	Hawai'i; Haleakala NP	1982–2010	Krushelnycky et al. 2013
	Herb and forb species of cool habitats declined	Oregon; Oregon Caves NM Pres.	1949–2007	Whittaker 1960; Damschen, Harrison, and Grace 2010
	Joshua tree (*Yucca brevifolia*) mortality due to wildfire in invasive grasses, drought, and gnawing by rodents	Joshua Tree NP	2000–2004	DeFalco et al. 2010
	Soft-leaved paintbrush (*Castilleja mollis*) growth reduced as growing-season temperatures increased	Channel Islands NP	1995–2006	McEachern et al. 2009
Shrubs, nonwoody plants	Herbaceous plants declined with increasing temperatures, but some shrubs increased	Arches NP, Canyonlands NP, Natural Bridges NM	1989–2009	Munson, Belnap, and Okin 2011; Munson et al. 2011
Trees	Cherry tree (*Prunus* × *yedoensis*) blooming advanced 7 days	Washington, DC; Baltimore, MD; National Capital Parks, Rock Creek Park	1970–1999	Abu-Asab et al. 2001
	Conifer growth increased since 1950 at high elevation	California; Yosemite NP	1000–1990	Bunn, Graumlich, and Urban 2005
		Lake Clark NP Pres.	1769–2003	Driscoll et al. 2005
	Large trees declined, small trees increased, and oak (*Quercus* spp.) increased as climate water deficit increased	Kings Canyon NP, Sequoia NP, Yosemite NP	1929–2010	McIntyre et al. 2015

	Quaking aspen (*Populus tremuloides*) recruitment decreased by increased browsing as snowpack decreased	Yellowstone NP	2007–2009	Brodie et al. 2012
	Tree dieback after years of drought	Southwestern United States; Bandelier NM	2002–2003	Breshears et al. 2005
	Tree dieback with 38% decrease of crown area after years of drought	Bandelier NM	2002–2011	Garrity et al. 2013
Wetlands	Wetland vegetation increased and dryland vegetation decreased as sea level rose	Gulf Islands NS	1978–2004	Lucas and Carter 2010
Wildfire	Burned area increased	Alaska NPs	1860–2009	Kasischke et al. 2010
	Burned area related to temperature increase and rainfall decrease	Yellowstone NP	1895–1989	Balling, Meyer, and Wells 1992
	Fire frequency and burned area increased with temperature	Western United States; NPs	1970–2003	Westerling et al. 2006
Animals				
Amphibians	Ranavirus outbreaks related to temperature and other factors	Acadia NP	1999–2005	Gahl and Calhoun 2010
	Salamander body sizes reduced as metabolic requirements increased	Catoctin Mountain Park, Great Smoky Mountains NP, Shenandoah NP	1950–2012	Caruso et al. 2014
	Species richness declined due to wetland desiccation	Yellowstone NP	1992–2008	McMenamin, Hadly, and Wright 2008
Birds	Elevation shift of ranges tracked temperature and precipitation	Lassen Volcanic NP, Sequoia NP, Yosemite NP	1911–2008	Tingley et al. 2009; Tingley et al. 2012
	Shift northward of ranges of 26 species 2.4 km year[-1]	Eastern United States; NPs	1967–2002	Hitch and Leberg 2007
	White-tailed Ptarmigan (*Lagopus leucurus*) hatching advanced 15 days	Rocky Mountain NP	1975–1999	Wang et al. 2002
Corals	Increased prevalence of white pox and other diseases after bleaching	Virgin Islands NP	2004–2006	Muller et al. 2008
Mammals	Belding's ground squirrel (*Urocitellus beldingi*) extirpated from 42% of sites as snow cover decreased	Lassen Volcanic NP, Yosemite NP	1902–2011	Morelli et al. 2012
	Harbor seal (*Phoca vitulina richardii*) population declined as glacial ice calved from tidewater glaciers (used for resting and raising pups) declined	Glacier Bay NP Pres.	1992–2008	Womble et al. 2010

(continued)

Table 6.2 *(continued)*

Resource	Change	National park	Years	Reference
Animals				
	Small mammals shifted ranges upslope or downslope in at least one of three regions for 25 of 34 species analyzed, with temperature the main factor explaining changes in high-elevation species range shifts	Kings Canyon NP, Lassen Volcanic NP, Sequoia NP, Yosemite NP	1911–2010	Rowe et al. 2015
	Squirrel body size increased as food plant growing season lengthened	Lassen Volcanic NP, Sequoia NP, Yosemite NP	1902–2008	Eastman et al. 2012
Mussels	Species richness declined as water warmed	California; Channel Islands NP	1968–2002	Smith, Fong, and Ambrose 2006
Reptiles	Loggerhead sea turtle (*Caretta caretta*) nesting advanced 1 week as water temperatures warmed	Canaveral National NS	1989–2003	Pike et al. 2006
Cultural				
Archaeological artifacts	Antler projectile points, wooden arrow shafts, a birch bark basket fragment, a caribou calf hide, and other artifacts from ca. 500 to 1770 exposed by melting glaciers	Lake Clark NP Pres., Wrangell–Saint Elias NP Pres.	2000–2010	Dixon, Manley, and Lee 2005; VanderHoek et al. 2012

Note: See table 6.1 abbreviations. NM Pres. = National Monument and Preserve.

unpublished data). Because of the limited skill of GCMs in projecting precipitation, GCMs disagree on the direction of projected precipitation change (increase or decrease) across much of the National Park System. The projections show more than 80% agreement for the system as a whole, although half of the GCMs project precipitation increases and half project decreases in some national parks in the southwestern United States, California, and Florida. Based on GCM ensemble averages, precipitation may decrease on ~5% of system area. In general, projected precipitation outside the tropics increases with distance from the equator (IPCC 2013).

GCMs project increased frequency and severity of extreme climate events. In North America, the maximum temperature of days so hot that they occur only once every 20 years (1981–2000) may increase by 2°C to 6°C by 2100 (IPCC 2012). In North America, the type of storm with precipitation so heavy that it has occurred only once in 20 years (1981–2000) may increase in frequency to once in 5 to 10 years by 2100 (IPCC 2012). Projections of North Atlantic hurricanes and Pacific tropical cyclones under climate change do not agree on the direction of future trends (IPCC 2013).

Vulnerabilities of Physical Resources

Analyses project potential future vulnerabilities to climate change of air quality, glaciers, permafrost, lake and groundwater levels, and river and stream flows in numerous national parks (table 6.3). In Glacier National Park, Hall and Fagre (2003) estimated that a temperature increase of 1°C could lead to complete melting of glaciers, which, at a rate of 3.3°C per century, could occur as early as 2030. Nineteen national parks on the Atlantic and Pacific coasts are vulnerable to inundation and coastal erosion from sea-level rise and storm surges (Pendleton, Thieler, and Williams 2010). Grand Canyon and Big Bend National Parks are vulnerable to lower river flows because of increased aridity and human water withdrawals.

Vulnerabilities of Plants

Analyses project potential future vulnerabilities to climate change of vegetation in numerous national parks (see table 6.3). National parks are vulnerable to northward and upslope vegetation shifts, with 16%–41% of National Park System area highly vulnerable to biome shifts (Gonzalez et al. 2010), and 4%–31% of system area highly vulnerable to the combination of biome shifts due to climate change and habitat fragmentation due to roads, urbanization, and agriculture (Eigenbrod et al. 2015) (fig. 6.3).

Table 6.3 Future vulnerabilities to projected climate change of resources in US national parks

Resource	Vulnerability	National park	Scenario	Reference
Physical				
Air quality	Nitrogen deposition from motor vehicles and other sources exceeds critical loads	17–25 NPs nationwide	RCP2.6, RCP8.5	Ellis et al. 2013
Coasts	Coastal and lakeshore parks moderately vulnerable overall, Atlantic and Gulf of Mexico parks highly vulnerable	22 NPs on ocean coasts and Great Lakes shores	Water level −57 to +37 cm century[−1]	Pendleton, Thieler, and Williams 2010
	Inundation of one-quarter of the area	Cape Cod NS	Sea level +1–2 m	Murdukhayeva et al. 2013
	Inundation of one-third of the area	Assateague Island NS	Sea level +0.6–2 m	Murdukhayeva et al. 2013
Glaciers	Complete disappearance as early as 2030	Glacier NP	CO_2 doubling	Hall and Fagre 2003
	Sperry Glacier, 80% volume reduction by 2100 (min.), disappearance by 2040 (max.)	Rocky Mountain NP	+1°C–10°C	Brown, Harper, and Humphrey 2010
	Toklat River glaciers melt 2–11 m	Denali NP	B2, A2	Crossman, Futter, Whitehead 2013
Lakes	Lake Mead decrease to 7 m above dead storage (water level of lowest intake) 2%–9% probability by 2035	Lake Mead NRA	B1, A1B, A2	Dawadi and Ahmad 2012
	Lakes Mead and Powell, 50% chance of loss of live storage (min. water level for hydroelectric production) by 2021	Glen Canyon NRA, Lake Mead NRA	Runoff −20%	Barnett and Pierce 2008
Permafrost	Reduction of area 40%–100%	Rocky Mountain NP	+0.5°C–4°C	Janke 2005
Rivers	Colorado River flow change of −42% to +18%	Arizona, California, Colorado, Mexico; Glen Canyon NRA, Grand Canyon NP, Lake Mead NRA	A2	US Bureau of Reclamation 2012; Vano et al. 2014
	Rio Grande flow reduction	Mexico, Texas; Big Bend NP, Rio Grande WSR	B1, A1B, A2	US Bureau of Reclamation 2013

Streams	Spring runoff decline 15%–27%, advance spring runoff 4–6 weeks by 2080	Glacier NP	B1, A1F1	Larson et al. 2011
Water table	Water table decrease up to 1.1 m	Everglades NP	+1.5°C, precipitation ±10%, sea level +46 cm	Nungesser et al. 2015
Plants				
Biomes	Shift northward and upslope, high vulnerability of 16%–41% of National Park System area	Global; US National Park System	B1, A1B, A2	Gonzalez et al. 2010; Eigenbrod et al. 2015
	Shift northward and upslope, together with habitat fragmentation, high vulnerability of 4%–31% of National Park System area	Global; US National Park System	B1, A1B, A2	Eigenbrod et al. 2015
	Shift northward of boreal forest into tundra	Alaska; Bering Land Bridge N. Pres.	+2°C–4°C	Rupp, Chapin, and Starfield 2000
	Shift northward of boreal forest into tundra	Alaska; Kobuk Valley NP, Noatak N. Pres.	+2°C–4°C	Rupp, Chapin, and Starfield 2001
	Shift upslope of temperate conifer forest into subalpine forest into alpine meadows	Olympic NP	+2°C, precipitation ±20%	Zolbrod and Peterson 1999
Plants, nonwoody	Hoffmann's slender-flowered gilia (*Gilia tenuiflora* ssp. *hoffmannii*), Northern Channel Island phacelia (*Phacelia insularis* var. *insularis*), and Santa Cruz Island chicory (*Malacothrix indecora*) germination reduction with higher temperatures after first rains	Channel Islands NP	HS	Levine, McEachern, and Cowan 2008
	Joshua tree (*Yucca brevifolia*) up to 90% loss of habitat, no refugia in Joshua Tree NP	Southwestern United States; Death Valley NP, Joshua Tree NP, Mojave N. Pres., Tule Springs Fossil Beds NM	A1B	Cole et al. 2011
	Joshua tree (*Yucca brevifolia*) up to 90% loss of habitat, some refugia in Joshua Tree NP	Joshua Tree NP	+3°C	Barrows and Murphy-Mariscal 2012

(continued)

Table 6.3 *(continued)*

Resource	Vulnerability	National park	Scenario	Reference
Plants				
Polar vegetation	Vegetation type change of 6%–17% of land area	Bering Land Bridge N. Pres., Cape Krusenstern NP, Gates of the Arctic NP Pres., Kobuk Valley NP, Noatak N. Pres.	A1B (6°C)	Jorgenson et al. 2015
Trees, invasive	Tree-of-heaven (*Ailanthus altissima*) habitat increase 48%	Appalachian Trail	RCP6.0	Clark, Wang, and August 2014
Trees	Bishop pine (*Pinus muricata*) requires water from fog and the cooling effect of overcast skies	Channel Islands NP	HS	Fischer, Still, and Williams 2009; Carbone et al. 2013
	Cherry tree (*Prunus* × *yedoensis*) peak bloom advance of 1 week to 1 month	National Capital Parks	A1B, A2	Chung et al. 2011
	Coastal scrub vegetation reduction	Point Reyes NS	A1B, A2	Hameed et al. 2013
	Douglas-fir (*Pseudotsuga menziesii*) growth reduction; mountain hemlock (*Tsuga mertensiana*) and subalpine fir (*Abies lasiocarpa*) growth increase	Mount Rainier NP, North Cascades NP, and Olympic NP	B1, A1B	Albright and Peterson 2013
	Eastern US trees, 134 species, habitat change for one-fourth to three-fourths	121 eastern US NPs	B1, A1FI	Fisichelli et al. 2014
	Foothills palo verde (*Parkinsonia microphylla*), ocotillo (*Fouquieria splendens*), and creosote bush (*Larrea tridentata*) increased mortality	Organ Pipe Cactus NM, Saguaro NP	HS	Munson et al. 2012
	Limber pine (*Pinus flexilis*) upslope range contraction	Rocky Mountain NP	RCP4.5, RCP8.5	Monahan et al. 2013
	Quaking aspen (*Populus tremuloides*) habitat reduction 46%–94%	Yellowstone NP	B1, B2, A2	Rehfeldt, Ferguson, and Crookston 2009
	Single-leaf piñon (*Pinus monophylla*) and California juniper (*Juniperus californica*) habitat reduction	Joshua Tree NP	+3°C	Barrows et al. 2014

	Torrey pine (*Pinus torreyana* ssp. *insularis*) requires water from fog and the cooling effect of overcast skies	Channel Islands NP	HS	Williams et al. 2008
	Tree dieback	Southwestern United States; Bandelier NM	A2	Williams et al. 2013
	Tropical montane cloud forests sensitive to drought	Hawai'i, Haleakala NP	HS	Loope and Giambelluca 1998
	Western white pine (*Pinus monticola*) and mountain hemlock (*Tsuga mertensiana*) increase in climate water deficit	Yosemite NP	B1 (1.5°C)	Lutz, Wagtendonk, and Franklin 2010
	Whitebark pine (*Pinus albicaulis*) habitat reduction 71%–99%	Greater Yellowstone Ecosystem; Grand Teton NP, John D. Rockefeller Jr. Memorial Parkway, Yellowstone NP	RCP4.5, RCP8.5	Chang, Hansen, and Piekielek 2014
Wetlands	Buttonwood (*Conocarpus erectus*), mahogany (*Swietenia mahagoni*), and other species killed by saltwater intrusion	Everglades NP	Sea level +10 cm	Saha et al. 2011
	Mangrove forests soil accumulation too low, eroded soil affects benthic habitats	Everglades NP	+1.5°C, precipitation ±10%, sea level +46 cm	Koch et al. 2015
	Sawgrass decrease with decreased precipitation	Everglades NP	B1, A1B, A2	Todd et al. 2012
	Tall sawgrass and pine savanna more strongly affected than sawgrass	Everglades NP	Inundation time +30% to −60%	Foti et al. 2013
	Wetland or upland vegetation decrease, depending on precipitation	Everglades NP	+1.5°C, precipitation ±10%, sea level +46 cm	van der Valk, Volin, and Wetzel 2015
Wildfire	Fire frequency increase 3 to 10 times historical frequencies	Yellowstone NP	A2	Westerling et al. 2011

(*continued*)

Table 6.3 (continued)

Resource	Vulnerability	National park	Scenario	Reference
Animals				
Amphibians	Mountain pond habitat reduction	Washington; Mount Rainier NP, North Cascades NP, Olympic NP	A1B	Ryan et al. 2014
Birds	Habitat decline for half of 162 species	Bering Land Bridge N. Pres., Cape Krusenstern NP, Gates of the Arctic NP Pres., Kobuk Valley NP, Noatak N. Pres.	A1B	Marcot et al. 2015
	Loggerhead Shrike (*Lanius ludovicianus*), Scaled Quail (*Callipepla squamata*), and Rock Wren (*Salpinctes obsoletus*) ranges shift from desert and grassland to shrubland	Big Bend NP	A1FI	White et al. 2011
	Northern Spotted Owl (*Strix occidentalis caurina*) survival decreases in areas of warmer, wetter winters and hotter, drier summers	Oregon, Washington; Olympic NP	HS	Glenn et al. 2011
	Rufa Red Knot (*Calidris canutus rufa*) endangered by habitat reduction from sea-level rise, prey reduction from heat mortality, phenology mismatch of Red Knot migration and prey availability	Atlantic Coast, Gulf of Mexico; NPs	HS	US Department of the Interior 2014
	Wading bird habitat reduced under drier scenarios	Everglades NP	+1.5°C, precipitation ±10%, sea level +31 cm	Catano et al. 2015
	White-tailed Ptarmigan (*Lagopus leucurus*) population reduction by half	Rocky Mountain NP	+2°C–3°C	Wang et al. 2002
Corals	Coral early life-phases particularly susceptible to ocean acidification	Dry Tortugas NP	HS	Kuffner, Hickey, and Morrison 2013
	Corals vulnerable to bleaching but show some tolerance and adaptive capacity	NP of American Samoa	HS	Craig, Birkeland, and Belliveau 2001; Oliver and Palumbi 2009; Palumbi et al. 2014

Fish	Devil's Hole pupfish (*Cyprinodon diabolis*) favorable spawning conditions reduction up to 2 weeks	Death Valley NP	RCP2.6, RCP4.5, RCP6.0, RCP8.5	Hausner et al. 2014
	Freshwater fish reduced abundance from advanced spring ice breakup and melting permafrost	Kobuk Valley NP	A1B	Durand et al. 2011
	Pinfish (*Lagodon rhomboides*) estuarine habitat decrease	Everglades NP	+1°C, precipitation ±10%, sea level +46 cm	Kearney et al. 2015
	Yellowstone cutthroat trout (*Oncorhynchus clarkii bouvieri*) habitat decrease in low-elevation streams, but growth increase at higher elevations	Yellowstone NP	A2	Al-Chokhachy et al. 2013
Insect pests	Argentine ant (*Linepithema humile*) habitat increase with temperature increase	Haleakala NP	HS	Hartley, Krushelnycky, and Lester 2010
Insects	Karner Blue butterfly (*Lycaeides melissa samuelis*) susceptible to higher temperatures, among other factors	Indiana Dunes NL	HS	Grundel and Pavlovic 2007
	Meltwater stonefly (*Lednia tumana*) habitat reduction 80%	Glacier NP	A1B	Muhlfeld et al. 2011
Mammals	For 213 mammal species, average park loss of 8% of species, gain 48%, rodents 40% of the species influx	Acadia NP, Big Bend NP, Glacier NP, Great Smoky Mountains NP, Shenandoah NP, Yellowstone NP, Yosemite NP, Zion NP	CO_2 doubling	Burns, Johnston, and Schmitz 2003
	For 39 mammal species, habitat decline for 62%	Bering Land Bridge N. Pres., Cape Krusenstern NP, Gates of the Arctic NP Pres., Kobuk Valley NP, Noatak N. Pres.	A1B	Marcot et al. 2015
	American bison (*Bison bison*) forage quality and animal weight reduction from hotter and drier conditions	Western United States; Badlands NP, Great Sand Dunes NP Pres.	HS	Craine 2013

(*continued*)

Table 6.3 (continued)

Resource	Vulnerability	National park	Scenario	Reference
Mammals				
	American marten (*Martes americana*), American pika (*Ochotona princeps*), Canada lynx (*Lynx canadensis*), hoary marmot (*Marmota caligata*), mountain goat (*Oreamnos americanus*), and wolverine (*Gulo gulo*) range reductions	Washington; Mount Rainier NP, North Cascades NP, Olympic NP	A1B, A2	Johnston, Freund, and Schmitz 2012
	American pika (*Ochotona princeps*) extirpation	Lassen NP, Sequoia NP, Yosemite NP	RCP4.5, RCP8.5	Stewart et al. 2015
	Belding's ground squirrel (*Urocitellus beldingi*) habitat reduction 52%–99%	Lassen NP, Sequoia NP, Yosemite NP	A2	Morelli et al. 2012
	Desert bighorn sheep (*Ovis canadensis nelsoni*) habitat upslope shift	California; Joshua Tree NP, Mojave N. Pres.	HS	Epps et al. 2006; Epps et al. 2007
	Indiana bat (*Myotis sodalis*) northward range shift away from the park	Eastern United States; Cumberland Gap NHP	B2, A1B	Loeb and Winters 2013
	Northern elephant seals (*Mirounga angustirostris*) haul-out habitat inundation up to 69%	Point Reyes NS	Sea level +1.4 m	Funayama et al. 2013
	Polar bear (*Ursus maritimus*) threatened by loss of sea ice habitat	Polar areas; NPs	HS	US Department of the Interior 2008
	Wolverine (*Gulo gulo*) habitat northward shift	Western United States; Glacier NP	RCP2.6, RCP4.5, RCP8.5	Peacock 2011
	Wolverine (*Gulo gulo*) habitat reduction	Western United States; Glacier NP	A1B	McKelvey et al. 2011
Reptiles	American alligator (*Alligator mississippiensis*) habitat reduction under drier scenarios	Everglades NP	+1.5°C, precipitation ±10%, sea level +31 cm	Catano et al. 2015
	Desert tortoise (*Gopherus agassizii*) habitat reduction 57%–93%, common chuckwalla (*Sauromalus ater*) habitat reduction 49%–74%	Joshua Tree NP	+1°C–3°C, precipitation −25 to −75%	Barrows 2011

	Green turtles (*Chelonia mydas*) nest flooding from tropical storms	Canaveral NS	HS	Pike and Stiner 2007
	Lizards (*Sceloporus* spp.) habitat reduction by half	Joshua Tree NP	+1°C–3°C, precipitation −25 to −75%	Barrows and Fisher 2014
	Loggerhead sea turtle (*Caretta caretta*) hatching success reduction up to 15%	Buck Island Reef NM, Padre Island NS	B1, A1B, A2	Pike 2014
Cultural				
Archaeological mounds	Oyster shell middens flooding and erosion from sea-level rise	Canaveral NS	HS	Stalter and Kincaid 2004
Historical monuments	Flooding from sea-level rise and storm surge	Washington, DC; Constitution Gardens, Korean War Veterans Memorial, Martin Luther King Jr. Memorial, National Capital Parks, National Mall, Pennsylvania Avenue NHS, Theodore Roosevelt Island Park, Thomas Jefferson Memorial, World War I Memorial, World War II Memorial	Sea level +0.1–5 m	Ayyub, Braileanu, and Qureshi 2012
Historical monuments	Flooding from sea-level rise and storm surge	UNESCO World Heritage Sites globally; Independence NHP, San Juan NHS, Statue of Liberty NM	Sea level +2.8–4.4 m	Marzeion and Levermann 2014

Note: Emissions scenarios and projections of global average temperature increase (mean ±standard deviation) from IPCC (2013): 1986–2005 to 2081–2100, RCP2.6 = 1°C ±0.4°C, RCP4.5 = 1.8°C ±0.5°C, RCP6.0 = 2.2°C ±0.5°C, RCP8.5 = 3.7°C ±0.7°C; IPCC (2007b): 1980–1999 to 2090–2099, B1 = 1.8°C (1.1°C–2.9°C), B2 = 2.4°C (1.4°C–3.8°C), A1B = 2.8°C (1.7°C–4.4°C), A2 = 3.4°C (2.0°C–5.4°C), A1FI = 4.0°C (2.4°C–6.4°C); IPCC (2001b): CO$_2$ doubling in 70 years = 3.5°C ±0.9°C. HS = sensitivity based on historical or experimental data. See table 6.1 abbreviations.

Areas of high vulnerability include parts of Acadia, Joshua Tree, Mount Rainier, Rocky Mountain, Saguaro, and Yosemite National Parks, while potential refugia include parts of Death Valley National Park, Organ Pipe Cactus National Monument, and White Sands National Monument.

Bandelier National Monument and the southwestern United States are vulnerable to tree dieback and possible conversion of some forest to grassland because of drought stress under climate change rising to its highest level in 1,000 years (Williams et al. 2013). Everglades National Park is vulnerable to inundation of extensive areas because of sea-level rise and alterations of upland vegetation due to changes in precipitation (see table 6.3). Joshua Tree National Park is vulnerable to nearly complete disappearance of suitable habitat for the Joshua tree (*Yucca brevifolia*) (Cole et al. 2011; Barrows and Murphy-Mariscal 2012).

Warmer and wetter conditions render many ecosystems vulnerable to increased spread of invasive species (Bellard et al. 2013). The Appalachian Trail is vulnerable to increased spread of the invasive tree-of-heaven (*Ailanthus altissima*) (Clark, Wang, and August 2014).

Although wildfire is a natural and necessary part of many forest ecosystems, climate change could increase fire frequencies far above levels to

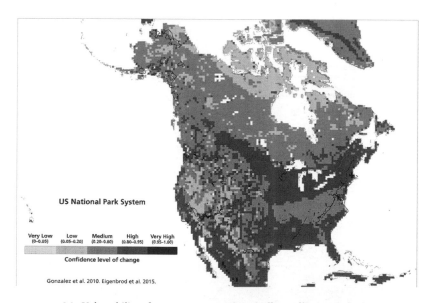

6.3. Vulnerability of ecosystems to combined effects of biome shifts due to climate change and habitat fragmentation due to land-cover change, based on 1901–2002 climate trends, 1990–2100 vegetation projections, and 2009 land cover (Gonzalez et al. 2010; Eigenbrod et al. 2015).

which current vegetation are adapted (Turner et al., this volume, ch. 5). Under high emissions, hotter temperatures could increase wildfire frequencies in Yellowstone and Grand Teton National Parks and across the Greater Yellowstone Ecosystem by 300% to 1,000% by 2100 (Westerling et al. 2011).

Vulnerability of Animals

Analyses project potential future vulnerabilities to climate change of corals, insects, amphibians, reptiles, birds, and mammals in many national parks (see table 6.3). Climate change renders coral reefs vulnerable to bleaching from warmer waters and to dissolving from ocean acidification (Wong et al. 2014). Corals in the National Park of American Samoa show some tolerance and adaptive capacity (Craig, Birkeland, and Belliveau 2001; Oliver and Palumbi 2009; Palumbi et al. 2014). In Dry Tortugas National Park, ocean acidification could especially affect early life-phases of coral (Kuffner, Hickey, and Morrison 2013).

Numerous wildlife species in US national parks that are listed as endangered under the US Endangered Species Act are vulnerable to increased mortality under climate change. In Indiana Dunes National Lakeshore, the Karner blue butterfly (*Lycaeides melissa samuelis*) is vulnerable to extirpation under hotter temperatures because of acceleration of larval development, decreased fitness, and a lack of wild lupines (*Lupinus perennis*), its food source (Grundel and Pavlovic 2007). The Devil's Hole pupfish (*Cyprinodon diabolis*), found in the world in only one small pool in Death Valley National Park, is vulnerable to a reduction of favorable spawning conditions from 74 days to 57 days under high emissions (Hausner et al. 2014). At Canaveral National Seashore, green turtles (*Chelonia mydas*) are vulnerable to potential flooding of nests from increases in storms (Pike and Stiner 2007). Using data from national parks and other areas, the US government has added two species to the US Endangered Species Act lists because of vulnerability to climate change. The polar bear (*Ursus maritimus*) is listed as endangered under the act because of the reduction of its sea ice habitat under climate change (US Department of the Interior 2008). The Rufa Red Knot (*Calidris canutus rufa*), a migratory shorebird found in Padre Island National Seashore and along the Atlantic coast, is listed as threatened under the act because of urban development, sea-level rise, and reductions in food species due to climate change (US Department of the Interior 2014).

Upslope and poleward shifting of cooler climates and biomes increases the vulnerability of high-elevation mammals. American pika (*Ochotona princeps*) is vulnerable to extirpation in Lassen, Sequoia, and Yosemite National Parks (Stewart et al. 2015). American pika, Canada lynx (*Lynx canadensis*), hoary marmot (*Marmota caligata*), and the wolverine (*Gulo gulo*) are vulnerable to range contractions in Mount Rainier, North Cascades, and Olympic National Parks (Johnston, Freund, and Schmitz 2012).

Vulnerability of Cultural Resources

Thawing and exposure of archaeological artifacts as glaciers melt in Wrangell–Saint Elias and Lake Clark National Parks and Preserves can cause organic objects to decompose and be lost forever (Dixon, Manley, and Lee 2005; VanderHoek et al. 2012) if they are not detected, secured, and protected. In addition, sea-level rise renders vulnerable cultural sites in national parks along the Atlantic and Pacific coasts, including oyster shell middens over a millennium old in Canaveral National Seashore (Stalter and Kincaid 2004), the National Mall and other monuments in Washington, DC (Ayyub, Braileanu, and Qureshi 2012), and UNESCO World Heritage Sites such as the Statue of Liberty National Monument (Marzeion and Levermann 2014).

Conclusions

Field evidence documents impacts of human climate change across the US National Park System. The alteration of ecosystems and physical and cultural resources in US national parks reflects the widespread impact of climate change around the world. If we do not reduce greenhouse gas emissions, vulnerability analyses project future damage to the irreplaceable and globally unique wonders of US national parks.

While dedicated park managers may make extraordinary efforts to protect the national parks, the most effective way to attack a problem is to eliminate its cause. Reducing greenhouse gas emissions will reduce the magnitude of future climate change and threats to national parks. Climate change projections are not predictions—they are not inevitable. Greenhouse gas mitigation analyses by the IPCC (2014b) show that it is within our power to avoid the most drastic impacts of climate change by improving energy efficiency, installing renewable energy systems, conserving forests with large carbon stocks, expanding public transit, and using other measures to reduce emissions. Billions of small actions caused the problem of climate change, so billions of small sustainable actions can help us solve it.

Literature Cited

Abu-Asab, M. S., P. M. Peterson, S. G. Shetler, and S. S. Orli. 2001. Earlier plant flowering in spring as a response to global warming in the Washington, DC, area. Biodiversity and Conservation 10:597–612.

Albright, W. L., and D. L. Peterson. 2013. Tree growth and climate in the Pacific Northwest, North America: a broad-scale analysis of changing growth environments. Journal of Biogeography 40:2119–2133.

Al-Chokhachy, R., J. Alder, S. Hostetler, R. Gresswell, and B. Shepard. 2013. Thermal controls of Yellowstone cutthroat trout and invasive fishes under climate change. Global Change Biology 19:3069–3081.

Allen, C. D., and D. D. Breshears. 1998. Drought-induced shift of a forest–woodland ecotone: rapid landscape response to climate variation. Proceedings of the National Academy of Sciences USA 95:14839–14842.

Arendt, A. A., S. B. Luthcke, C. F. Larsen, W. Abdalati, W. B. Krabill, and M. J. Beedle. 2008. Validation of high-resolution GRACE mascon estimates of glacier mass changes in the St Elias Mountains, Alaska, USA, using aircraft laser altimetry. Journal of Glaciology 54:778–787.

Ault, T. R., A. K. Macalady, G. T. Pederson, J. L. Betancourt, and M. D. Schwartz. 2011. Northern Hemisphere modes of variability and the timing of spring in western North America. Journal of Climate 24:1003–1014.

Ayyub, B. M., H. G. Braileanu, and N. Qureshi. 2012. Prediction and impact of sea level rise on properties and infrastructure of Washington, DC. Risk Analysis 32:1901–1918.

Balling, R. C., G. A. Meyer, and S. G. Wells. 1992. Climate change in Yellowstone National Park—is the drought-related risk of wildfires increasing? Climatic Change 22:35–45.

Balser, A. W., J. B. Jones, and R. Gens. 2014. Timing of retrogressive thaw slump initiation in the Noatak Basin, northwest Alaska, USA. Journal of Geophysical Research: Earth Surface 119:1106–1120.

Barnett, T. P., and D. W. Pierce. 2008. When will Lake Mead go dry? Water Resources Research 44:W03201.

Barnett, T. P., D. W. Pierce, H. G. Hidalgo, C. Bonfils, B. D. Santer, T. Das, G. Bala, et al. 2008. Human-induced changes in the hydrology of the western United States. Science 319:1080–1083.

Baron, J. S., L. Gunderson, C. D. Allen, E. Fleishman, D. McKenzie, L. A. Meyerson, J. Oropeza, and N. Stephenson. 2009. Options for national parks and reserves for adapting to climate change. Environmental Management 44:1033–1042.

Barrows, C. W. 2011. Sensitivity to climate change for two reptiles at the Mojave-Sonoran Desert interface. Journal of Arid Environments 75:629–635.

Barrows, C. W., and M. Fisher. 2014. Past, present and future distributions of a local assemblage of congeneric lizards in southern California. Biological Conservation 180:97–107.

Barrows, C. W., J. Hoines, K. D. Fleming, M. S. Vamstad, M. Murphy-Mariscal, K. Lalumiere, and M. Harding. 2014. Designing a sustainable monitoring framework for assessing impacts of climate change at Joshua Tree National Park, USA. Biodiversity and Conservation 23:3263–3285.

Barrows, C. W., and M. L. Murphy-Mariscal. 2012. Modeling impacts of climate change on Joshua trees at their southern boundary: how scale impacts predictions. Biological Conservation 152:29–36.

Bearup, L. A., R. M. Maxwell, D. W. Clow, J. E. McCray. 2014. Hydrological effects of forest transpiration loss in bark beetle-impacted watersheds. Nature Climate Change 4:481–486.

Bellard, C., W. Thuiller, B. Leroy, P. Genovesi, M. Bakkenes, and F. Courchamp. 2013. Will climate change promote future invasions? Global Change Biology 19:3740-3748.

Bindoff, N. L., P. A. Stott, K. M. AchutaRao, M. R. Allen, N. Gillett, D. Gutzler, K. Hansingo, et al. 2013. Detection and attribution of climate change: from global to regional. Pages 867–952 in T. F. Stocker, D. Qin, G. K. Plattner, M. Tignor, S. K. Allen, J. Boschung, A. Nauels, et al., eds. Climate change 2013: the physical science basis. Contribution of Working Group I to the Fifth Assessment Report of the Intergovernmental Panel on Climate Change. Cambridge University Press, Cambridge, United Kingdom, and New York, New York.

Bonfils, C., B. D. Santer, D. W. Pierce, H. G. Hidalgo, G. Bala, T. Das, T. P. Barnett, et al. 2008. Detection and attribution of temperature changes in the mountainous western United States. Journal of Climate 21:6404-6424.

Breshears, D. D., N. S. Cobb, P. M. Rich, K. P. Price, C. D. Allen, R. G. Balice, W. H. Romme, et al. 2005. Regional vegetation die-off in response to global-change-type drought. Proceedings of the National Academy of Sciences USA 102:15144-15148.

Brodie, J., E. Post, F. Watson, and J. Berger. 2012. Climate change intensification of herbivore impacts on tree recruitment. Proceedings of the Royal Society B 279:1366-1370.

Brown, J., J. Harper, and N. Humphrey. 2010. Cirque glacier sensitivity to 21st century warming: Sperry Glacier, Rocky Mountains, USA. Global and Planetary Change 74: 91–98.

Bunn, A. G, L. J. Graumlich, and D. L. Urban. 2005. Trends in twentieth-century tree growth at high elevations in the Sierra Nevada and White Mountains, USA. The Holocene 15:481–488.

Burns, C. E., K. M. Johnston, and O. J. Schmitz. 2003. Global climate change and mammalian species diversity in US national parks. Proceedings of the National Academy of Sciences USA 100:11474-11477.

Carbone, M. S., A. P. Williams, A. R. Ambrose, C. M. Boot, E. S. Bradley, T. E. Dawson, S. M. Schaeffer, J. P. Schimel, and C. J. Still. 2013. Cloud shading and fog drip influence the metabolism of a coastal pine ecosystem. Global Change Biology 19:484-497.

Caruso, N. M., M. W. Sears, D. C. Adams, and K. R. Lips. 2014. Widespread rapid reductions in body size of adult salamanders in response to climate change. Global Change Biology 20:1751-1759.

Catano, C. P., S. S. Romañach, J. M. Beerens, L. G. Pearlstine, L. A. Brandt, K. M. Hart, F. J. Mazzotti, and J. C. Trexler. 2015. Using scenario planning to evaluate the impacts of climate change on wildlife populations and communities in the Florida Everglades. Environmental Management 55:807–823.

Cavanaugh, K. C., J. R. Kellner, A. J. Forde, D. S. Gruner, J. D. Parker, W. Rodriguez, and I. C. Feller. 2014. Poleward expansion of mangroves is a threshold response to decreased frequency of extreme cold events. Proceedings of the National Academy of Sciences USA 111:723-727.

Chang, T., A. J. Hansen, and N. Piekielek. 2014. Patterns and variability of projected bioclimatic habitat for Pinus albicaulis in the greater Yellowstone area. PLoS ONE 9:e111669.

Chung, U., L. Mack, J. Yun, and S. H. Kim. 2011. Predicting the timing of cherry blossoms in Washington, DC and Mid-Atlantic States in response to climate change. PLoS ONE 6:e27439.

Church, J. A., P. U. Clark, A. Cazenave, J. M. Gregory, S. Jevrejeva, A. Levermann, M. A.

Merrifield, et al. 2013. Sea level change. Pages 1137–1216 *in* T. F. Stocker, D. Qin, G. K. Plattner, M. Tignor, S. K. Allen, J. Boschung, A. Nauels, et al., eds. Climate change 2013: the physical science basis. Contribution of Working Group I to the Fifth Assessment Report of the Intergovernmental Panel on Climate Change. Cambridge University Press, Cambridge, United Kingdom, and New York, New York.

Church, J. A., and N. J. White. 2011. Sea-level rise from the late 19th to the early 21st century. Surveys in Geophysics 32:585–602.

Clark, J., Y. Wang, and P. V. August. 2014. Assessing current and projected suitable habitats for tree-of-heaven along the Appalachian Trail. Philosophical Transactions of the Royal Society B 369:20130192.

Cole, K. L., K. Ironside, J. Eischeid, G. Garfin, P. B. Duffy, and C. Toney. 2011. Past and ongoing shifts in Joshua tree distribution support future modeled range contraction. Ecological Applications 21:137–149.

Craig, P., C. Birkeland, and S. Belliveau. 2001. High temperatures tolerated by a diverse assemblage of shallow-water corals in American Samoa. Coral Reefs 20:185–189.

Craine, J. M. 2013. Long-term climate sensitivity of grazer performance: a cross-site study. PLoS ONE 8:e67065.

Crossman, J., M. N. Futter, P. G. Whitehead. 2013. The significance of shifts in precipitation patterns: modelling the impacts of climate change and glacier retreat on extreme flood events in Denali National Park, Alaska. PLoS ONE 8:e74054.

Daly, C., M. Halbleib, J. I. Smith, W. P. Gibson, M. K. Doggett, G. H. Taylor, J. Curtis, and P. P. Pasteris. 2008. Physiographically sensitive mapping of climatological temperature and precipitation across the conterminous United States. International Journal of Climatology 28:2031–2064.

Damschen, E. I., S. Harrison, and J. B. Grace. 2010. Climate change effects on an endemic-rich edaphic flora: resurveying Robert H. Whittaker's Siskiyou sites (Oregon, USA). Ecology 91:3609–3619.

Dawadi, S., and S. Ahmad. 2012. Changing climatic conditions in the Colorado River Basin: implications for water resources management. Journal of Hydrology 430–431: 127–141.

DeFalco, L. A., T. C. Esque, S. J. Scoles-Sciulla, and J. Rodgers. 2010 Desert wildfire and severe drought diminish survivorship of the long-lived Joshua tree (*Yucca brevifolia*; Agavaceae). American Journal of Botany 97:243–250.

Dixon, E. J., W. F. Manley, and C. M. Lee. 2005. Glacier artifacts the emerging archaeology of glaciers and ice patches: examples from Alaska's Wrangell–St. Elias National Park and Preserve. American Antiquity 70:129–143.

Dolanc, C. R., J. H. Thorne, and H. D. Safford. 2013. Widespread shifts in the demographic structure of subalpine forests in the Sierra Nevada, California, 1934 to 2007. Global Ecology and Biogeography 22:264–276.

Driscoll, W. W., G. C. Wiles, R. D. D'Arrigo, and M. Wilmking. 2005. Divergent tree growth response to recent climatic warming, Lake Clark National Park and Preserve, Alaska. Geophysical Research Letters 32:L20703.

Durand, J. R., R. A. Lusardi, D. M. Nover, R. J. Suddeth, G. Carmona-Catot, C. R. Connell-Buck, S. E. Gatzke, et al. 2011. Environmental heterogeneity and community structure of the Kobuk River, Alaska, in response to climate change. Ecosphere 2:art44.

Eakin, C. M., J. A. Morgan, S. F. Heron, T. B. Smith, G. Liu, L. Alvarez-Filip, B. Baca, et al. 2010. Caribbean corals in crisis: record thermal stress, bleaching, and mortality in 2005. PLoS ONE 5:e13969.

Eastman, L. M., T. L. Morelli, K. C. Rowe, C. J. Conroy, and C. Moritz. 2012. Size increase in high elevation ground squirrels over the last century. Global Change Biology 18: 1499–1508.

Eigenbrod, F., P. Gonzalez, J. Dash, and I. Steyl. 2015. Vulnerability of ecosystems to climate change moderated by habitat intactness. Global Change Biology 21:275–286.

Ellis, R. A., D. J. Jacob, M. P. Sulprizio, L. Zhang, C. D. Holmes, B. A. Schichtel, T. Blett, et al. 2013. Present and future nitrogen deposition to national parks in the United States: critical load exceedances. Atmospheric Chemistry and Physics 13:9083–9095.

Epps, C. W., P. J. Palsbøll, J. D. Wehausen, G. K. Roderick, and D. R. McCullough. 2006. Elevation and connectivity define genetic refugia for mountain sheep as climate warms. Molecular Ecology 15:4295–4302.

Epps, C. W., J. D. Wehausen, V. C. Bleich, S. G. Torres, and J. S. Brashares. 2007. Optimizing dispersal and corridor models using landscape genetics. Journal of Applied Ecology 44:714–724.

Fischer, D. T., C. J. Still, and A. P. Williams. 2009. Significance of summer fog and overcast for drought stress and ecological functioning of coastal California endemic plant species. Journal of Biogeography 36:783–799.

Fisichelli, N. A., S. R. Abella, M. Peters, and F. J. Krist. 2014. Climate, trees, pests, and weeds: change, uncertainty, and biotic stressors in eastern US national park forests. Forest Ecology and Management 327:31–39.

Foti, R., M. del Jesus, A. Rinaldo, and I. Rodriguez-Iturbe. 2013. Signs of critical transition in the Everglades wetlands in response to climate and anthropogenic changes. Proceedings of the National Academy of Sciences USA 110:6296–6300.

Franklin, J. F., W. H. Moir, G. W. Douglas, and C. Wiberg. 1971. Invasion of subalpine meadows by trees in the Cascade Range, Washington and Oregon. Arctic and Alpine Research 3:215–224.

Funayama, K., E. Hines, J. Davis, and S. Allen. 2013. Effects of sea-level rise on northern elephant seal breeding habitat at Point Reyes Peninsula, California. Aquatic Conservation: Marine and Freshwater Ecosystems 23:233–245.

Gahl, M. K., and A. J. K. Calhoun. 2010. The role of multiple stressors in ranavirus-caused amphibian mortalities in Acadia National Park wetlands. Canadian Journal of Zoology 88:108–121.

Garrity, S. R., C. D. Allen, S. P. Brumby, C. Gangodagamage, N. G. McDowell, D. M. Cai. 2013. Quantifying tree mortality in a mixed species woodland using multitemporal high spatial resolution satellite imagery. Remote Sensing of Environment 129:54–65.

Glenn, E. M., R. G. Anthony, E. D. Forsman, and G. S. Olson. 2011. Local weather, regional climate, and annual survival of the northern spotted owl. The Condor 113:159–176.

Gonzalez, P. 2011. Science for natural resource management under climate change. Issues in Science and Technology 27(4):65–74.

Gonzalez, P., R. P. Neilson, J. M. Lenihan, and R. J. Drapek. 2010. Global patterns in the vulnerability of ecosystems to vegetation shifts due to climate change. Global Ecology and Biogeography 19:755–768.

Granshaw, F. D., and A. G. Fountain. 2006. Glacier change (1958–1998) in the North Cascades National Park Complex, Washington, USA. Journal of Glaciology 52:251–256.

Grundel, R., and N. B. Pavlovic. 2007. Resource availability, matrix quality, microclimate, and spatial pattern as predictors of patch use by the Karner blue butterfly. Biological Conservation 135:135–144.

Hall, M. H. P., and D. B. Fagre. 2003. Modeled climate-induced glacier change in Glacier National Park, 1850–2100. BioScience 53:131–140.

Hameed, S. O., K. A. Holzer, A. N. Doerr, J. H. Baty, and M. W. Schwartz. 2013. The value of a multi-faceted climate change vulnerability assessment to managing protected lands: lessons from a case study in Point Reyes National Seashore. Journal of Environmental Management 121:37–47.

Hansen, A. J., N. Piekielek, C. Davis, J. Haas, D. M. Theobald, J. E. Gross, W. B. Monahan, T. Olliff, and S. W. Running. 2014. Exposure of US national parks to land use and climate change 1900–2100. Ecological Applications 24:484–502.

Hartley, S., P. D. Krushelnycky, and P. J. Lester. 2010. Integrating physiology, population dynamics and climate to make multi-scale predictions for the spread of an invasive insect: the Argentine ant at Haleakala National Park, Hawaii. Ecography 33:83–94.

Hausner, M. B., K. P. Wilson, D. B. Gaines, F. Suárez, G. G. Scoppettone, and S. W. Tyler. 2014. Life in a fishbowl: prospects for the endangered Devils Hole pupfish (*Cyprinodon diabolis*) in a changing climate. Water Resources Research 50:7020–7034.

Hessl, A. E., and W. L. Baker. 1997. Spruce and fir regeneration and climate in the forest-tundra ecotone of Rocky Mountain National Park, Colorado, USA. Arctic and Alpine Research 29:173–183.

Hitch, A. T., and P. L. Leberg. 2007. Breeding distributions of North American bird species moving north as a result of climate change. Conservation Biology 21:534–539.

Hoffman, M. J., A. G. Fountain, and J. M. Achuff. 2007. 20th-century variations in area of cirque glaciers and glacierets, Rocky Mountain National Park, Rocky Mountains, Colorado, USA. Annals of Glaciology 46:349–354.

Intergovernmental Panel on Climate Change (IPCC). 2001a. Climate change 2001: impacts, adaptation, and vulnerability. Cambridge University Press, Cambridge, United Kingdom.

———. 2001b. Climate change 2001: the scientific basis. Cambridge University Press, Cambridge, United Kingdom.

———. 2007a. Climate change 2007: impacts, adaptation, and vulnerability. Cambridge University Press, Cambridge, United Kingdom.

———. 2007b. Climate change 2007: the physical science basis. Cambridge University Press, Cambridge, United Kingdom.

———. 2012. Managing the risks of extreme events and disasters to advance climate change adaptation. Cambridge University Press, Cambridge, United Kingdom.

———. 2013. Climate change 2013: the physical science basis. Cambridge University Press, Cambridge, United Kingdom.

———. 2014a. Climate change 2014: impacts, adaptation, and vulnerability. Cambridge University Press, Cambridge, United Kingdom.

———. 2014b. Climate change 2014: mitigation of climate change. Cambridge University Press, Cambridge, United Kingdom.

Jakubos, B., and W. H. Romme. 1993. Invasion of subalpine meadows by lodgepole pine in Yellowstone National Park, Wyoming, USA. Arctic and Alpine Research 25:382–390.

Janke, J. R. 2005. Modeling past and future alpine permafrost distribution in the Colorado Front Range. Earth Surface Processes and Landforms 30:1495–1508.

Johnston, K. M., K. A. Freund, and O. J. Schmitz. 2012. Projected range shifting by montane mammals under climate change: implications for Cascadia's national parks. Ecosphere 3:art97.

Jones, B. M., G. Grosse, C. D. Arp, M. C. Jones, K. M. Walter Anthony, and V. E. Romanovsky. 2011. Modern thermokarst lake dynamics in the continuous permafrost zone, northern Seward Peninsula, Alaska. Journal of Geophysical Research 116:G00M03.

Jorgenson, M. T., B. G. Marcot, D. K. Swanson, J. C. Jorgenson, and A. R. DeGange. 2015.

Projected changes in diverse ecosystems from climate warming and biophysical drivers in northwest Alaska. Climatic Change 130:131–144.

Kasischke, E. S., D. L. Verbyla, T. S. Rupp, A. D. McGuire, K. A. Murphy, R. Jandt, J. L. Barnes, et al. 2010. Alaska's changing fire regime—implications for the vulnerability of its boreal forests. Canadian Journal of Forest Research 40:1313–1324.

Kearney, K. A., M. Butler, R. Glazer, C. R. Kelble, J. E. Serafy, and E. Stabenau. 2015. Quantifying Florida Bay habitat suitability for fishes and invertebrates under climate change scenarios. Environmental Management 55:836–856.

Klasner, F. L., and D. B. Fagre. 2002. A half century of change in alpine treeline patterns at Glacier National Park, Montana, USA. Arctic, Antarctic, and Alpine Research 34:49–56.

Koch, M. S., C. Coronado, M. W. Miller, D. T. Rudnick, E. Stabenau, R. B. Halley, and F. H. Sklar. 2015. Climate change projected effects on coastal foundation communities of the greater Everglades using a 2060 scenario: need for a new management paradigm. Environmental Management 55:857–875.

Krushelnycky, P. D., L. L. Loope, T. W. Giambelluca, F. Starr, K. Starr, D. R. Drake, A. D. Taylor, and R. H. Robichaux. 2013. Climate-associated population declines reverse recovery and threaten future of an iconic high-elevation plant. Global Change Biology 19:911–922.

Kuffner, I. B., T. D. Hickey, and J. M. Morrison. 2013. Calcification rates of the massive coral Siderastrea siderea and crustose coralline algae along the Florida Keys (USA) outer-reef tract. Coral Reefs 32:987–997.

Kuffner, I. B., B. H. Lidz, J. H. Hudson, and J. S. Anderson. 2015. A century of ocean warming on Florida Keys coral reefs: historic in situ observations. Estuaries and Coasts 38:1085–1096.

Larsen, C. F., R. J. Motyka, A. A. Arendt, K. A. Echelmeyer, and P. E. Geissler. 2007. Glacier changes in southeast Alaska and northwest British Columbia and contribution to sea level rise. Journal of Geophysical Research 112:F01007.

Larson, R. P., J. M. Byrne, D. L. Johnson, S. W. Kienzie, and M. G. Letts. 2011. Modelling climate change impacts on spring runoff for the Rocky Mountains of Montana and Alberta II: runoff change projections using future scenarios. Canadian Water Resources Journal 36:35–52.

La Sorte, F. A., and F. R. Thompson. 2007. Poleward shifts in winter ranges of North American birds. Ecology 88:1803–1812.

Lesica, P. 2014. Arctic-alpine plants decline over two decades in Glacier National Park, Montana, USA. Arctic, Antarctic, and Alpine Research 46:327–332.

Lesica, P., and B. McCune. 2004. Decline of arctic-alpine plants at the southern margin of their range following a decade of climatic warming. Journal of Vegetation Science 15:679–690.

Levine, J. M., A. K. McEachern, and C. Cowan. 2008. Rainfall effects on rare annual plants. Journal of Ecology 96:795–806.

Littell, J. S., D. McKenzie, D. L. Peterson, and A. L. Westerling. 2009. Climate and wildfire area burned in western US ecoprovinces, 1916–2003. Ecological Applications 19:1003–1021.

Loeb, S. C., and E. A. Winters. 2013. Indiana bat summer maternity distribution: effects of current and future climates. Ecology and Evolution 3:103–114.

Logan, J. A., W. W. Macfarlane, and L. Willcox. 2010. Whitebark pine vulnerability to climate-driven mountain pine beetle disturbance in the Greater Yellowstone Ecosystem. Ecological Applications 20:895–902.

Loope, L. L., and T. W. Giambelluca. 1998. Vulnerability of island tropical montane cloud forests to climate change, with special reference to East Maui, Hawaii. Climatic Change 39:503–517.

Lucas, K. L., and G. A. Carter. 2010. Decadal changes in habitat-type coverage on Horn Island, Mississippi, USA. Journal of Coastal Research 26:1142–1148.

Lutz, J. A., J. W. van Wagtendonk, and J. F. Franklin. 2010. Climatic water deficit, tree species ranges, and climate change in Yosemite National Park. Journal of Biogeography 37:936–950.

Macfarlane, W. W., J. A. Logan, and W. R. Kern. 2013. An innovative aerial assessment of Greater Yellowstone Ecosystem mountain pine beetle-caused whitebark pine mortality. Ecological Applications 23:421–437.

Marcot, B. G., M. T. Jorgenson, J. P. Lawler, C. M. Handel, and A. R. DeGange. 2015. Projected changes in wildlife habitats in Arctic natural areas of northwest Alaska. Climatic Change 130:145–154.

Marlon, J. R., P. J. Bartlein, D. G. Gavin, C. J. Long, R. S. Anderson, C. E. Briles, K. J. Brown, et al. 2012. Long-term perspective on wildfires in the western USA. Proceedings of the National Academy of Sciences USA 109:E535–E543.

Marzeion, B., J. G. Cogley, K. Richter, and D. Parkes. 2014. Attribution of global glacier mass loss to anthropogenic and natural causes. Science 345:919–921.

Marzeion, B., and A. Levermann. 2014. Loss of cultural world heritage and currently inhabited places to sea level rise. Environmental Research Letters 9:034001.

McEachern, A. K., D. M. Thomson, and K. A. Chess. 2009. Climate alters response of an endemic island plant to removal of invasive herbivores. Ecological Applications 19:1574–1584.

McIntyre, P. J., J. H. Thorne, C. R. Dolanc, A. L. Flint, L. E. Flint, M. Kelly, and D. D. Ackerly. 2015. Twentieth-century shifts in forest structure in California: denser forests, smaller trees, and increased dominance of oaks. Proceedings of the National Academy of Sciences USA 112:1458–1463.

McKelvey, K. S., J. P. Copeland, M. K. Schwartz, J. S. Littell, K. B. Aubry, J. R. Squires, S. A. Parks, M. M. Elsner, and G. S. Mauger. 2011. Climate change predicted to shift wolverine distributions, connectivity, and dispersal corridors. Ecological Applications 21. 2882–2897.

McMenamin, S. K., E. A. Hadly, and C. K. Wright. 2008. Climatic change and wetland desiccation cause amphibian decline in Yellowstone National Park. Proceedings of the National Academy of Sciences USA 105:16988–16993.

Millar, C. I., R. D. Westfall, D. L. Delany, J. C. King, and L. J. Graumlich. 2004. Response of subalpine conifers in the Sierra Nevada, California, USA, to 20th-Century warming and decadal climate variability. Arctic, Antarctic, and Alpine Research 36:181–200.

Monahan, W. B., T. Cook, F. Melton, J. Connor, and B. Bobowski. 2013. Forecasting distributional responses of limber pine to climate change at management-relevant scales in Rocky Mountain National Park. PLoS ONE 8:e83163.

Monahan, W. B., and N. A. Fisichelli. 2014. Climate exposure of US national parks in a new era of change. PLoS ONE 9:e101302.

Morelli, T. L., A. B. Smith, C. R. Kastely, I. Mastroserio, C. Moritz, and S. R. Beissinger. 2012. Anthropogenic refugia ameliorate the severe climate-related decline of a montane mammal along its trailing edge. Proceedings of the Royal Society B 279:4279–4286.

Moritz, C., J. L. Patton, C. J. Conroy, J. L. Parra, G. C. White, and S. R. Beissinger. 2008. Impact of a century of climate change on small-mammal communities in Yosemite National Park, USA. Science 322:261–264.

Moss, R. H., J. A. Edmonds, K. A. Hibbard, M. R. Manning, S. K. Rose, D. P. van Vuuren, T. R. Carter, et al. 2010. The next generation of scenarios for climate change research and assessment. Nature 463:747–756.

Muhlfeld, C. C., J. J. Giersch, F. R. Hauer, G. T. Pederson, G. Luikart, D. P. Peterson, C. C. Downs, and D. B. Fagre. 2011. Climate change links fate of glaciers and an endemic alpine invertebrate. Climatic Change 106:337–345.

Muller, E. M., C. S. Rogers, A. S. Spitzack, and R. van Woesik. 2008. Bleaching increases likelihood of disease on Acropora palmata (Lamarck) in Hawksnest Bay, St John, US Virgin Islands. Coral Reefs 27:191–195.

Munson, S. M., J. Belnap, and G. S. Okin. 2011. Responses of wind erosion to climate-induced vegetation changes on the Colorado Plateau. Proceedings of the National Academy of Sciences USA 108:3854–3859.

Munson, S. M., J. Belnap, C. D. Schelz, M. Moran, and T. W. Carolin. 2011. On the brink of change: plant responses to climate on the Colorado Plateau. Ecosphere 2:art68.

Munson, S. M., R. H. Webb, J. Belnap, J. A. Hubbard, D. E. Swann, and S. Rutman. 2012. Forecasting climate change impacts to plant community composition in the Sonoran Desert region. Global Change Biology 18:1083–1095.

Murdukhayeva, A., P. August, M. Bradley, C. LaBash, and N. Shaw. 2013. Assessment of inundation risk from sea level rise and storm surge in northeastern coastal national parks. Journal of Coastal Research 29:1–16.

Necsoiu, M., C. L Dinwiddie, G. R Walter, A. Larsen, and S. A Stothoff. 2013. Multi-temporal image analysis of historical aerial photographs and recent satellite imagery reveals evolution of water body surface area and polygonal terrain morphology in Kobuk Valley National Park, Alaska. Environmental Research Letters 8:025007.

Nungesser, M., C. Saunders, C. Coronado-Molina, J. Obeysekera, J. Johnson, C. McVoy, and B. Benscoter. 2015. Potential effects of climate change on Florida's Everglades. Environmental Management 55:824–835.

Oliver, T. A., and S. R. Palumbi. 2009. Distributions of stress-resistant coral symbionts match environmental patterns at local but not regional scales. Marine Ecology Progress Series 378:93–103.

Palumbi, S. R., D. J. Barshis, N. Traylor-Knowles, and R. A. Bay. 2014. Mechanisms of reef coral resistance to future climate change. Science 344:895–898.

Peacock, S. 2011. Projected 21st century climate change for wolverine habitats within the contiguous United States. Environmental Research Letters 6:014007.

Pederson, G. T., D. B. Fagre, S. T. Gray, and L. J. Graumlich. 2004. Decadal-scale climate drivers for glacial dynamics in Glacier National Park, Montana, USA. Geophysical Research Letters 31:L12203.

Pederson, G. T., S. T. Gray, C. A. Woodhouse, J. L. Betancourt, D. B. Fagre, J. S. Littell, E. Watson, B. H. Luckman, and L. J. Graumlich. 2011. The unusual nature of recent snowpack declines in the North American Cordillera. Science 333:332–335.

Pelto, M. S. 2006. The current disequilibrium of North Cascade glaciers. Hydrological Processes 20:769–779.

Pendleton, E. A., E. R. Thieler, and S. J. Williams. 2010. Importance of coastal change variables in determining vulnerability to sea- and lake-level change. Journal of Coastal Research 26:176–183.

Pierce, D. W., T. P., Barnett, H. G. Hidalgo. T. Das, C. Bonfils, B. D. Santer, G. Bala, et al. 2008. Attribution of declining western US snowpack to human effects. Journal of Climate 21:6425–6444.

Pike, D. A. 2014. Forecasting the viability of sea turtle eggs in a warming world. Global Change Biology 20:7–15.

Pike, D. A., R. L. Antworth, and J. C. Stiner. 2006. Earlier nesting contributes to shorter nesting seasons for the Loggerhead Seaturtle, *Caretta caretta*. Journal of Herpetology 40:91–94.

Pike, D. A., and J. C. Stiner. 2007. Sea turtle species vary in their susceptibility to tropical cyclones. Oecologia 153:471–478.

Portmann, R. W., S. Solomon, and G. C. Hegerl. 2009. Spatial and seasonal patterns in climate change, temperatures, and precipitation across the United States. Proceedings of the National Academy of Sciences USA 106:7324–7329.

Raffa, K. F., B. H. Aukema, B. J. Bentz, A. L. Carroll, J. A. Hicke, M. G. Turner, and W. H. Romme. 2008. Cross-scale drivers of natural disturbances prone to anthropogenic amplification: the dynamics of bark beetle eruptions. BioScience 58:501–517.

Rehfeldt, G. E., D. E. Ferguson, and N. L. Crookston. 2009. Aspen, climate, and sudden decline in western USA. Forest Ecology and Management 258:2353–2364.

Rhein, M., S. R. Rintoul, S. Aoki, E. Campos, D. Chambers, R. A. Feely, S. Gulev, et al. 2013. Observations: ocean. Pages 255–316 *in* T. F. Stocker, D. Qin, G. K. Plattner, M. Tignor, S. K. Allen, J. Boschung, A. Nauels, et al., eds. Climate change 2013: the physical science basis. Contribution of Working Group I to the Fifth Assessment Report of the Intergovernmental Panel on Climate Change. Cambridge University Press, Cambridge, United Kingdom, and New York, New York.

Riordan, B., D. Verbyla, and A. D. McGuire. 2006. Shrinking ponds in subarctic Alaska based on 1950–2002 remotely sensed images. Journal of Geophysical Research 111:G04002.

Rochefort, R. M., and D. L. Peterson. 1996. Temporal and spatial distribution of trees in subalpine meadows of Mount Rainier National Park, Washington, USA. Arctic and Alpine Research 28:52–59.

Roush, W., J. S. Munroe, and D. B. Fagre. 2007. Development of a spatial analysis method using ground based repeat photography to detect changes in the alpine treeline ecotone, Glacier National Park, Montana, USA. Arctic, Alpine, and Antarctic Research 39:297–308.

Rowe, K. C., K. M. C. Rowe, M. W. Tingley, M. S. Koo, J. L. Patton, C. J. Conroy, J. D. Perrine, S. R. Beissinger, and C. Moritz. 2015. Spatially heterogeneous impact of climate change on small mammals of montane California. Proceedings of the Royal Society B 282:20141857.

Rupp, T. S., F. S. Chapin, and A. M. Starfield. 2000. Response of subarctic vegetation to transient climatic change on the Seward Peninsula in north-west Alaska. Global Change Biology 6:541–555.

———. 2001. Modeling the influence of topographic barriers on treeline advance at the forest-tundra ecotone in northwestern Alaska. Climatic Change 48:399–416.

Ryan, M. E., W. J. Palen, M. J. Adams, and R. M. Rochefort. 2014. Amphibians in the climate vise: loss and restoration of resilience of montane wetland ecosystems in the western US. Frontiers in Ecology and the Environment 12:232–240.

Saha, A. K., S. Saha, J. Sadle, J. Jiang, M. S. Ross, R. M. Price, L. S. L. O. Sternberg, and K. S. Wendelberger. 2011. Sea level rise and South Florida coastal forests. Climatic Change 107:81–108.

Settele, J., R. Scholes, R. Betts, S. Bunn, P. Leadley, D. Nepstad, J. T. Overpeck, and M. A. Taboada. 2014. Terrestrial and inland water systems. Pages 217–360 *in* C. B. Field,

V. R. Barros, D. J. Dokken, K. J. Mach, M. D. Mastrandrea, T. E. Bilir, M. Chatterjee, et al., eds. Climate change 2014: impacts, adaptation, and vulnerability. Part A: global and sectoral aspects. Contribution of Working Group II to the Fifth Assessment Report of the Intergovernmental Panel on Climate Change. Cambridge University Press, Cambridge, United Kingdom, and New York, New York.

Smith, J. R., P. Fong, and R. F. Ambrose. 2006. Dramatic declines in mussel bed community diversity: response to climate change? Ecology 87:1153–1161.

Stalter, R., and D. Kincaid. 2004. The vascular flora of five Florida shell middens. Journal of the Torrey Botanical Society 131:93–103.

Stewart, J. A. E., J. D. Perrine, L. B. Nichols, J. H. Thorne, C. I. Millar, K. E. Goehring, C. P. Massing, and D. H. Wright. 2015. Revisiting the past to foretell the future: summer temperature and habitat area predict pika extirpations in California. Journal of Biogeography 42:880–890.

Suarez, F., D. Binkley, M. W. Kaye, and R. Stottlemyer. 1999. Expansion of forest stands into tundra in the Noatak National Preserve, northwest Alaska. Ecoscience 6:465–470.

Swetnam, T. W. 1993. Fire history and climate change in giant sequoia groves. Science 262:885–889.

Swetnam, T. W., C. H. Baisan, A. C. Caprio, P. M. Brown, R. Touchan, R. S. Anderson, and D. J. Hallett. 2009. Multi-millennial fire history of the Giant Forest, Sequoia National Park, California, USA. Fire Ecology 5:120–150.

Taylor, A. H. 1995. Forest expansion and climate-change in the mountain hemlock (*Tsuga mertensiana*) zone, Lassen Volcanic National Park, California, USA. Arctic and Alpine Research 27:207–216.

Taylor, A. H., and A. E. Scholl. 2012. Climatic and human influences on fire regimes in mixed conifer forests in Yosemite National Park, USA. Forest Ecology and Management 267:144–156.

Tingley, M. W., M. S. Koo, C. Moritz, A. C. Rush, and S. R. Beissinger. 2012. The push and pull of climate change causes heterogeneous shifts in avian elevational ranges. Global Change Biology 18:3279–3290.

Tingley, M. W., W. B. Monahan, S. R. Beissinger, and C. Moritz. 2009. Birds track their Grinnellian niche through a century of climate change. Proceedings of the National Academy of Sciences USA 106:19637–19643.

Todd, M. J., R. Muneepeerakul, F. Miralles-Wilhelm, A. Rinaldo, and I. Rodriguez-Iturbe. 2012. Possible climate change impacts on the hydrological and vegetative character of Everglades National Park, Florida. Ecohydrology 5:326–336.

Trouet, V., A. H. Taylor, E. R. Wahl, C. N. Skinner, and S. L. Stephens. 2010. Fire-climate interactions in the American West since 1400 CE. Geophysical Research Letters 37:L04702.

US Bureau of Reclamation. 2012. Colorado River Basin Water Supply and Demand Study. US Bureau of Reclamation, Washington, DC.

———. 2013. Lower Rio Grande Basin Study. US Bureau of Reclamation, Denver, Colorado.

US Department of the Interior. 2008. Endangered and threatened wildlife and plants; determination of threatened status for the polar bear (*Ursus maritimus*) throughout its range. Federal Register 73:28212–28303.

———. 2014. Endangered and threatened wildlife and plants; threatened species status for the rufa red knot. Federal Register 79:73706–73748.

VanderHoek, R., E. J. Dixon, N. L. Jarman, and R. M. Tedor. 2012. Ice patch archaeology in Alaska: 2000–10. Arctic 65(S1):153–164.

van der Valk, A. G., J. C. Volin, and P. R. Wetzel. 2015. Predicted changes in interannual water-level fluctuations due to climate change and its implications for the vegetation of the Florida Everglades. Environmental Management 55:799–806.

van Mantgem, P. J., N. L. Stephenson, J. C. Byrne, L. D. Daniels, J. F. Franklin, P. Z. Fule, M. E. Harmon, et al. 2009. Widespread increase of tree mortality rates in the western United States. Science 323:521–524.

Vano, J. A., B. Udall, D. R. Cayan, J. T. Overpeck, L. D. Brekke, T. Das, H. C. Hartmann, et al. 2014. Understanding uncertainties in future Colorado River streamflow. Bulletin of the American Meteorological Society 95:59–78.

Vaughan, D. G., J. C. Comiso, I. Allison, J. Carrasco, G. Kaser, R. Kwok, P. Mote, et al. 2013. Observations: cryosphere. Pages 317–382 in T. F. Stocker, D. Qin, G. K. Plattner, M. Tignor, S. K. Allen, J. Boschung, A. Nauels, et al., eds. Climate change 2013: the physical science basis. Contribution of Working Group I to the Fifth Assessment Report of the Intergovernmental Panel on Climate Change. Cambridge University Press, Cambridge, United Kingdom, and New York, New York.

von Storch, H., and F. W. Zwiers. 1999. Statistical Analysis in Climate Research. Cambridge University Press, Cambridge, United Kingdom.

Wang, G. M., N. T. Hobbs, K. M. Giesen, H. Galbraith, D. S. Ojima, and C. E. Braun. 2002. Relationships between climate and population dynamics of white-tailed ptarmigan Lagopus leucurus in Rocky Mountain National Park, Colorado, USA. Climate Research 23:81–87.

Westerling, A., H. G. Hidalgo, D. R. Cayan, and T. W. Swetnam. 2006. Warming and earlier Spring increase western US forest wildfire activity. Science 313:940–943.

Westerling, A. L., M. G. Turner, E. A. H. Smithwick, W. H. Romme, and M. G. Ryan. 2011. Continued warming could transform Greater Yellowstone fire regimes by mid-21st century. Proceedings of the National Academy of Sciences USA 108:13165–13170.

White, J. D., K. J. Gutzwiller, W. C. Barrow, L. Johnson-Randall, L. Zygo, and P. Swint. 2011. Understanding interaction effects of climate change and fire management on bird distributions through combined process and habitat models. Conservation Biology 25:536–546.

Whittaker, R. H. 1960. Vegetation of the Siskiyou Mountains, Oregon and California. Ecological Monographs 30:279–338.

Williams, A. P., C. D. Allen, A. K. Macalady, D. Griffin, C. A. Woodhouse, D. M. Meko, T. W. Swetnam, et al. 2013. Temperature as a potent driver of regional forest drought stress and tree mortality. Nature Climate Change 3:292–297.

Williams, A. P., C. J. Still, D. T. Fischer, and S. W. Leavitt. 2008. The influence of summertime fog and overcast clouds on the growth of a coastal Californian pine: a tree-ring study. Oecologia 156:601–611.

Womble, J. N., G. W. Pendleton, E. A. Mathews, G. M. Blundell, N. M. Bool, and S. M. Gende. 2010. Harbor seal (Phoca vitulina richardii) decline continues in the rapidly changing landscape of Glacier Bay National Park, Alaska 1992–2008. Marine Mammal Science 26:686–697.

Wong, P. P., I. J. Losada, J. P. Gattuso, J. Hinkel, A. Khattabi, K. L. McInnes, Y. Saito, and A. Sallenger. 2014. Coastal systems and low-lying areas. Pages 361–410 in C. B. Field, V. R. Barros, D. J. Dokken, K. J. Mach, M. D. Mastrandrea, T. E. Bilir, M. Chatterjee, et al., eds. Climate change 2014: impacts, adaptation, and vulnerability. Part A: global and sectoral aspects. Contribution of Working Group II to the Fifth Assessment Report of the Intergovernmental Panel on Climate Change. Cambridge University Press, Cambridge, United Kingdom, and New York, New York.

Woodward, A., E. G. Schreiner, and D. G. Silsbee. 1995. Climate, geography, and tree establishment in subalpine meadows of the Olympic Mountains, Washington, USA. Arctic and Alpine Research 27:217–225.

Zolbrod, N., and D. L. Peterson. 1999. Response of high-elevation forests in the Olympic Mountains to climatic change. Canadian Journal of Forest Research 29:1966–1978.

SEVEN

Protecting National Parks from Air Pollution Effects: Making Sausage from Science and Policy

JILL S. BARON, TAMARA BLETT, WILLIAM C. MALM, RUTH M. ALEXANDER, AND HOLLY DOREMUS

Introduction

The story of air pollution research, policy development, and management in national parks is a fascinating blend of cultural change, vision, interdisciplinary and interagency collaboration, and science-policy-management-stakeholder collaborations. Unable to ignore the loss of iconic vistas from regional haze and loss of fish from acid rain in the 1980s, the National Park Service (NPS) embraced an obligation to protect resources from threats originating outside park boundaries. Upholding the Organic Act requirement for parks to remain "unimpaired" for the enjoyment of future generations, and using the Clean Air Act statement that the NPS has an "affirmative responsibility" to protect park resources, the NPS has supported, and effectively used, research as a means to protect lands, waters, and vistas from a mostly unseen threat. Using visibility and atmospheric nitrogen deposition as examples, we will illustrate some success stories where the NPS led the way to benefit not only parks but the nation.

Recent scholarship by scientists and environmental historians documents a transition in the management practices of national parks in the latter decades of the 20th century. From the founding of the US National Park Service well into the 1960s, park management focused on recreational tourism, rather than on the preservation of natural resources. Resource management was generally uninformed by science, partly because of a lack of capacity but also partly because the science of natural resource management was itself developing (National Research Council 1992). In the 1970s, however, the NPS began to move toward ecological management founded on scientific understanding in order to protect and preserve its natural resources (Leopold et al. 1963; Sellars 1999), and the NPS is now moving to protect ecological integrity under conditions of continuous environmental and social change (Colwell et al. 2012; Stephenson 2014). Today, science-based management is accepted by the NPS, although it was not always so (National Research Council 1992; Jarvis 2008).

As resource management in the parks has become more scientifically based, it has also widened its geographic scope. Until the late 20th century, while there was acknowledgment in the literature of external threats to parks—especially air pollution, water pollution, and incompatible outside land use—these threats were rarely, if at all, addressed in park management (National Park Service 1980; Shafer 2012). Similarly, visitors to parks were never treated as an outside threat, in spite of increasing numbers that strained infrastructure and staff. Enabling legislation for most national parks explicitly acknowledged their role in providing for visitor enjoyment, with specific language in some park documents (including those for Rocky Mountain National Park) identifying them as "pleasuring grounds." Visitor enjoyment was persistently prioritized over the goal of conservation, though the latter was also identified as a key purpose of the National Park System in the Organic Act of 1916.

Management policies through the late 20th century focused almost exclusively on actions within national parks. The 10-year NPS infrastructure project titled Mission 66 attempted to manage visitor crowding through internally focused policies (National Park Service 1956). So, too, the Leopold Report of 1963—although it highlighted ecological processes—relied on an inward focus as it set the boundaries for active management of the parks' natural resources for years to come. The boundaries were based on wildlife and their habitat, and they included on-the-ground implementation of policies within the political boundaries of parks (Leopold et al. 1963). Thus, NPS resource management culture supported tackling natural resource issues with local actions. Parks were managed as islands.

Parks are not islands, of course. Political boundaries are quite porous to air pollution and other human-caused phenomena, including climate change. Pressure from the scientific community, from scientifically oriented NPS policy analysts and managers, and from stakeholders provoked new attentiveness over time to external threats to resource conservation throughout the NPS.

Air pollution, one of the first issues to be recognized as an external threat, involved both harm to visitor enjoyment and harm to natural resources. In its various forms, air pollution degraded viewshed visibility, a priority for park visitors, as well as ecosystems. A 1979 paper reported a significant deterioration in visibility over the period of 1950–1975 in the southwest United States, affecting the views of visitors to desert national parks such as Grand Canyon and Canyonlands (Trijonis 1979). Visibility degradation is one of the most obvious effects of air pollution on the environment. Five years previously, another paper on a related subject, acid rain, had documented the widespread occurrence of acid precipitation and its damaging effects to lakes, streams, and possibly forests (Likens and Bormann 1974). *The New York Times* captured the impressions of the time: "not so gentle rain . . . the acidity of orange juice . . . damaging crops, trees, buildings, statues and car finishes . . . 200 lakes and ponds in the Adirondacks officially dead—devoid of both brook trout and trash fish" (quoted in Ogden 1983).

The Clean Air Act (CAA), first passed in 1963, established a legal foundation for research, monitoring, and control of air pollution. The CAA Amendments of 1977[1] specifically mandated the protection of visual air quality in large national parks and wilderness areas, and required that the NPS work with the Environmental Protection Agency to identify the sources of visibility impairment. Responding to scientific and popular concern about acid rain a few years later, Congress passed the Acid Precipitation Act of 1980,[2] establishing a comprehensive 10-year multiagency federal research plan. The resulting National Acid Precipitation Assessment Program was tasked with conducting research on the causes, extent, and effects of acid rain nationwide (Likens and Bormann 1974; Burns 2012).

NPS program managers in the Washington office recognized early the need for rigorous inquiry into the effects of acid rain on park air, lands, and waters. The NPS used acid rain as justification for developing a quantifiable foundation and cumulative body of knowledge about natural processes and human influences on the parks beginning in the 1970s and 1980s. Perhaps because of the legal tools available for improving visibility through the CAA and the national attention on acid rain, air quality research in the NPS steadily advanced and was put to use to protect national park resources even as the adoption of other monitoring and experimental approaches to science-based management practices in parks took longer to implement (National Research Council 1992; Shaver and Malm 1996).

We illustrate the development of research into the causes and consequences of air pollution in parks with two examples, visibility and nitrogen deposition, and describe the advancement and expansion of these valuable

1. P.L. 95-95.
2. P.L. 96-294, Title VII.

research programs today. Since the 1970s, the NPS has promoted, funded, and catalyzed study of many other air pollutants, including ozone, mercury, acid rain, nitrogen, and other organic and inorganic contaminants (table 7.1). The NPS has supported research into sources, deposition, and environmental effects of air pollution in the national parks. That research has supported the implementation of policies that improved air quality and associated natural resources within parks. The science-based policies promoted by the NPS, in fact, have been a powerful force for cleaning up the air all over the nation.

Visibility Research

Fly ash and sulfate particles produced as by-products of energy production on the Colorado Plateau in the 1960s and 1970s reduced visibility, compromising the ability to see and enjoy the unique scenic resources of the region (fig. 7.1). Possible sources of visibility impairment included the large coal-fired Navajo Generating Station and Four Corners Power Plant, coal mines in the region, the town of Page (created to house workers building Glen Canyon Dam), and sources farther away (W. C. Malm, unpublished data).

Table 7.1 Air quality science in US National Parks: NPS clean air accomplishments since 1990.

What	Purpose	For whom
National-scale air quality monitoring data	To develop air pollution risk and air quality condition and trends assessments for air chemistry and visibility in and near parks over the past 30 years	Parks, which use data to see whether air standards are being met and to communicate air quality impacts to the public; researchers and teachers, who use data to characterize air quality in the United States; NPS managers, who use data to target air improvements in parks and evaluate effectiveness of existing air pollution regulations
Web cameras at 18 parks	To characterize visibility conditions and views from park vistas	Public, who have real-time web access to park visibility and other visual information
Special air quality studies	To provide high-quality data and peer-reviewed science through atmospheric and ecological studies assessing pollution sources, source types, and impacts to sensitive park resources	NPS managers, who use results to assess potential impacts to visitor health, visibility, and ecosystems; scientists, who synthesize results for stakeholders and policymakers to understand parks in larger-scale contexts

Table 7.1 (*continued*)

What	Purpose	For whom
FLAG: Federal Land Managers Air Quality Guidance	To synthesize air pollution effects science and air quality modeling protocols into a standardized set of recommendations for analyses needed to quantify where emissions go, how much ends up in parks, and how much pollution it takes to exceed visibility and ecosystem thresholds	Stakeholders (industry), who develop estimates for potential future emissions impacts at parks; air regulators, who use these estimates to set allowable emissions levels for industrial facilities and projects; land managers, who utilize a consistent, science-based process to assess effects in parks of projected new emissions
Nitrogen Deposition Reduction Plan for Rocky Mountain National Park	To use science-based air pollution effects thresholds to communicate concerns about nitrogen deposition impacts in the park; to identify pollution sources affecting the park and set expectations for improvement; to serve as a model for developing similar work in other parks	Park managers and air regulators, who set collaborative goals for what emissions improvements are needed and how quickly; stakeholders (agricultural producers in Colorado), who understand nitrogen emissions issues and voluntarily reduce emissions using best management practices
Public communication products	To share with the public via kiosks, interpretive panels, and evening programs air quality stories based on park research and monitoring data	NPS staff, who utilize the best available science to communicate to the public that clean, clear air is a valuable resource in parks
Air quality dispersion models	To develop new models and methods to understand pollution transport and transformation from sources to receptors in parks and to determine cost-effective emissions reductions	Land managers, air regulators, and researchers, who create and use large-scale, modeling approaches to predict future park air pollution impacts and assess where emissions reductions are needed—a process resulting in millions of tons of pollutants removed from park airsheds
Science-based policy recommendations	To disseminate information through journal articles, reports, position papers	Land managers, who articulate policy options that use best available science as a basis to protect resources from air pollution (e.g., Shaver and Malm 1996; Porter et al. 2005) so that air regulators and stakeholders can make informed recommendations and decisions
Citizen science air quality projects	To engage the public in collection of data (e.g., mercury in dragonfly larvae) in large-scale efforts to advance park science	Public, who develop increased awareness of air pollution issues; students and teachers, who get excited about field science; land managers, who acquire data at large spatial scales

7.1. Sulfate haze in Grand Canyon. Photo courtesy of W. C. Malm.

Visibility measurements, or measurements of atmospheric clarity, were initiated by researchers at Northern Arizona University in the 1970s in response to general concern over regional haze and its causes. These measurements included the transmission properties of the atmosphere, yielding a measure of visibility (O'Dell and Layton 1974), and particulate and gas concentrations using high-volume air samplers and gas bubblers (Malm 1974). Subsequent studies culminated in a report titled *The Excellent but Deteriorating Air Quality in the Lake Powell Region* (Walther, Malm, and Cudney. 1977).

These studies were more than ordinary scientific research and reporting. They represented an important partnership between private citizens, businesses, local governments, public leaders, and the scientific community that led to new policy (Bishop 1994, 1996). The research results, coupled with a vigorous educational and public relations campaign led by Friends of the Earth, convinced Congress to include the Visibility Protection Amendment[3] in the CAA Amendments of 1977, which was specifically written to reduce sulfur dioxide (SO_2) emissions at the Navajo Generating Station and other emitting sources. Soon after the enactment of the Visibility Protection Amendment, the NPS began to hire policy and legal experts as well as scientists to address the new air quality policy.

The amendment codified visibility protection, an aesthetic value, for certain federal lands, referred to as federal Class I areas. Congress declared

3. CAA § 169A.

a national goal of correcting past visibility impairment in these areas from human-made air pollution and preventing any future impairment.[4] It directed the EPA to develop regulations to ensure progress toward this goal, requiring the states (which are the primary air pollution regulators under the CAA's "cooperative federalism" structure) to control emissions from sources affecting visibility in federal Class I areas. The amendment identified large national parks and wilderness areas as among the mandatory federal Class I areas;[5] 16 national parks and wildernesses on the Colorado Plateau fell within the federal Class I category.

As is the case with so much environmental legislation in the United States, special interests lined up in support of, and in opposition to, implementation of the visibility regulations the EPA promulgated under the amendment. Nongovernmental organizations (NGOs) played an important watchdog role in enforcing implementation of the new legislation, building on their sizeable role in shaping US social values and public policy. The visibility provision required states to make reasonable progress toward natural conditions by reducing emissions from sources that may cause or contribute to visibility impairment in Class I areas.[6] The EPA's implementing regulations required states with such sources to include "regional haze programs" in their State Implementation Plans for compliance with the CAA.[7] If those programs did not adequately address visibility in federal Class I areas, the EPA was required to disapprove them and impose its own visibility requirements.[8] By 1982, when neither states nor the EPA had produced implementation plans for regional haze, the Environmental Defense Fund sued the EPA to require it to fulfill its statutory mandate to protect visibility.[9] That lawsuit forced the EPA to act. From 1984 to 1991, the EPA developed regional haze plans for 35 states.

The EPA's regulations established the legal basis for federal land manager review of new pollution sources and major modifications to large industrial sources under the New Source Review program.[10] The EPA also established a cooperative federal visibility monitoring program between the EPA and the federal land management agencies. The monitoring program was the beginning of the Interagency Monitoring of Protected Visual En-

4. CAA § 169A.
5. CAA § 162(a).
6. CAA § 169A(b)(2).
7. 40 CFR § 51.308.
8. CAA § 110(c).
9. Environmental Defense Fund v. Reilly, No. C82-6850-RPA, N.D. Cal. 1984.
10. CAA § 162.

vironments (IMPROVE) network. On 24 March 1986, Grand Canyon and Canyonlands National Parks certified visibility impairment, likely associated with emissions from the Navajo Generating Station.

Studies beginning in 1978 had already measured visibility across southwestern national parks (Snelling, Pitchford, and Pitchford 1984; W. C. Malm, unpublished data). Data on visibility were important for documenting the problem but could not be used to identify the sources of impairment. For that, transport models and tracer studies were needed. The NPS in collaboration with the EPA developed and tested a series of increasingly sophisticated models over more than 10 years. That work culminated in the Winter Haze Intensive Tracer Experiment (WHITEX) (Malm, Pitchford, and Iyer 1989), carried out from 7 January through 18 February 1987. In December 1989, the NPS released its final WHITEX report, affirming that the Navajo Generating Station emissions contributed to visibility impairment in Grand Canyon National Park. The results of the WHITEX study were contested by the energy industry and particularly the owners of the Navajo Generating Station, who did not want to install costly emissions reduction scrubbers. A National Academy of Sciences review found "at some times during the study period, [Navajo Generating Station] contributed significantly to haze in the Grand Canyon National Park" (National Research Council 1990). The National Academy review and further studies led to a negotiated agreement in 1991 to reduce SO_2 emissions from the power plant by 90%. More recently, in 2014, the EPA issued a final Regional Haze Rule under the CAA that provides for a nitrogen oxide (NO_x) emissions reduction plan for Navajo Generating Station.

While the CAA was the mechanism by which the NPS protected and improved visibility, it was the rigorous monitoring, models, and dedicated work of many scientists and policymakers over many years that produced a remarkable success story: "The collection of good data and the performance of many analyses provided an impetus for the regulatory process. More was needed, however, to achieve a successful resolution. First, an internal and external support network needed to be established for both the science and most park-protective solution. Constant consultation between NPS scientists and air-quality regulatory experts promoted an understanding and trust in the science on one hand, and appreciation of the regulatory context and responsibility on the other" (Shaver and Malm 1996). Today, visibility is significantly better at many parks. Of the 157 national park units where visibility is monitored routinely, improvements have been seen at 49, while visibility has not declined at another 14 (National Park Service 2010). Many of these parks are in the western United States, including California, where visibility improvements have occurred in spite of increased population growth, energy consumption, vehicle miles traveled, and gross domestic product (Bachmann 2007).

Atmospheric Deposition Research

There are currently more than 90 NPS employees working in the Water Resources and Air Resources Divisions, but there were only about a dozen in their joint predecessor, the NPS Washington Office Air and Water Resource Division, in 1977. Division Chief Raymond Herrmann assigned one of his three water resource specialists (J. S. Baron) to research the magnitude and effects of acid rain on national parks. At that time the science of acid rain sources, transport, deposition, and ecological effects was in the very early stages of discovery (Likens and Bormann 1974). Few, if any, national parks had even basic knowledge of what biological resources could be vulnerable to acid rain. Most parks did have geologic maps, however; these were used in a coarse separation of parks that might or might not be sensitive to acid inputs. Parks located in regions underlain by granite, sandstone, or ba salts could be sensitive to acid deposition because these rock types weather slowly and provide little acid neutralizing capacity.

With funding from the National Acid Precipitation Assessment Program, three parks underlain by granitic rocks were selected to initiate acid rain research, but the underlying vision was much broader. The need for acid rain research served to justify long-term studies designed to build understanding of park ecosystems and the impacts of external threats. An instrumented watershed approach was initially adopted for long-term monitoring and research in Isle Royale, Sequoia, and Rocky Mountain National Parks; many other parks were added later (Herrmann and Stottlemyer 1991). Watershed studies allow for quantification of chemical and hydrologic budgets and, if monitored over time, provide records of change. Within the confines of a watershed, they can also support ecosystem studies to quantify biogeochemical and biological processes and the flow of acids, nutrients, and other chemical compounds through soils, vegetation, or surface waters (Herrmann 1997).

The watershed studies, begun in 1980–1982, preceded by many years the National Research Council's recommendation "that accomplishing the mission of the National Park Service requires far more than passive protection; it requires sound understanding of park resources, their status and trends, and the measures needed to correct or prevent problems in these dynamic ecosystems" (National Research Council 1992). The watershed studies were used by the National Research Council committee, in fact, as examples of the kind of basic data needed for effective park management. Other acid rain effects research, particularly in eastern parks, contributed to the larger body of knowledge that led to passage of amendments to the CAA in 1990 that directly resulted in reduction of power plant SO_2 emissions.

The watershed atmospheric deposition research in Rocky Mountain National Park described below had the ultimate goal of informing and improving park management but began with basic discovery. What researchers found, thanks to the low-sulfur coal that produced far lower SO_2 emissions in the western than in the eastern United States, was not acid rain but high inputs of nitrogen in rain and snow. Many years of monitoring and research built a body of knowledge that was used by a creative coalition of scientists, resource agencies, and NGOs to develop strategies for park protection.

The legal and policy leverage for application of watershed research results was the Prevention of Significant Deterioration program, a product of the 1977 CAA Amendments. One of the congressionally declared purposes of those amendments was "to preserve, protect, and enhance the air quality in national parks, national wilderness areas, national monuments, national seashores, and other areas of special national or regional natural, recreational, scenic, or historic value."[11] That purpose is implemented primarily through the states, which must have permit review programs for proposed new or modified sources of air pollutants. Federal land managers must be notified of applications for sources whose emissions may affect federal Class I areas under their supervision. Federal officials "have an affirmative responsibility to protect the air quality related values," including but not limited to visibility, of Class I lands.[12] No permit may be issued if federal officials demonstrate that emissions from the facility will have an adverse impact on air quality–related values in a Class I area.[13] These provisions provide the NPS (and other federal land agencies) with incentives to understand the air quality–related values of their lands and the threats to those values. The ability to demonstrate adverse impacts on air quality–related values carries with it the ability to block new sources.

Early atmospheric deposition measurements from Rocky Mountain National Park and elsewhere in Colorado found sulfate concentrations in

11. CAA § 160(2).
12. CAA § 165(d).
13. CAA § 165(d).

rain and snow that were 10 times lower than in the eastern United States, but nitrate values that were similar to those in the East (Gibson and Baron 1984). Precipitation was not very acidic, and reconstructions of past atmospheric deposition from proxies in lake sediment cores did not find the signature increase in lead, copper, zinc, and vanadium concentrations that accompanied coal- or oil-fired power plants in the East and in Europe after the industrial revolution. Assemblages of diatoms, algae that are well preserved in lake sediments, were used to reconstruct lake water chemistry over time. The metals analysis and pH profiles inferred from diatom stratigraphy did not suggest a history of acidic atmospheric deposition (Baron et al. 1986).

While subsequent work confirmed this conclusion, further examination of diatoms preserved in lake sediments found a profound shift in assemblages beginning about 1950. This was roughly coincident with an increase in metropolitan growth, industrial cattle feedlots, and widespread application of synthetic nitrogen fertilizers in the region east of Rocky Mountain National Park. Enhanced nitrogen deposition in the park was a by-product of these changes. Instead of evidence of acid rain, there was evidence of eutrophication (Wolfe, Baron, and Cornett 2001). Monitoring, field experiments, modeling studies, and comparative regional analyses discovered biogeochemical and biological changes in alpine and forest soils, vegetation, lakes, and streams; the changes were attributable to atmospheric nitrogen deposition (Baron et al. 2000; Bowman et al. 2006). Endorsement of the scientific validity and importance of the results came from an external review of the published research (Burns 2004). While the effects from nitrogen deposition were subtle, the results of the research, coupled with strong inference from the history of atmospheric deposition effects in the eastern United States, pointed to the early stages of a trajectory that would ultimately lead to ecosystem acidification if nitrogen deposition increased over time (Porter and Johnson 2007) (fig. 7.2).

A final piece fell into place with a scientific publication defining the critical nitrogen load for alpine lakes of the park (Baron 2006). A critical load is defined as the amount of atmospheric deposition below which harmful effects are not known to occur (Nilsson and Grennfelt 1988). For Rocky Mountain National Park, the alpine lake critical nitrogen load was determined by estimating the amount of atmospheric deposition in the 1950s when diatom assemblages shifted from those characteristic of nutrient-poor waters to those typical of nutrient-rich waters (Baron 2006). As information accumulated, studies began to point toward Colorado's own power plants, transportation corridors, and especially agriculture as

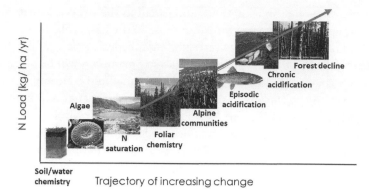

N Load (kg/ ha /yr)

Forest decline

Chronic
acidification

Episodic
acidification

Algae

Alpine
communities

Foliar
chemistry

N
saturation

Soil/water
chemistry   Trajectory of increasing change

7.2. Trajectory of change in alpine mountain ecosystems from increasing atmospheric nitrogen deposition. Figure modified from Porter and Johnson (2007).

major sources of atmospheric nitrogen deposition to Rocky Mountain National Park (Baron and Denning 1993; Day et al. 2012).

As was the case with visibility, stakeholder groups and NGOs became important to policy implementation. The Environmental Defense Fund (EDF) produced a detailed report in 2004 describing the effects and probable sources of nitrogen pollution and, with Trout Unlimited, urged Rocky Mountain National Park and the Department of the Interior to exercise their affirmative responsibility to protect the air quality–related values of the park.[14] Superintendent Vaughn Baker adopted the published critical load, setting the baseline for emissions reductions. At a 2004 meeting, with help from the EDF, the EPA, Colorado Department of Air Quality Control, and NPS, the Colorado Air Quality Control Commission proposed to prepare a plan for regional reductions in nitrogen emissions. The Colorado Livestock Association, Colorado Farm Bureau, and Colorado Corn Growers agreed to participate in plan development.

A strategy to facilitate interagency coordination, the Rocky Mountain National Park Initiative resulted in a signed memorandum of understanding (MOU) between the NPS, Colorado Department of Public Health and Environment (CDPHE), and EPA (Porter et al. 2005). The MOU agencies issued the Nitrogen Deposition Reduction Plan in 2007;[15] the Colorado Air

14. Environmental Defense Fund, Groups petition Interior Department to protect Rocky Mountain National Park, 1 September 2004, accessed 7 March 2016, http://www.edf.org/news/groups-petition-interior-department-protect-rocky-mountain-national-park.

15. CDPHE, Rocky Mountain National Park Initiative, accessed 7 March 2016, https://www.colorado.gov/pacific/cdphe/rocky-mountain-national-park-initiative.

Quality Control Commission added its endorsement (Morris et al. 2014). This plan is currently in place and represents a remarkable collaboration of public land managers (especially Rocky Mountain National Park staff), stakeholders, and NGOs with a common objective of protecting and restoring the natural resources of Rocky Mountain National Park. It would not have happened without the scientific underpinnings based on scientific research.

Sausage Recipes and Outcomes

The partnership of scientists, NPS Air Resources Division and EPA policymakers, resource managers, and environmental stakeholders spawned pollution controls on power plants throughout the country, shaped more stringent and new air quality standards (e.g., for ozone and fine particles, including sulfur and nitrogen subsets), initiated research and policy discussions on toxic pollutants, and premiered the adoption of critical loads for management of ecological endpoints. Christine Shaver was an environmental lawyer who served as division chief of the NPS Air Resources Division for more than 20 years, during which the NPS played a critical role in national air quality policy. Coming to the NPS from the EDF, she was instrumental in gaining singular air quality attention for national parks in the CAA Amendments and EPA regulations. In 2015, several years after retirement, she noted that because of the strong science underpinning the policy evaluation, dozens of new power plants planned in the late 1990s were never built (C. Shaver, pers. comm.). Many more power plants were required to install enhanced emissions controls.

The unprecedented MOU with the CDPHE and EPA to reverse a trend of increasing nitrogen deposition has been in place since 2007. Progress toward the resource management goal of achieving the critical load by 2032 is measured by whether there is a gradual decrease in the amount of wet nitrogen deposition measured. Milestones at five-year intervals beginning in 2012 are meant to evaluate progress toward the goal. A contingency plan was developed to put corrective measures in place should the interim milestones not be achieved. The 2012 milestone was not met but the weight of evidence suggested deposition trends had stabilized, and the MOU agencies in 2013 agreed to not trigger the contingency plan (Morris et al. 2014).

The MOU requires the CDPHE to develop an air management strategy that will help meet park goals. The Colorado Air Quality Control Commission has also established a Rocky Mountain National Park Initiative subcommittee to involve stakeholders, review the research, identify information needs, and discuss options for improving conditions in the park (Burns 2004).

Certainly none of this would have come about without long-lasting partnerships of scientists with managers and policy analysts. Frequent communication developed a mutual understanding of the implications of basic research, the tools available for action, and the language to convey the importance of unseen management threats to the public. In the case of visibility, there were direct legal avenues for using the knowledge gained to protect resources. For atmospheric deposition, policy followed the scientific discovery process: Was there acid rain or atmospheric nitrogen deposition to national parks, and if there was, were there measurable effects? Initially results were mainly informative and the links to policy came later. While initial research in both the visibility and atmospheric deposition examples was scientist initiated, in recent years research questions have also been posed to the scientists by resource managers and policy offices. For the air resources of national parks, a true symbiosis between science and management has evolved. In all cases, the sharing of information without sensationalizing the outcomes was important to the credibility of both scientists and managers.

Environmental stakeholders, including NGOs, provided a critical partner for air quality research. For visibility, environmental stakeholders played a strong role in calling for research on the front end, and for atmospheric deposition, environmental stakeholders were equally important to action on the back end. Friends of the Earth exposed the loss of iconic vistas in national parks of the Southwest with presentations to regulators and draft language for a Visibility Protection Amendment beginning in 1975 and a PBS documentary in 1982. The EDF and Trout Unlimited acted on the published body of information related to changes caused by nitrogen deposition to alpine and subalpine ecosystems by notifying the Secretary of the Interior that these "changes" violated the NPS's affirmative responsibility to protect Class I areas, including Rocky Mountain National Park. With goals of protecting natural systems on which life depends (EDF) and protecting wild trout from harm (Trout Unlimited), these two organizations had common interests in preventing environmental damage from atmospheric deposition. Since the Nitrogen Deposition Reduction Plan MOU was signed by the NPS, CDPHE, and EPA in 2007, another group of stakeholders has become vital to achieving nitrogen reduction goals: Colorado crop, livestock, and dairy producers. A devoted group of agricultural producers meets regularly with representatives from state and federal agencies to develop and test ways of reducing emissions from fields, feedlots and farms.[16] An alternative to litigation, this forum, built on science and trust, is part of the solution.

The NPS is not a regulatory agency. In order to meet NPS air quality goals, it must work collaboratively with states, the EPA, and neighboring land management partners. It does so as a full partner by bringing years of rigorous monitoring and analytical data to the table, along with prodigious legal assessments. By understanding and sharing information about air quality conditions and trends in parks with regulatory agencies and the public, the NPS has effectively managed its resources and helped shape many federal and state air pollution control programs. The NPS has provided guidance for other public management agencies, promoted citizen science and public awareness around air quality topics, and sponsored research into other pollutants (see table 7.1). Information on air quality conditions and trends in parks has provided the impetus for a number of collaborative efforts with states, tribes, the EPA, the private sector, and the public to protect and improve air quality in parks (National Park Service 2010). In addition to the examples described previously, research is underway to assess the effects of nitrogen, sulfur, or mercury deposition on plants, soils, or waters in many national parks. Excess nitrogen effects on plant communities and soil nutrient cycling have been documented at more than 25 national parks (National Park Service 2010). Ongoing research continues to investigate acidification of soils and streams from sulfur and nitrogen deposition at Great Smoky Mountains, Shenandoah, Acadia, and Isle Royale National Parks and the Appalachian National Scenic Trail. A citizen science initiative to collect dragonfly larvae for mercury bioaccumulation measures is being considered at Acadia, Big Cypress, Cape Cod, Channel Islands, Denali, Great Smoky Mountains, Marsh-Billings, Mammoth Cave, North Cascades, Rocky Mountain, Saint Croix, and Saguaro National Parks. NPS-sponsored research on mercury burdens in fish tissues at 19 other parks quantified the threat of this heavy metal to the fish and their consumers, including humans (National Park Service 2010).

Findings from air quality and effects research in national parks have had benefits far beyond the parks themselves. As mentioned above, the NPS Air Resources Division had a role in preventing construction of some coal-fired power plants and imposing stringent emissions controls on others. As a result, carbon dioxide (CO_2) and other pollutants have been kept out of the atmosphere. Regulatory changes developed collaboratively with the

16. CDPHE, Rocky Mountain National Park Initiative, accessed 7 March 2016, https://www.colorado.gov/pacific/cdphe/rocky-mountain-national-park-initiative.

EPA to improve visibility and reductions in ozone in parks have also increased air quality for surrounding regions. The converse is also true. In the eastern United States, where ozone concentrations in national parks like Great Smoky Mountains, Mammoth Cave, and Shenandoah sometimes exceed health-based ozone standards, ozone trends have improved over the past 10 years because of reductions in emissions from power plants, industry, and vehicles (National Park Service 2010) (fig. 7.3). Improvements in visibility in the eastern United States are influenced by reductions in SO_2

7.3. Scenery photos of the Great Smoky Mountains on a good (*top*) and hazy (*bottom*) visibility day. Photo pair courtesy of Great Smoky Mountains National Park.

and NO_x emissions from electric utilities and industrial boilers, as required by the Acid Precipitation Act and State Implementation Plans for nitrogen oxides; these reductions benefit entire regions by reducing particulates that contribute to respiratory disease. In Colorado, where a conversion from coal-fired energy production to natural gas and renewable energy is ongoing owing to market forces and the State Climate Action Plan, reduction of nitrogen emissions to benefit Rocky Mountain National Park is recognized as a co-benefit of reducing greenhouse gas emissions.[17]

The Frontiers Ahead

The air quality successes from strong scientific research in national parks will be increasingly challenged by the interactions of air pollution with climate change. Warmer temperatures interact with air pollutants to produce more ozone and haze; both warming and nitrogen inputs diminish native biodiversity; and increased energy use for cooling may produce a positive feedback loop, resulting in more emissions and increased warming (Hobbs et al. 2010; Suddick et al. 2013). Site specific research will be important for producing local information that can aid adaptation. Maintenance and expansion of monitoring efforts within parks will be essential for future action and for bearing witness to the many large-scale changes taking place. As scientists continue their research in parks, finding solutions to air quality problems will require larger coalitions between science, management, and policy.

Air quality science is conducted in national parks by a diverse mixture of academic and federal scientists. Fostering and supporting a symbiotic relationship among scientists, policy analysts, managers, and stakeholders is critical to "making sausage" from the complex ingredients inherent in both ecosystem sciences and policy. This mixture must be encouraged and supported with financial and intellectual resources. Air pollution is an excellent example of the permeability of park boundaries to threats; these threats affect many other public (and private) lands. Public land management agencies are increasingly sharing scientific knowledge, developing common policies, and developing common management practices. This encouraging trend bodes well for developing regional problem-solving capabilities for environmental problems, such as climate change and air pollution, that do not respect agency boundaries (Pardo et al. 2012; Blett et al. 2014).

17. See http://www.xcelenergy.com/Environment/Programs/Colorado_Clean_Air-Clean _Jobs_Plan (accessed 24 March 2016).

Acknowledgments

We are indebted to the many scientists and policymakers who have worked over the years to protect natural resources and air quality from air pollution. Barbara Brown, Phil Wondra, Dave Shaver, John Christiano, and Chris Shaver developed the highly successful NPS Air Resources Division that so successfully protects air resources of national parks. Many managers of Rocky Mountain National Park have worked tirelessly on nitrogen deposition issues, particularly Vaughn Baker, Ben Bobowski, Judy Visty, Jim Cheatham, and Paul McLaughlin. We especially thank Ray Hermann, who was ahead of his time with the vision to promote long-term watershed studies in national parks. The manuscript was greatly improved by reviews from Ellen Porter, Christine Shaver, and Will Wright.

Literature Cited

Bachman, J. 2007. Will the circle be unbroken: a history of the US National Ambient Air Quality Standards. Journal of Air and Waste Management Association 57:652–697.

Baron, J., and A. S. Denning. 1993. The influence of mountain meteorology on precipitation chemistry at low and high elevations of the Colorado Front Range, USA. Atmospheric Environment 27A:2337–2349.

Baron, J., S. A. Norton, D. R. Beeson, and R. Herrmann. 1986. Sediment diatom and metal stratigraphy from Rocky Mountain lakes with special reference to atmospheric deposition. Canadian Journal of Fisheries and Aquatic Sciences 43:1350–1362.

Baron, J. S. 2006. Hindcasting nitrogen deposition to determine an ecological critical load. Ecological Applications 16:433–439.

Baron, J. S., H. M. Rueth, A. P. Wolfe, K. R. Nydick, E. J. Allstott, J. T. Minear, B. Moraska. 2000. Ecosystem responses to nitrogen deposition in the Colorado Front Range. Ecosystems 3:352–368.

Bishop, J., Jr. 1994. Who speaks for the Colorado Plateau? High Country News, 4 April.

———. 1996. Pact promises cleaner air. High Country News, 24 June.

Blett, T. F., J. A. Lynch, L. H. Pardo, C. Huber, R. Haeuber, and R. Pouyat. 2014. FOCUS: a pilot study for national-scale critical loads development in the United States. Environmental Science and Policy 38:225–236.

Bowman, W. D., J. R. Gartner, K. Holland, and M. Wiedermann. 2006. Nitrogen critical loads for alpine vegetation and terrestrial ecosystem response: are we there yet? Ecological Applications 16:1183–1193.

Burns, D. A. 2004. The effects of atmospheric nitrogen deposition in the Rocky Mountains of Colorado and southern Wyoming, USA—a critical review. Environmental Pollution 127:257–269.

———. 2012. National Acid Precipitation Assessment Program Report to Congress 2011: an integrated assessment. Office of Science and Technology Policy, Washington, DC.

Colwell, R., S. Avery, J. Berger, G. E. Davis, H. Hamilton, T. Lovejoy, S. Malcolm, et al. 2012. Revisiting Leopold: resource stewardship in the national parks. A report of the National Park System Advisory Board Science Committee. http://www.nps.gov/calltoaction/PDF/LeopoldReport_2012.pdf.

Day, D. E., X. Chen, K. A. Gebhart, C. M. Carrico, F. M. Schwandner, K. B. Benedict, B. A. Schichtel, and J. L. Collett. 2012. Spatial and temporal variability of ammonia and other inorganic aerosol species. Atmospheric Environment 61:490–498.

Gibson, J. H., and J. S. Baron. 1984. Acidic deposition in the Rocky Mountain Region. Pages 29–43 in T. A. Colbert and R. L. Cuany, eds. Proceedings: high altitude revegetation workshop no. 6, Colorado Water Resources Research Institute information series no. 53. Colorado State University, Fort Collins, Colorado.

Herrmann, R. 1997. Long-term watershed research and monitoring to understand ecosystem change in parks and equivalent reserves. Journal of the American Water Resources Association 33:747–753.

Herrmann, R., and R. Stottlemyer. 1991. Long-term monitoring for environmental change in US national parks a watershed approach. Environmental Monitoring and Assessment 17:51–65.

Hobbs, W. O., R. J. Telford, H. J. B. Birks, J. E. Saros, R. R. O. Hazewinkel, B. B. Perren, E. Saulnier-Talbot, and A. P. Wolfe. 2010. Quantifying recent ecological changes in remote lakes of North America and Greenland using sediment diatom assemblages. PLoS ONE 5:e10026.

Jarvis, J. B. 2008. The natural resource challenge: a retrospective and view to the future. US National Park Service Publications and Papers. Paper 27. http://digitalcommons.unl.edu/natlpark/27.

Leopold, A. S., S. A Cain, C. M. Cottam, I. N. Gabrielson, and T. L. Kimball. 1963. Wildlife management in the national parks: the Leopold Report. Unpublished report to the US Secretary of the Interior.

Likens, G. E and F. H. Bormann. 1974. Acid rain: a serious regional environmental problem. Science 184:1176–1179.

Malm, W. C. 1974. Air movement in the Grand Canyon. Plateau 46:135–132.

Malm, W. C., M. Pitchford, and H. L. Iyer. 1989. Design and implementation of the winter haze intensive tracer experiment. Pages 432–458 in Proceedings of the 81st annual meeting and exhibition of the Air Pollution Control Association. Air Pollution Control Association, Pittsburgh, Pennsylvania.

Morris, K. A, M. A. Mast, G. Wetherbee, J. S. Baron, C. Taipale, T. Blett, D. Gay, and J. Heath. 2014. 2012 Monitoring and tracking wet nitrogen deposition at Rocky Mountain National Park. Natural Resource Report NPS/NRSS/ARD/NRR-2014-757. National Park Service, Denver, Colorado.

National Park Service. 1956. Mission 66: to provide adequate protection and development of the National Park System for human use. Unpublished report to the National Park Service.

———. 1980. State of the parks, 1980. A report to the Congress. Office of Science and Technology, National Park Service, Washington, DC.

National Park Service, Air Resources Division. 2010. Air quality in national parks: 2009 annual performance and progress report. Natural Resource Report NPS/NRPC/ARD/NRR—2010/266. National Park Service, Denver, Colorado.

National Research Council. 1990. Haze in the Grand Canyon: an evaluation of the winter haze intensive tracer experiment. National Academies Press, Washington, DC.

———. 1992. Science and the national parks. National Academies Press, Washington, DC.

Nilsson, J., and P. Grennfelt, eds. 1988. Critical loads for sulphur and nitrogen. Nordic Council of Ministers, Copenhagen, Denmark.

O'Dell, K. D., and R. G. Layton. 1974. Visibility studies in the Grand Canyon. Plateau 46:133–134.

Ogden, M. D. 1983. Stealthy destruction from the sky. New York Times, 22 May.

Pardo, L. H., M. E. Fenn, C. L. Goodale, L. H. Geiser, C. T. Driscoll, E. B., Allen, J. S. Baron, et al. 2012. Effects of nitrogen deposition and empirical nitrogen critical loads for ecoregions of the United States. Ecological Applications 8:3049–3082.

Porter, E., T. Blett, D. U. Potter, and C. Huber. 2005. Protecting resources on federal lands: implications of critical loads for atmospheric deposition of nitrogen and sulfur. BioScience 55:603–612.

Porter, E., and S. Johnson. 2007. Translating science into policy: using ecosystem thresholds to protect resources in Rocky Mountain National Park. Environmental Pollution 149:268–280.

Sellars, R. W. 1999. Preserving nature in the national parks: a history. Yale University Press, New Haven, Connecticut.

Shafer, C. L. 2012. Chronology of awareness about US national park external threats. Environmental Management 50:1098–1110.

Shaver, C. L., and W. C. Malm. 1996. Air quality in the Grand Canyon. Pages 229–249 in W. L. Halvorsen and G. E. Davis, eds. Science and ecosystem management in the national parks. University of Arizona Press, Tucson, Arizona.

Snelling, R. N., M. Pitchford, and A. Pitchford. 1984. Visibility investigative experiment in the west (VIEW). EPA-600/S4-84-060. Environmental Protection Agency, Environmental Monitoring Systems Laboratory, Las Vegas, Nevada.

Stephenson, N. L. 2014. Making the transition to the third era of natural resources management. The George Wright Forum 31:227–235.

Suddick, E. C., P. Whitney, A. R. Townsend, and E. A. Davidson. 2013. The role of nitrogen in climate change and the impacts of nitrogen—climate interactions in the United States: foreword to thematic issue. Biogeochemistry 114:1–10.

Trijonis, J. 1979. Visibility in the southwest—an exploration of the historical data base. Atmospheric Environment 13:833–843.

Walther, E. G., W. C. Malm, and R. A. Cudney. 1977. The excellent but deteriorating air quality in the Lake Powell Region. Journal of the Air Pollution Control Association 29:378–380.

Wolfe, A. P., J. S. Baron, and R. J. Cornett. 2001. Unprecedented changes in alpine ecosystems related to anthropogenic nitrogen deposition. Journal of Paleolimnology 25:1–7.

EIGHT

Biological Invasions in the National Parks and in *Park Science*

DANIEL SIMBERLOFF

Introduction

Although particular biological invasions (e.g., that of the gypsy moth, *Lymantria dispar*, into North America) had been noted by the mid-19th century, the scope of the problem and the great variety of inimical impacts were not widely appreciated until the mid-1980s (Simberloff 2010). The first 15 years of modern invasion biology were largely dominated by research on impacts at the population level—how a particular nonnative species affects one native species or a limited group of native species, usually by one of a few mechanisms: trampling or browsing by introduced herbivores, predation by introduced predators, competition for resources, spread of disease, and hybridization. More recently, many other impacts have been recognized, some affecting entire ecosystems (Simberloff 2013). Introduction of nonnative species has increasingly been recognized as one of the major global changes affecting native species in ecosystems in myriad ways worldwide (Drake et al. 1989; National Research Council 2000), but when the earliest US national parks were established, the issue was almost totally unrecognized.

It is therefore unsurprising that the US National Park System was not initially concerned with biological invasions. Nor was science a focus. The 1872 congressional act establishing the first national park, Yellowstone, did "provide for the preservation, from injury or spoliation, of all timber, mineral deposits, natural curiosities, or wonders within said park, and their retention in their natural condition," which can loosely be construed to support conservation, and possibly action against nonnative species. However, the key motivation supporting establishment of the early large parks, such as Yellowstone, Sequioa, and Yosemite, was tourism (Sellars 1997). Nature was primarily construed as scenery for tourism or a source of raw materials such as minerals and timber, although the concept of wilderness was prominent in John Muir's activism in support of the 1890 establishment of Yosemite National Park (Sellars 1997).

With a different impetus—protection of archaeological sites—Congress in 1906 passed the Antiquities Act, which allowed the president to establish national monuments. By including among worthy targets of this status "historic structures" and "other objects of historic interest," this act led the subsequently established National Park Service (NPS) to have a somewhat schizophrenic mission with respect to nonnative species. National monuments were brought under the umbrella of the newly formed service, and many historic structures and landscapes contained nonnative plants established during associated historic eras that were to be preserved or emulated. However, the Antiquities Act was also used to set aside as national monuments huge tracts of land largely revered for scenic reasons that subsequently became national parks, such as the Grand Canyon and Mount Olympus.

The 1916 National Parks Organic Act establishing the NPS furthered the schizophrenia—it mandated the conservation of scenery, natural objects, and wildlife, but also "historic objects." In addition, it allowed the director to "grant the privilege to graze live stock within any national park, monument, or reservation . . . when in his judgment such use is not detrimental to the primary purpose for which such park, monument, or reservation was created."

Nonnative Species Introductions and Management in the Parks

Many activities brought nonnatives to the US national parks in the early years, particularly stocking for sportfishing, which began in Yellowstone in 1881 with the introduction of regionally native cutthroat trout (*Oncorhynchus clarkii*) to fishless lakes (Sellars 1997). In 1889, nonnative brook trout (*Salvelinus fontinalis*) and rainbow trout (*O. mykiss*) were introduced in Yellowstone, and by the 1920s Yellowstone had nonnative rainbow trout, brook trout, lake trout (*S. namaycush*), and brown trout (*Salmo trutta*) (Sellars 1997). Rainbow trout were introduced to previously fishless Crater Lake in 1888, 14 years before it became a national park, and stocking continued after it became a park (Sellars 1997). Similarly, in the 1890s stocking of nonnative fish began in Yosemite, and Sequoia and Glacier also developed stocking programs (Sellars 1997). Hatcheries for some nonnatives were soon established within national parks.

Neither of the first policy documents of the new NPS (the 1918 Lane letter and the 1925 Work letter) specifically banned nonnative species, and in the 1920s the NPS was still heavily promoting sportfishing, relying on the Bureau of Fisheries for stocking and running hatcheries, and collaborating with state fish and game agencies (Sellars 1997). Chinese Pheasant (*Phasianus colchicus*) and nonnative subspecies of Wild Turkey (*Meleagris gallopavo*) were introduced to Sequoia National Park (Adams 1925), and various nonnative shrubs and trees were planted in several national parks (Lien 1991). In 1921, the Ecological Society of America passed a resolution opposing all nonnative animal and plant species introductions in any national park (Lien 1991), and the American Association for the Advancement of Science passed a similar resolution the same year (Moore 1925). In 1922, Horace Albright, field assistant to the director of the NPS, responded, stating that NPS policy was that "foreign plant and animal life are not to be brought in" (Adams 1925). Nevertheless, the NPS continued introducing nonnative trout in Yellowstone, saying this was only in streams where they had already been introduced (Sellars 1997). In fact, the NPS continued to stock fish on a great scale, as well as many nonnative plant species to landscape developed areas (Sellars 1997). Nonnative grasses were also planted in irrigated fields in Yellowstone to provide hay for winter.

In 1929, Albright, now the NPS director, claimed that "exotic plants, animals, and birds are excluded from the parks," but there were many exceptions, including continued introduction of nonnative fish (Sellars 1997). However, in the 1920s the NPS had begun trying to remove some nonnative animals because they were harming native plants and animals—for example, many feral burros (*Equus africanus asinus*) were killed in 1924 in the Grand Canyon, and in the 1920s Hawai'i Volcanoes National Park initiated efforts to eradicate goats (*Capra aegagrus hircus*) (Sellars 1997).

In 1933, NPS biologists George Wright, Joseph Dixon, and Ben Thompson submitted a report—*Fauna of the National Parks of the United States* (*Fauna No. 1*)—warning that nonnative species in the parks posed threats, including transmission of disease to native species. They particularly stressed hybridization with natives, exemplified by hybridization of introduced Siberian reindeer (*Rangifer tarandus sibericus*) with native caribou (*R. t. caribou*) in Mount McKinley National Park. This report gave new impetus to efforts to control nonnative species, but it also highlighted the obvious disconnect between stated policy and actions on the ground. The emphasis by critics had been heavily on problems caused by nonnative animals, but in 1935 Wright's assistant complained that no part of Glacier National Park had pristine areas worthy of being research reserves because all streams had nonnative fish and "exotic plants have been carried to practically every corner of the park" (Sellars 1997). Nevertheless, and despite the fact that their fish management specialist agreed that introducing nonnative fishes had adversely affected natural conditions of park waters, the NPS under the leadership of Albright and his successor, Arno Cammerer, continued to promote sportfishing of nonnative fish (Sellars 1997).

However, park biologists in the 1930s inspired a slight improvement in policies, with the recommendation of Wright, Dixon, and Thompson (1933) to reduce populations of nonnative species already present, not to introduce new ones, and to set aside one watershed in each park to reflect a natural state, with no introductions allowed. In 1936, NPS Director Cammerer announced a policy that prohibited wider distribution of nonnative fishes and forbade introduction of nonnative fishes into waters that did not already contain them. However, the NPS gave park managers great leeway in managing nonnatives. In waters where nonnative species were "best suited to the environment and have proven of higher value for fishing purposes than native species," stocking of nonnatives could continue if approved by the park superintendent and NPS director, so substantial stocking continued (e.g., in Yellowstone, including in fishless waters) (Sellars 1997). Carl Hubbs objected repeatedly, citing threats posed by nonnative fishes to native fishes (Hubbs 1940; Hubbs and Wallis 1948; Hubbs and Lagler 1949), but the practice persisted.

In the 1960s, two reports on the national parks—the Leopold Report on wildlife management in the parks (Leopold et al. 1963) and the Robbins Report on research in the parks (Robbins et al. 1963)—reiterated that areas set aside to preserve natural objects and wildlife, as stated in the Organic Act, should not be the locus of nonnative species introductions (Drees 2004). Several informal statements by NPS officials served as responses, such as one by NPS scientist Lowell Sumner in 1964 that "nonnative species are to be eradicated, or held to a minimum if complete eradication is impossible" (quoted in Drees 2004). Finally in 1968, the NPS issued *Administrative Policies for Natural Areas of the National Park System*, stating that "nonnative species may not be introduced into natural areas. Where they have become established or threaten invasion of a natural area, an appropriate management plan should be developed to control them, where feasible," and further that "nonnative species of plants and animals will be eliminated where it is possible to do so by approved methods which will preserve wilderness qualities" (National Park Service [1968] quoted in Drees 2004).

And, in fact, the NPS did increase its efforts to reduce or eradicate populations of nonnative species. A 1967 report listed 30 parks with programs to manage nonnative plants and 9 parks with programs to manage nonnative animals (Sellars 1997). However, attempts to eradicate nonnative mammals sometimes elicited heated objections from either animal rights groups (e.g., burros in Grand Canyon National Park) (Dodge 1951) or hunting organizations (e.g., wild boar, *Sus scrofa*, in Great Smoky Mountains National Park) (Sellars 1997). The resultant controversies led to scientific research to demonstrate the inimical effects of the nonnatives, particularly with respect to habitat destruction. Such research was instrumental in the NPS's successful defense against a lawsuit to prevent shooting burros in Bandelier National Monument, which in turn allowed the NPS to eliminate remaining burros in Grand Canyon National Park after live-trapping and removal by the Fund for Animals (Sellars 1997). However, even substantial research on boar impact in Great Smoky Mountains National Park (e.g., Singer 1981; Singer, Swank, and Clebsch 1984) did not carry the day, as opposition from North Carolina hunters led to an odd policy: shooting boar on the Tennessee side of the park but not on the North Carolina side (Sellars 1997). As of 1981, at least four US national parks allowed recreational hunting of boar (Singer 1981).

In Hawai'i Volcanoes National Park in the 1970s, public pressure led the NPS to forestall fencing and to allow hunters to participate in reducing goat numbers, with no demonstrated effect on populations and complaints from a ranger that this policy constituted sustained-yield recreation for hunters (Sellars 1997). The stated policy of wanting to control but not eliminate goats in Hawai'i Volcanoes obviously contradicted the official NPS policy of eliminating nonnative species where possible (Sellars 1997), and led to an attempted delicate balancing act that has proven difficult if not impossible to maintain with highly fecund ungulates. Even against complaints from conservation-minded citizens and against the advice of a park biologist, this policy was maintained, and objections to it by the park superintendent even led to his removal (Sellars 1997). However, beginning in 1971, a fencing program was instituted along with the killing of goats near the fences, and it had largely succeeded by 1980 in keeping large areas nearly goat-free (Sellars 1997). Native vegetation subsequently recovered.

Biological Invasions in *Park Science*

Park Science arrived on the scene in 1980, slightly before the explosive rise of modern invasion biology in the mid-1980s in the wake of the Scientific Committee on Problems of the Environment (SCOPE) program on biological invasions, which was inspired by a SCOPE workshop that same year, 1980 (Simberloff 2010). *Park Science* began as *Pacific Park Science* but after one year became a national NPS publication. Its stated purpose was to help the process established 10 years earlier with the advent of the Cooperative Park Study Units (CPSUs)—that is, bringing university scientists and park scientists and managers together to bring the best science to bear on NPS management matters (Dickenson 1980).

From the outset, nonnative species issues figured heavily in *Park Science*. For instance, volume 1, number 2 included: a notice that Charles van Riper III, whose main focus at the University of Hawai'i CPSU had been avian diseases introduced with nonnative birds (van Riper et al. 1986), had been appointed unit leader at the new CPSU at the University of California, Davis; a report on an initiative to incorporate an integrated pest management approach in NPS activities, primarily with respect to targeted nonnative species; and a report on the feasibility of a restoration project at John Day Fossil Beds National Monument entailing redressing the impact of grazing livestock and invasive plants such as cheatgrass (*Bromus tectorum*) and Russian thistle (*Kali tragus*). Volume 1, number 3 had more detailed and broadly ranging articles on the utility of experimental management of nonnative herbivores like goats in Hawai'i, mountain goats (*Oreamnos americanus*) in Olympic National Park, and wild boar in the Great Smoky Mountains (Houston 1981); the threat to native plants in Hawai'i from the combined assault of feral goats and pigs and several nonnative plants (Loope 1981); and work at the University of Hawai'i CPSU on threats and management of exotic animals, not only herbivores but also rats (*Rattus* spp.) and the small Indian mongoose (*Herpestes auropunctatus*) (Stone 1981). Through 2014, *Park Science* published 222 articles on invasive nonnative species, plus several notices of relevant meetings and literature; the number varies greatly along with the great vicissitudes in frequency and size of issues (fig. 8.1). But, from the second year of publication, the journal has almost always had several articles on nonnative species.

Park scientists and other researchers working in the NPS were among the vanguard in the explosive growth of invasion biology in the late 1980s and 1990s. For instance, the two SCOPE books that spurred the field (Mooney and Drake 1986; Drake et al. 1989) had several contributions focused on invasion impacts in US national parks, notably Peter Vitousek's research on the impact of fire tree, *Morella faya* (formerly *Myrica faya*), on nutrient cycling in Hawai'i Volcanoes National Park (Vitousek 1986), John Ewel's work on the impact of Australian paperbark (*Melaleuca quinquenervia*) and Brazilian pepper (*Schinus terebinthifolius*) in Everglades National Park and

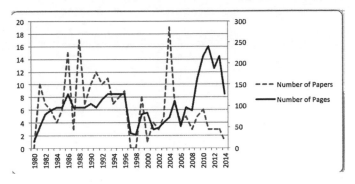

8.1. Number of articles in *Park Science* on biological invasions (*dashed line*), and total number of pages in *Park Science* (*solid line*), by year.

Big Cypress National Preserve (Ewel 1986), and Lloyd Loope and Dieter Mueller-Dombois's review of impacts of introduced ungulates and plants on native vegetation in Hawai'i's national parks (Loope and Mueller-Dombois 1989).

Although several NPS scientists published substantial research on invasions in international journals and proceedings, as did many academic researchers working in the national parks, the "house science organ," *Park Science*, reflects the early growth and evolution of the field. The goal of the SCOPE program that launched modern invasion biology was to bring science to bear on management problems caused by invasions. However, the focus both within the program and in the myriad publications that quickly followed it was not on management, but instead it centered largely on the impacts of invasions and questions about why some species are particularly invasive and some habitats particularly prone to invasion impacts (Simberloff 2013). Because the raison d'être of *Park Science* is to facilitate management, articles there tended from the outset to focus somewhat more on specific management issues and methods associated with invasions than did those in academic journals and the SCOPE volumes themselves.

Nevertheless, many articles in *Park Science* have been about invasion processes and impacts, and not primarily about management. During the first years of the journal, which nearly coincided with the beginning of modern invasion biology, *Park Science* reports abounded on the impacts of particular nonnative species on individual native species or small groups of them (table 8.1). Several NPS scientists published repeatedly on nonnative species in *Park Science* during this period and raised the profile of the issue within the NPS. Susan Bratton, working in several southeastern parks, and

Table 8.1 Sample of papers from the first 15 years of Park Science describing impacts on nonnative species in the US National Park System

Species	Impact	Location	Reference
Goats, pigs	Trampling, browsing, rooting	Hawai'i Volcanoes NP, Haleakala NP	Loope 1981; Tunison, Stone, and Cuddihy 1986
Rats, small Indian mongoose	Predation on birds	Hawai'i Volcanoes NP	Stone 1981
Mountain goats	Trampling	Olympic NP	Moorhead 1981, 1989; Schreiner and Woodward 1994
Salt cedar	Competition	Grand Canyon NP	Johnson 1981
Dutch elm disease, European elm bark beetle	Disease, disease vector	National Capital Region	Sherald and Hammerschlag 1982
Fire tree, banana poka	Competition	Hawai'i Volcanoes NP, Haleakala NP	Gardner 1982; Gardner and Smith 1985
Kudzu	Competition	Several	Bratton 1983
Cheatgrass	Competition	Whitman Mission NHS	Herrera 1988
Water hyacinth	Competition	Jean Lafitte NHP and Pres.	Anon. 1991
Purple loosestrife	Competition	Voyageurs NP	Benedict 1990
Musk thistle, Canada thistle	Competition	Mesa Verde NP	Floyd-Hanna et al. 1993
Mosquitofish	Hybridization	Big Bend NP	Hoddenbach 1982
European rabbit	Grazing	San Juan Island NHP	Agee 1984
Ferret	Predation	San Juan Island NHP	Agee 1984
Dogwood anthracnose	Disease	Catoctin Mountain Park	Mielke and Langdon 1986
Feral horse	Grazing	Several	Bratton 1986
Nutmeg Mannikin, House Finch, Japanese White-eye, House Sparrow	Vector disease	Hawai'i Volcanoes NP, Haleakala NP	Anon. 1988
Gypsy moth	Defoliation	Shenandoah NP	Haskell and Teetor 1988; Vaughan and Karish 1991
Hemlock woolly adelgid	Defoliation	Shenandoah NP	Watson 1992; Hayes 1992
Cutthroat trout, rainbow trout	Predation, competition	North Cascades NP	Liss and Larson 1991

Note: Scientific names for species in the table that are not mentioned in the text: European elm bark beetle (*Scolytus multistriatus*), banana poka (*Passiflora mollissima*), water hyacinth (*Eichhornia crassipes*), milk thistle (*Silybum marianum*), Canada thistle (*Cirsium arvense*), mosquitofish (*Gambusia affinis*), European rabbit (*Oryctolagus cuniculalus*), ferret (*Mustela putorius*), dogwood anthracnose (*Discula destructiva*), Nutmeg Mannikin (*Lonchura punctulata*), House Finch (*Haemorhous mexicanus*), Japanese White-eye (*Zosterops japonicus*), and House Sparrow (*Passer domesticus*). National park abbreviations: NHP = National Historical Park; NHP and Pres. = National Historical Park and Preserve; NHS = National Historic Site; NP = National Park

Lloyd Loope, working in Hawai'i, were particularly persistent and notable for the scope of their invasion publications, and both became well known among invasion biologists generally.

In 1992, *Park Science* maintained its mission (to provide "a report to park managers of recent and on-going research in parks with emphasis on its implications for planning and management") but shifted its form somewhat. Whereas previous numbers had each been smorgasbords of all sorts of habitats and many kinds of science, many numbers from this point on focused wholly or heavily on single issues or regions, such as climate change, soundscapes, or Caribbean-area research. Whereas almost every previous number had one or (usually) several articles or notices relating to biological invasions, some subsequent numbers, because of their foci (e.g., paleontology, soundscapes) had none. However, most continued to have at least some such articles, either because a theme topic (climate change) interacted with biological invasions or because a number had material in addition to the pages devoted to the theme.

One such number (vol. ??, no. ?) was an entire issue, 71 pages, devoted to invasive species, guest-edited by Ron Hiebert, a park scientist who had long focused on invasive plants (fig. 8.2). Both in scope and depth, this number departed from previous ones, with considerable material not closely related to management, including historical material, book reviews, and many articles on ecological impacts of particular invaders. It could almost serve as a primer in invasion biology.

From 1996 through 2006, the NPS published a second series, the annual *Natural Resource Year in Review*, which summarized the application of science to resource management in the NPS. This was a glossy, beautifully illustrated journal with volumes of about 100 pages each featuring shorter, snappier articles—mostly one or two pages. Many articles summarized or updated longer reports in *Park Science*, but others treated new subjects. Nonnative species were a major focus in each volume, with 88 articles throughout the life of the series and entire sections of several volumes devoted to nonnative species or to restoration that entailed management of nonnatives.

Invasion Biology and *Park Science* Evolve

By 2000, as the science of invasion biology matured and expanded, research in the "one-on-one" vein that dominated the first 15 years continued to be important—impacts of the great majority of introduced species had not been studied, new invasions continued to occur, and it was increasingly

ARE WE DOING ENOUGH?

ASSESSING THE INVASIVE
PLANT ISSUE

INVASIONS IN THE SEA

HEMLOCK WOOLLY ADELGID
AND THE DISINTEGRATION OF
EASTERN HEMLOCK FORESTS

ECOLOGICAL EFFECTS OF ANIMAL INTRODUCTIONS
AT CHANNEL ISLANDS

IMPACTS OF
WEST NILE VIRUS
ON WILDLIFE

PLUS, THE CHALLENGE OF PRESERVING
NATIVE MUSSEL, SALAMANDER,
AND FISH SPECIES

8.2. Special issue of *Park Science* (vol. 22, no. 2) wholly
devoted to biological invasions in US national parks.

apparent that some nonnative species that had initially been restricted and
innocuous could, after a substantial lag time, abruptly spread and become
highly damaging invaders (Crooks 2011). However, two new foci came to
be the leading edge of the field (Simberloff 2013).

First was invasion impacts at the ecosystem level rather than at the population level. Vitousek had pointed to the importance and variety of such
impacts during the SCOPE project, particularly with his research on fire tree

in Hawai'i Volcanoes National Park (Vitousek 1986). However, his arguments did not lead to many other studies of this phenomenon for at least a decade. But beginning circa 2000, a flurry of publications reported such ecosystem impacts caused by a variety of mechanisms, including modification of nutrient cycling, hydrology, fire regimes, and physical structure (Ehrenfeld 2010; Simberloff 2011a). Some of this research was conducted in national parks: such as Vitousek's continuing work on impacts of fire tree (Vitousek et al. 1987; Asner and Vitousek 2005); research on impacts of another nonnative nitrogen-fixer, black locust (*Robinia pseudoacacia*), in Cape Cod National Seashore (Von Holle et al. 2006; Von Holle et al. 2013); and studies of impacts of changing fire regimes caused by Australian paperbark in Everglades National Park (Serbesoff-King 2003) and by nonnative grasses in Hawai'i Volcanoes National Park (D'Antonio, Tunison, and Loh 2000).

Several articles in *Park Science* reflected this new research thrust. Notably, Hayes (1992), Mahan (1999), and Evans (2004, 2005) detailed a number of impacts, including some at the ecosystem level, that would likely follow the invasion of eastern parks by the hemlock woolly adelgid (*Adelges tsugae*). Biggam (2004) pointed to ways in which impacts by nonnative plants, earthworms, and boar can affect ecosystems through modification of nutrient cycles and hydrology. Esque et al. (2006) described how changed fire cycles induced by buffelgrass (*Cenchrus ciliaris*) invasion at Saguaro National Park affect ecosystem structure and function. Sturm (2008) described research showing ecosystem impacts of horses (*Equus ferus caballus*) at Assateague Island National Seashore.

The second new focus of modern invasion biology was the incorporation of genetics and evolution (Simberloff 2013). For reasons that have not been explored, the initial burst of invasion research following the SCOPE project was almost wholly ecological. Although evolutionists and geneticists were engaged in the SCOPE project, no uptick in such research ensued for about 15 years (Simberloff 2010); the first monograph on invasion and evolution was by Cox in 2004. However, in the new century, and particularly with the advent of increasingly accessible and inexpensive tools of molecular genetics, papers on genetics and evolution flooded journals, heralding the arrival of an entire subfield, termed invasion genetics (Barrett 2015). This research addressed such topics as the rapid evolution of nonnative species in their new range (e.g., Huey et al. 2000), the evolution of native species in response to nonnative invaders (e.g., Strauss, Lau, and Carroll 2006), the role of multiple propagules in the establishment and spread of nonnative species (e.g., Lavergne and Molofsky 2007), and hybridization between native and nonnative species and between different populations of single nonnative species (e.g. Schierenbeck and Ellstrand 2009).

Not many papers in *Park Science* have tracked the explosion of research in invasion genetics, even though recent molecular genetic research has cast light on the causes and trajectories of invasions by many nonnative species of great concern to the NPS, such as Brazilian pepper (Mukherjee et al. 2012), *Phragmites* (Meyerson and Cronin 2013), Dutch elm disease (*Ophiostoma* spp.) (Brasier 2001), salt cedar (*Tamarix* spp.) (Gaskin and Schaal 2002), purple loosestrife (*Lythrum salicaria*) (Chun, Nason, and Moloney 2009), and trout (Allendorf et al. 2004). Sakai (2004) noted the likelihood that Barred Owl (*Strix varia*), facilitated in spreading to the West by landscaping practices across the Great Plains, hybridizes with the threatened Northern Spotted Owl (*S. occidentalis caurina*), and Halbert et al. (2006) reported that American bison (*Bison bison*) herds in national parks manifested little or no genetic introgression from cattle. A recent *Park Science* report by Marburger and Travis (2013) describes how hybridization between a native cattail (*Typha*) species and one that has directly or indirectly been moved by humans into its range has produced a hybrid that is invasive in several national parks. Perhaps the dearth of *Park Science* papers on invasion genetics and evolution relative to those on ecosystem impacts is because the research usually seems more academic and less immediately useful to policymakers and especially managers than ecological research. It is noteworthy that Marburger and Travis (2013) are at pains to show how their molecular approach can aid in distinguishing hybrids from the parental species and also to detail management approaches.

Controversies in the Parks, the Science, and *Park Science*

The NPS has long been beset by conflicts regarding nonnative species, as noted above with respect to fish introductions and mammal control or eradication. *Park Science* has generally ignored or soft-pedaled these conflicts. For instance, removing goats from Olympic National Park has generated persistent controversy from animal rights advocates; several articles mention extraordinary measures to remove them without shooting them but do not explain the rationale for these measures (e.g., Moorhead 1981, 1989). Remarkably, Tuler and Janda (1991) reported on great risks to personnel involved in this effort to remove goats alive (fig. 8.3) without mentioning the underlying controversy that motivated the effort. Finally, Crawford (1993) described the controversy as a matter of animal rights and as "a major test of NPS policies on exotic species management," and noted

8.3. Captured mountain goats arriving at the staging area in Olympic
National Park for transfer elsewhere by state agency. Photo by Janis Burger,
courtesy of the National Park Service, from Tuler and Janda (1991).

extensive national press coverage of the issue. Fencing out and finally shooting goats in Hawai'i Volcanoes National Park was at least as controversial because of pressure from hunters wanting to maintain populations and conservationists and park scientists wanting to eradicate them (Bonsey 2011), but a description of the fencing project (Loope 1981) failed to mention any conflict.

Burro control in western parks was also controversial because of objections from animal rights groups, but Douglas (1981) described the ecological problem and potential solutions without mentioning opposition to either the methods or the goals. Others did not shy away from featuring the controversy. Fletcher (1983) detailed the dispute between the NPS and the Fund for Animals over removing burros from Bandelier National Monument and its resolution, which was similar to that for the burros of Death Valley National Monument (Anon. 1984).

Scientists documented similar problems caused by horses on eastern islands, but measures similar to those applied to burros have never been attempted. Bratton (1986) described major ecological problems caused by feral horses at Cape Lookout National Seashore and Cumberland Island National Seashore, asserting that the main reason the herds are not eliminated is their popularity with visitors. Wild horses were declared a "desirable exotic species" when Assateague Island National Seashore was founded, but they are ecologically damaging (Anon. 1996). Sturm (2008) reports on research finding substantial ecological damage from horses on Assateague, where they were again declared by the NPS in 1982 to be a "desirable species" managed as wildlife.

African oryx (*Oryx beisa*) were removed from White Sands National Monument by expensive nonlethal methods—primarily helicopters and slings—after fencing had failed (Conrod 2004). The public objected to the NPS's preferred alternative, shooting, after a critical newspaper article, but the last few animals were shot anyway.

Perhaps the most remarkable downplaying of such controversies concerns *Park Science* reportage on the 1991–1992 eradication of black rats (*Rattus rattus*) from Anacapa Island in the Channel Islands National Park, which elicited massive objections from animal rights advocates and a concerted (but failed) sabotage attempt (Simberloff 2011b). McEachern (2004) described in some detail the Anacapa rat eradication with no word of the controversy surrounding it, and *Park Science* editor Selleck (2005) declared it a tremendous victory, again without mentioning the controversy.

Animal rights conflicts have long plagued invasion biology and management outside the national parks, a frequent issue when control, management, or eradication of nonnative mammals or birds is attempted (Simberloff 2012). Four other controversies have arisen with respect to managing nonnative species (Simberloff 2013), but none have surfaced with respect to nonnative species in the NPS, nor have they resonated in the pages of *Park Science*. A small number of critics among ecologists have argued that the harmful impacts of invasions by nonnatives are overblown, pointing to the fact that most introduced species are not known to cause ecological problems (e.g., Davis et al. 2011). This argument is weakened by the facts that most nonnative populations have not been studied, that some introduced species undergo a long quiescent lag before abruptly spreading and wreaking ecological havoc (Crooks 2011), and that some substantial impacts, such as those described above caused by nutrient cycle changes, are initially subtle and difficult to recognize (Simberloff 2013; Simberloff et al. 2013). Sax, Gaines, and Brown (2002) and Thompson (2014) have observed that, in some locations, the number of established introduced species outweighs the number of recently extinct ones, so that local biodiversity is increased. However, many global species extinctions, especially but not exclusively on islands, are ascribed to invasive species, and many native species that persist do so in greatly reduced numbers (indeed, many are even on various lists of imperiled species) wholly or partly because of invasions (Simberloff 2013), so the net global impact of invasions on biodiversity is negative. The entire enterprise of managing nonnative species has been condemned as a form of displaced xenophobia, primarily by critics in the humanities and social sciences rather than by biologists (Simberloff 2003, 2012). These critics uniformly ignore or downplay the ecological impacts of many nonnative invaders and instead engage in a social construction (Brown 2001) of the field of invasion biology and management, based on the perceived psychology and power relationships among the participants rather than data and phenomena from nature.

Perhaps the most potentially damaging criticism of invasion management is the argument that, even if impacts are substantial, little can be done to prevent them in the face of the ongoing globalization and economic forces that cause the great majority of invasions. Consider this statement by Mark Gardener, former director of the Charles Darwin Research Station, about nonnative *Rubus niveus* in the Galapagos: "Blackberries now cover more than 30,000 hectares here, and our studies show that island biodiversity is reduced by at least 50% when it's present. But as far as I am concerned, it's now a Galápagos native, and it's time we accepted it as such" (quoted in Vince 2011, 1383). This pessimistic view, which has recently been echoed by advocates of abandoning traditional ecological restoration in favor of fashioning novel ecosystems that provide ecosystem services for humans (e.g., Hobbs, Higgs, and Hall 2013), is particularly pernicious because it gives license to policymakers to decrease or eliminate funding for invasive species prevention and management on the grounds that the effort is expensive and futile (Murcia et al. 2014).

In fact, invasion prevention and management have achieved many successes, both in eradication and in lessening populations of persistent invaders by physical and mechanical means, herbicides and pesticides, biological control, and various novel methods (Simberloff 2013, 2014). Furthermore, ongoing improvements in efficacy and in minimizing nontarget impacts characterize the recent history of all these approaches. Though no article in *Park Science* has explicitly responded to the claim that the whole enterprise is misguided and hopeless, from its inception, numerous articles tout progress in managing damaging nonnatives and in restoring ecosystems in the wake of such management. For instance, Consolo (1986), Loope (1991), Conrod (2004), and Selleck (2005) report small-scale eradications of invaders, while Syphax and Hammerschlag (1995), Whitworth, Carter, and Koepke (2005), and Wheeler, Thiet, and Smith (2013) report native species recovery or successful ecological restoration following reduction or elimination of nonnatives. Several successful projects entail persistence and incremental improvement in techniques. A good example is the striking success in reducing Australian paperbark in Everglades National Park after two decades of frustration and pessimism (e.g., Myers 1991), achieved by a combination of biological, chemical, and mechanical control (National Research Council 2014). *Park Science* is rife with reports of invasive species management using all these techniques. Beginning in 1980, the NPS increasingly incorporated integrated pest management (Norris 2011) in its management of nonnative species, particularly in the use of herbicides and pesticides (Ruggiero and Johnston 1984; Drees 2004). The aegis was concern about the amount of chemicals being used on park land (Drees 2004). Concomitant with the focus on integrated pest management has been an ongoing engagement with biological control (Anon. 1983), including striking successes (Anon. 1981; Holden 1985; Schreiner 2007).

If one asks why *Park Science* reports so few controversies regarding management of nonnative species in the parks, and none regarding the controversies swirling around invasion biology and management in academia, two answers come quickly to mind. First, *Park Science* is a federal government document, and such documents, except for National Research Council reports, are well known for dodging controversy, particularly controversies over federal agency policy and actions. As *Park Science* has become a glossy, more "professional" journal clearly aimed at a wider audience than just the managers who were its original target, it is not likely to feature conflicts. Second, the editors of and most authors published in *Park Science* are NPS personnel governed by a stated policy directing them not to introduce nonnative species and to lessen or eliminate existing populations of nonnative species where possible, with a few exceptions. They have a specific job to do on the ground, and they are focused on whatever can help them do that job. They are unlikely to be distracted by academic arguments about some aspects of the controversy, such as xenophobia, and their entire mission is antithetical to the claim that managing invasions is a largely futile endeavor.

This is not to say that NPS personnel and documents do not address controversial nonnative species issues in other publications. For instance, an NPS monograph on mountain goats in Olympic National Park (Houston, Schreiner, and Moorhead 1994) is explicit about controversies not only regarding this species but also several other mammals in various national parks. An NPS symposium on exotic pest plants addressed a controversy over whether Australian paperbark management would harm the nursery or apiary industries (Balciunas and Center 1991). However, these are monographic documents meant for scientific audiences.

Ongoing Invasive Species Problems

Despite heroic and sometimes successful efforts by NPS personnel, our national parks continue to be plagued by invasive nonnative species. Over 6,500 nonnative invasive species have been documented on park lands, of which ~70% are plants; 5% of park lands are dominated by invasive plants.[1] With increasingly sophisticated research and increasing time since some of the earlier introductions, more impacts are being detected even for longstanding nonnative populations. For example, in naturally fishless Crater Lake, from 1888 through 1941 nearly two million trout and landlocked salmon of five species were introduced, of which rainbow trout and kokanee (*O. nerka*) persist in large populations (Buktenica et al. 2007). Substantial study of the impact of such a drastic biotic change did not begin until the 1980s, and preliminary evidence suggests major effects on the entire food web (Buktenica et al. 2007; Larson et al. 2007; Urbach

---

1. National Park Service, Invasive species. . . . What are they and why are they a problem?, last updated 12 August 2009, accessed 8 March 2016, http://www.nature.nps.gov/biology/invasivespecies/.

et al. 2007). In 1915, ~20,000 signal crayfish (*Pacifastacus leniusculus*) were introduced to the lake as food for the introduced fish, as were nonnative amphipods (Brode 1938; Buktenica et al. 2015; M. W. Buktenica, pers. comm.). It is now apparent that the crayfish are spreading in the lake, increasing in number, and threatening the existence of a genetically distinct salamander found only in the lake, the Mazama newt (*Taricha granulosa mazamae*), which was first formally described only in the 1940s (Buktenica et al. 2015).

Many of the most prominent invaders are found in national parks: zebra mussel (*Dreissena polymorpha*), kudzu (*Pueraria lobata*), gypsy moth, cheatgrass, small Indian mongoose, garlic mustard (*Alliaria petiolata*), hemlock woolly adelgid, emerald ash borer (*Agrilus planipennis*), New Zealand mud snail (*Potamopyrgus antipodarum*), purple loosestrife, and Brazilian pepper. The one that has drawn the most recent attention is the Burmese python (*Python bivittatus*), which now numbers in the thousands in Everglades National Park and has been noted in *Park Science* (Blumberg 2009), popular books (e.g., Dorcas and Willson 2011), many reports in the scientific literature (e.g., Dorcas et al. 2012), and hundreds of newspaper and television reports. The python, whose arrival coincided with dramatic population crashes of all medium-sized and large mammals in the park (Dorcas et al. 2012), has occasioned a massive controversy between the federal government and snake hobbyists who object to its being added to the Lacey Act list of prohibited species.

However, the python is just one of myriad invaders that have transformed large parts of the Everglades and pose enormous challenges to the Everglades restoration project (National Research Council 2014). It joins Brazilian pepper, Australian paperbark, Australian pine (*Casuarina* spp.), Old World climbing fern (*Lygodium microphyllum*), the Argentine black and white tegu (*Salvator merianae*), the Cuban treefrog (*Osteopilus septentrionalis*), the Purple Swamphen (*Porphyrio porphyrio*), the island applesnail (*Pomacea maculata*), the pike killifish (*Belonesox belizanus*), the redbay ambrosia beetle (*Xyleborus glabratus*), the Mexican bromeliad weevil (*Metamasius callizona*), and at least 450 other nonnative species present in or very near the park (National Research Council 2014). Management is a Sisyphean task, as success with some species (e.g., paperbark and Australian pine) is more than balanced by the spread of other invaders and the arrival of new ones like the python, the tegu, and the ambrosia beetle.

Everglades National Park is perhaps the most striking example of the threats of nonnative species, but it exemplifies a principle that underlies the unique problems they pose to parks in general. Parks are islands, but introduced species do not recognize their boundaries, so parks are constantly besieged by species arriving from the outside—parks cannot legislate the nonnative species policies of the United States (Stohlgren, Loope, and Makarick 2013; National Research Council 2014). Thus, they will be faced with this threat in perpetuity. I close with an example from the Great Smoky Mountains National Park, close to my home. Oriental bittersweet (*Celastrus orbiculatus*), one of the most detested invasive plants in the United States, was first brought to the region by Frederick Law Olmsted for landscaping of George Washington Vanderbilt's Biltmore Estate near Asheville, North Carolina (Browder 2011). Olmsted created the Biltmore Nursery, which was a distribution center for plants of the estate and beyond (Alexander 2007). From there, bittersweet almost certainly reached the park from Fontana Lake, created by the Fontana Dam of the Tennessee Valley Authority in the early 1940s (K. Johnson, pers. comm.). The lake has many permanent and vacation homes at various locations on or near its shores, and many of these are extensively landscaped with nonnative plants. Oriental bittersweet was first found in the park in 1994 at four small sites near the Fontana Dam and has now spread to the furthest reaches of the park, despite significant effort by park managers to limit it (K. Johnson, pers. comm.). Containment of bittersweet will probably be an effort requiring some personpower in perpetuity unless a biological control agent is found.

Nevertheless, despite the fact that Great Smoky Mountains National Park personnel record about two new nonnative plant species a year, they have been managing nonnative plants since the 1950s with increasing sophistication, and the great majority of the park does not contain more than a smattering of invaders. No one is optimistic that the problem will be largely resolved soon, but neither is there a sense that the battle cannot be won. That seems also to be the thrust of most articles in *Park Science* and *Natural Resource Year in Review*: nonnative species are a challenge, but one that is well worth taking up and not hopeless (e.g., Snyder, Pernas, and Burch 2004; Pannebaker and Zimmerman 2005).

Conclusions

In light of the accelerating influx of nonnative species into the national parks (Stohlgren, Loope, and Makarick 2013), the inability of the parks to control invasive species in the surrounding landscape, and the straitened budgetary situation for federal resource agencies, one might question whether the apparent resolve of park scientists and managers to confront the problem is misplaced. An assessment in the 1990s of funding needs for nonnative species management in the parks determined that $80 million would be required annually for a fully adequate program (Drees 2004). This figure is of course far out of line with available funding. In 2002, the amount available from the NPS and external sources for invasive plant control in the parks was about $4 million. This disparity does not mean the situation is hopeless, as witnessed by the successes described above, but neither can every nonnative species be battled everywhere it is detected. Obviously a triage approach is needed, based on risk assessment of potential targets (Stohlgren, Loope, and Makarick 2013), which is improving in response to a major research thrust (Lonsdale 2011). Many—perhaps most—park managers employ such a risk-based triage system already, focusing particularly on nonnatives known to be highly invasive elsewhere, and especially on new infestations (K. Johnson, pers. comm.). A key component in such a system would be an early warning system, which can be greatly facilitated by engaging and training citizens as volunteers (Stohlgren, Loope, and Makarick 2013; Simberloff 2014). The NPS already engages trained volunteers in many nonnative plant removal projects (e.g., Blumberg 2004; Rapp 2006; Travaglini 2006).

Many of the most damaging nonnative species in national parks come from other continents: Burmese python, Australian paperbark, and Brazilian pepper. However, the impacts of nonnative fishes and crayfish in Crater Lake, Oregon, exemplify the fact that great distance to the native range is not a prerequisite for invasive threat. Rainbow trout, kokanee, and signal crayfish are all native nearby in Oregon. However, their arrival in a previously fishless lake was as great an ecological upheaval as if they had come from the Old World. "Nonnative" is defined as having arrived with deliberate or inadvertent human assistance at a site geographically discrete from the native range. Thus a species undergoing a continuous, incremental range expansion is not nonnative in each new area colonized. The propensity to cause damage to the native denizens is likely largely due to the absence of long periods of coevolution between the invader and the natives, as was recognized long ago by Aldo Leopold (1939).

In an era of rapid climate change, a frequent suggestion has been "managed relocation"—deliberately moving a population into an area currently outside, and not contiguous with, the geographic range of the species in order to avoid the possibility that it will go extinct because it will be unable to move quickly enough in the face of changing climate, especially in a highly anthropogenic landscape (Schwartz et al. 2012). This proposal is highly controversial, and a key part of the controversy is the contention that any such introduction, even over relatively short distances, poses the many and varied risks of any nonnative species (Ricciardi and Simberloff 2009, 2014).

Very similar suggestions have been contested by the NPS. For instance, the endangered Bolson tortoise (*Gopherus flavomarginatus*) was proposed by Donlan et al. (2005) for introduction to Big Bend National Park as part of a massive rewilding project on the grounds that, though restricted to a small part of northern Mexico today, it was very widely distributed in the Chihuahuan Desert until the late Pleistocene. The NPS, however, ruled that it would be nonnative in the park today (Houston and Schreiner 1995). Mountain goat removal in Olympic National Park was controversial not only because of animal rights concerns discussed previously, but also because it is native in nearby parts of Washington and removal opponents claimed that the species occupied the Olympic Peninsula during the late Quaternary. However, subsequent examination of evidence led the NPS to reject this claim and to view the species as nonnative in the park and the cause of substantial negative impacts on native species (Houston, Schreiner, and Moorhead 1994; Houston and Schreiner 1995). In cases such as these, in which a species or very similar relative currently living nearby may have occupied a park in the distant past, Houston and Schreiner (1995) advocate a conservative interpretation of the NPS policy forbidding introduction of nonnative species. They defend this view on the grounds of our general ignorance of ecosystem dynamics and processes, which hinders our prediction of the interactions of introduced species with abiotic forces (such as fires or climate change) and native species. This is the appropriate lens with which to view suggestions to prepare for climate change with managed relocations into the national parks.

Acknowledgments

I thank Kristine Johnson (Great Smoky Mountains National Park) and Mark Buktenica (Crater Lake National Park) for important unpublished information, Jeff Selleck (editor in chief of *Park Science*) for much informative discussion, David Ackerly and Mary Tebo for important comments on an early draft of this manuscript, and Tander Simberloff for preparing a figure.

Literature Cited

Adams, C. C. 1925. Ecological conditions in national forests and in national parks. Scientific Monthly 20:561–593.

Agee, J. K. 1984. Scientists follow rabbits in third wave "invasion." Park Science 5(1): 23–24.

Alexander, B. 2007. The Biltmore nursery. A botanical legacy. Natural History Press, Charleston, South Carolina.

Allendorf, F. W., R. F. Leary, N. P. Hitt, K. L. Knudsen, L. L. Lundquist, and P. Spruell. 2004. Intercrosses and the US Endangered Species Act: should hybridized populations be included as westslope cutthroat trout? Conservation Biology 18:1203–1213.

Anonymous. 1981. Biological control of Klamath weed in Yosemite NP. Park Science 2(1):10.

Anonymous. 1983. Regional highlights. Park Science 3(2):8–9.

Anonymous. 1984. BLM in the saddle-burros headed for last roundup. Park Science 4(2):16.

Anonymous. 1988. Regional highlights. Park Science 8(4):12–13.

Anonymous. 1991. Regional highlights. Park Science 11(1):16–18.

Anonymous. 1996. Information crossfile. Park Science 16(2):8

Asner, G. P., and P. M. Vitousek. 2005. Remote analysis of biological invasion and biogeochemical change. Proceedings of the National Academy of Sciences USA 102: 4383–4386.

Balciunas, J. K., and T. D. Center. 1991. Biological control of *Melaleuca quinquenervia*: prospects and conflicts. Pages 1–22 *in* T. D. Center, R. F. Doren, R. L. Hofstetter, R. L. Myers, and L. D. Whiteaker, eds. Proceedings of the symposium on exotic pest plants. National Park Service, Washington, DC.

Barrett, S. C. 2015. Foundations of invasion genetics: the Baker and Stebbins legacy. Molecular Ecology 24:1927–1941.

Benedict, J. 1990. Purple loosestrife control in voyageurs National Park. Park Science 10(3):21–22.

Biggam, P. 2004. Understanding relationships among invasive species and soils. Park Science 22(2):61–62.

Blumberg, B. 2004. Small Saint-Gaudens managing exotic invasives. Page 68 *in* Natural Resource Year in Review 2003. Natural Resource Program Division, Natural Resource Information Division, Denver, Colorado, and National Park Service, US Department of the Interior, Washington, DC.

———. 2009. Burmese pythons in southern Florida's Everglades. Park Science 26(2):19.

Bonsey, W. E. 2011. Goats in Hawai'i Volcanoes National Park: a story to be remembered. Unpublished report to the National Park Service.

Brasier, C. M. 2001. Rapid evolution of introduced plant pathogens via interspecific hybridization. BioScience 51(2):123–133.

Bratton, S. P. 1983. Kudzu eradication in Southeast parks. Park Science 3(4):11–12.

———. 1986. Vegetation management course emphasizes field projects. Park Science 6(4):3.

Brode, J. S. 1938. The denizens of Crater Lake. Northwest Science 12:50–57.

Browder, J. R. 2011. The effect of *Celastrus orbiculatus*, oriental bittersweet, on the herbaceous layer along a western North Carolina creek. MS thesis, Western Carolina University, Cullowhee, North Carolina.

Brown, J. R. 2001. Who rules science? Harvard University Press, Cambridge, Massachusetts.

Buktenica, M. W., S. F. Girdner, G. L. Larson, and C. D. McIntire. 2007. Variability of kokanee and rainbow trout food habits, distribution, and population dynamics in an ultraoligotrophic lake with no manipulative management. Hydrobiologia 574:235–264.

Buktenica, M. W., S. F. Girdner, A. M. Ray, D. K. Herring, and J. Umek. 2015. The impact of introduced crayfish on a unique population of salamander in Crater Lake, Oregon. Park Science 31(2):5–12.

Chun, Y. J., J. D. Nason, and K. A. Moloney. 2009. Comparison of quantitative and molecular genetic variation of native vs. invasive populations of purple loosestrife (*Lythrum salicaria* L., Lythraceae). Molecular Ecology 18:3020–3035.

Conrod, B. 2004. Last African oryx removed from White Sands National Monument. Park Science 22(2):6.

Consolo, S. 1986. Yellowstone takes action to avert ecological crisis. Park Science 6(3):7–8.

Cox, G. W. 2004. Alien species and evolution. Island Press, Washington, DC.

Crawford, P. 1993. Olympic mountain goat update. Park Science 13(3):15.

Crooks, J. A. 2011. Lag times and exotic species: the ecology and management of biological invasions in slow-motion. Pages 404–410 *in* D. Simberloff and M. Rejmánek, eds. Encyclopedia of biological invasions. University of California Press, Berkeley, California.

D'Antonio, C. M., J. T. Tunison, and R. K. Loh. 2000. Variation in the impact of exotic grasses on native plant composition in relation to fire across and elevation gradient in Hawaii. Austral Ecology 25:507–522.

Davis, M. A., M. K. Chew, R. J. Hobbs, A. E. Lugo, J. J. Ewel, G. J. Vermeij, J. H. Brown, et al. 2011. Don't judge species on their origins. Nature 474:153–154.

Dickenson, R. 1980. From the director. Pacific Park Science 1(1):1.

Dodge, N. N. 1951. Running wild. National Parks Magazine 25:10–15.

Donlan, J., H. W. Greene, J. Berger, C. E. Bock, J. H. Bock, D. A. Burney, J. A. Estes, et al. 2005. Re-wilding North America. Nature 436:913–914.

Dorcas, M. E., and J. D. Willson. 2011. Invasive pythons in the United States. University of Georgia Press, Athens, Georgia.

Dorcas, M. E., J. D. Willson, R. N. Reed, R. W. Snow, M. R. Rochford, M. A. Miller, W. E. Meshaka Jr., et al. 2012. Severe mammal declines coincide with proliferation of invasive Burmese pythons in Everglades National Park. Proceedings of the National Academy of Sciences USA 109:2418–2422.

Douglas, C. L. 1981. Burro impact on bighorn habitat studied. Pacific Park Science 1(3):15–16.

Drake, J. A., H. A. Mooney, F. diCastri, R. H. Groves, F. J. Kruger, M. Rejmánek, and M. Williamson, eds. 1989. Biological invasions. A global perspective. Wiley, Chichester, United Kingdom.

Drees, L. 2004. A retrospective on NPS policy and management. Park Science 22(2):21–26.

Ehrenfeld, J. G. 2010. Ecosystem consequences of biological invasions. Annual Review of Ecology, Evolution, and Systematics 41:59–80.

Esque, T. C., C. R. Schwalbe, J. A. Lissow, D. F. Haines, D. Foster, and M. C. Garnett. 2006. Buffelgrass fuel loads in Saguaro National Park, Arizona, increase fire danger and threaten native species. Park Science 24(2):33–37,56.

Evans, R. A. 2004. Hemlock woolly adelgid and the disintegration of eastern hemlock ecosystems. Park Science 22(2):53–56.

———. 2005. Update on hemlock woolly adelgid and the management of hemlock decline at Delaware Water Gap. Park Science 23(1):8.

Ewel, J. J. 1986. Invasibility: lessons from south Florida. Pages 214–230 in H. A. Mooney and J. A. Drake, eds. Ecology of biological invasions of North America and Hawaii. Springer, New York, New York.

Fletcher, M. R. 1983. Turtles and burros make SWR headlines. Park Science 4(1):8.

Floyd-Hanna, L., W. Romme, D. Kendall, A. Loy, and M. Colyer. 1993. Succession and biological invasion at Mesa Verde NP. Park Science 13(4):16–18.

Gardner, D. E. 1982. Exotic plants in Hawaii's national parks: a major challenge. Park Science 2(2):18–19.

Gardner, D. E., and C. W. Smith. 1985. Plant biocontrol quarantine facility at Hawaii Volcanoes. Park Science 6(1):3–4.

Gaskin, J. F., and B. A. Schaal. 2002. Hybrid Tamarix widespread in US invasion and undetected in native Asian range. Proceedings of the National Academy of Sciences USA 99:11256–11259.

Halbert, N. D., P. J. P. Gogan, R. Hiebert, and J. N. Derr. 2006. Where the buffalo roam: the role of history and genetics in the conservation of bison on federal lands. Park Science 24(2):22–29.

Haskell, D., and A. Teetor. 1988. Predicting gypsy moth defoliation patterns in Shenandoah National Park. Park Science 8(2):5.

Hayes, D. 1992. Eastern hemlock: the next American chestnut? Park Science 12(4):12.

Herrera, D. 1988. Mission strives to deserve its Indian name—Waiilatpu. Park Science 8(3):6–7.

Hobbs, R. J., E. S. Higgs, and C. M. Hall, eds. 2013. Novel ecosystems: intervening in the new ecological world order. Wiley-Blackwell, New York, New York.

Hoddenbach, G. A. 1982. Three strikes but NOT OUT—an endangered species continues to hang on in Big Bend. Park Science 2(3):12.

Holden, L. J. 1985. Tansy flea beetle wins ragwort sweepstakes at Redwood NP. Park Science 5(4):10–11.

Houston, D. B. 1981. Yellowstone elk: some thoughts on experimental management. Pacific Park Science 1(3):4–6.

Houston, D. B., and E. G. Schreiner. 1995. Alien species in national parks: drawing lines in space and time. Conservation Biology 9:204–209.

Houston, D. B., E. G. Schreiner, and B. B. Moorhead, eds. 1994. Mountain goats in Olympic National Park: biology and management of an introduced species. National Park Service, Washington, DC.

Hubbs, C. L. 1940. Fishes from the Big Bend region of Texas. Transactions of the Texas Academy of Science 23:3–12.

Hubbs, C. L., and K. F. Lagler. 1949. Fishes of Isle Royale, Lake Superior, Michigan. Papers of the Michigan Academy of Science, Arts, and Letters 33:73–133.

Hubbs, C. L., and O. L. Wallis. 1948. The native fish fauna of Yosemite National Park and its preservation. Yosemite Nature Notes 27(12):131–144.

Huey, R. B., G. W. Gilchrist, M. L. Carlsen, D. Berrigan, and L. Serra. 2000. Rapid evolution of a geographic size cline in an introduced fly. Science 287:308–309.

Johnson, R. R. 1981. Riparian management and the Colorado River. Park Science 2(1):6–7.

Larson, G. L., C. D. McIntire, M. W. Buktenica, S. F. Girdner, and R. E. Truitt. 2007. Distribution and abundance of zooplankton populations in Crater Lake, Oregon. Hydrobiologia 574:217–233.

Lavergne S., and J. Molofsky. 2007. Increased genetic variation and evolutionary potential drive the success of an invasive grass. Proceedings of the National Academy of Sciences USA 104:3883–3888.

Leopold, A. 1939. A biotic view of the land. Journal of Forestry 37:727–730.

Leopold, A. S., S. A. Cain, C. M. Cottam, I. N. Gabrielson, and T. L. Kimball. 1963. Wildlife management in the national parks: the Leopold Report. Unpublished report to the US Secretary of the Interior.

Lien, C. 1991. Olympic battleground. Sierra Club Books, San Francisco, California.

Liss, W. J., and G. L. Larson 1991. Ecological effects of stocked trout on North Cascades naturally fishless lakes. Park Science 11(3):22–23.

Lonsdale, W. M. 2011. Risk assessment and prioritization. Pages 604–609 *in* D. Simberloff and M. Rejmánek, eds. Encyclopedia of biological invasions. University of California Press, Berkeley, California.

Loope, L. L. 1981. Plant research at Haleakala. Pacific Park Science 1(3):9–10.

———. 1991. Haleakala rabbits declared eradicated (for now). Park Science 11(4):17.

Loope, L. L., and D. Mueller-Dombois. 1989. Characteristics of invaded islands, with special reference to Hawaii. Pages 257–280 *in* J. A. Drake, H. A. Mooney, F. diCastri, R. H. Groves, F. J. Kruger, M. Rejmánek, and M. Williamson, eds. Biological invasions. A global perspective. Wiley, Chichester, United Kingdom.

Mahan, G. C. 1999. Ecosystem-based assessment of biodiversity associated with eastern hemlock forests. Park Science 19(1):37–39.

Marburger, J., and S. Travis. 2013. Cattail hybridization in national parks: an example of cryptic plant invasion. Park Science 30(2):58–68.

McEachern, K. 2004. Ecological effects of animal introductions at Channel Islands National Park. Park Science 22(2):46–52.

Meyerson, L. A., and J. T. Cronin. 2013. Evidence for multiple introductions of *Phragmites australis* to North America: detection of a new non-native haplotype. Biological Invasions 15:2605–2608.

Mielke, M., and K. Langdon. 1986. Dogwood anthracnose fungus threatens Catoctin Mountain Park. Park Science 6(2):6–8.

Mooney, H. A., and J. A. Drake, eds. 1986. Ecology of biological invasions of North America and Hawaii. Springer, New York, New York.

Moore, B. 1925. Importance of natural conditions in national parks. Pages 340–355 *in* G. B. Grinnell and C. Sheldon, eds. Hunting and Conservation. Yale University Press, New Haven, CT.

Moorhead, B. B. 1981. Olympic National Park stages well-run removal of exotic goats. Park Science 2(1):5

———. 1989. Non-native mountain goat management undertaken at Olympic National Park. Park Science 9(3):10–11.

Mukherjee, A., D. A. Williams, G. S. Wheeler, J. P. Cuda, S. Pal, and W. A. Overholt. 2012. Brazilian peppertree (*Schinus terebinthifolius*) in Florida and South America: evidence of a possible niche shift driven by hybridization. Biological Invasions 14:1415–1430.

Murcia, C., J. Aronson, G. H. Kattan, D. Moreno-Mateos, K. Dixon, and D. Simberloff. 2014. A critique of the 'novel ecosystem' concept. Trends in Ecology and Evolution 29:548–553.

Myers, R. L. 1991. Are we serious about exotic species control? A symposium conclusion. Pages 255–259 *in* T. D. Center, R. F. Doren, R. L. Hofstetter, R. L. Myers, and L. D.

Whiteaker, eds. Proceedings of the symposium on exotic pest plants. National Park Service, Washington, DC.

National Park Service. 1968. Administrative policies for natural areas of the National Park System. US Government Printing Office, Washington, DC.

National Research Council. 2000. Global change ecosystems research. National Academies Press, Washington, DC.

———. 2014. Progress toward restoring the Everglades. The fifth biennial review. National Academies Press, Washington, DC.

Norris, R. F. 2011. Integrated pest management. Pages 353–356 in D. Simberloff and M. Rejmánek, eds. Encyclopedia of biological invasions. University of California Press, Berkeley, California.

Pannebaker, F., and K. Zimmerman. 2005. Park celebrates removal of last tamarisk. Pages 38–39 in Natural Resource Year in Review 2004. Natural Resource Program Division, Natural Resource Information Division, Denver, Colorado, and National Park Service, US Department of the Interior, Washington, DC.

Rapp, W. 2006. Maintaining the integrity of native plant communities in Glacier Bay. Pages 110–111 in Natural Resource Year in Review 2005. Natural Resource Program Division, Natural Resource Information Division, Denver, Colorado, and National Park Service, US Department of the Interior, Washington, DC.

Ricciardi, A., and D. Simberloff. 2009. Assisted colonization is not a viable conservation strategy. Trends in Ecology and Evolution 24:248–253.

———. 2014. Fauna in decline: first do no harm. Science 345:884.

Robbins, W. J., E. A. Ackerman, M. Bates, S. A. Cain, F. D. Darling, J. M. Fogg Jr., T. Gill, et al. 1963. National Academy of Sciences Advisory Committee on research in the national parks: the Robbins Report. A report by the Advisory Committee to the National Park Service on research. National Academy of Sciences, National Research Council, Washington, DC.

Ruggiero, M., and G. Johnston. 1984. Pest management: the IPM approach. Park Science 4(2):22.

Sakai, H. 2004. Range expansion of barred owls into Redwood National and State Parks: management implications and consequences for threatened northern spotted owls. Park Science 23(1):24–27,50.

Sax, D. F., S. D. Gaines, and J. H. Brown. 2002. Species invasions exceed extinctions on islands worldwide: a comparative study of plants and birds. American Naturalist 160:766–783.

Schierenbeck, K. A., and N. C. Ellstrand. 2009. Hybridization and the evolution of invasiveness in plants and other organisms. Biological Invasions 11:1093–1105.

Schreiner, E., and A. Woodward. 1994. Study documents mountain goat impacts at Olympic National Park. Park Science 14(2):23–25.

Schreiner, J. 2007. Beetles overcoming purple loosestrife infestations at Delaware Water Gap National Recreation Area. Park Science 24(2):7.

Schwartz, M. W., J. J. Hellmann, J. M. McLachlan, D. F. Sax, J. O. Borevitz, J. Brennan, A. E. Camacho, et al. 2012. Managed relocation: integrating the scientific, regulatory, and ethical challenges. BioScience 62:732–743.

Sellars, R. W. 1997. Preserving nature in the national parks. Yale University Press, New Haven, Connecticut.

Selleck, J. 2005. Celebrating the victories. Park Science 23(2):2.

Serbesoff-King, K. 2003. Melaleuca in Florida: a literature review on the taxonomy, distribution, biology, ecology, economic importance and control measures. Journal of Aquatic Plant Management 41:98–112.

Sherald, J. L., and R. S. Hammerschlag. 1982. Science and management work together in controlling Dutch elm disease in NCR. Park Science 2(2):10–11.

Simberloff, D. 2003. Confronting introduced species: a form of xenophobia? Biological Invasions 5:179–192.

———. 2010. Charles Elton—neither founder nor siren, but prophet. Pages 11–24 *in* D. M. Richardson, ed. Fifty years of invasion ecology. Wiley, New York, New York.

———. 2011a. How common are invasion-induced ecosystem impacts? Biological Invasions 13:1255–1268.

———. 2011b. Review of *When the killing's done*. Biological Invasions 13:2163–2166.

———. 2012. Nature, natives, nativism, and management: worldviews underlying controversies in invasion biology. Environmental Ethics 34:5–25.

———. 2013. Biological invasions: much progress plus several controversies. Contributions to Science 9:7–16.

———. 2014. Biological invasions: what's worth fighting and what can be won? Ecological Engineering 65:112–121.

Simberloff, D., J. L. Martin, P. Genovesi, V. Maris, D. A. Wardle, J. Aronson, F. Courchamp, et al. 2013. Impacts of biological invasions—what's what and the way forward. Trends in Ecology and Evolution 28:58–66.

Singer, F. J. 1981. Wild pig populations in the national parks. Environmental Management 5:263–270.

Singer, F. J., W. T. Swank, and E. E. C. Clebsch. 1984. Effects of wild pig rooting in a deciduous forest. Journal of Wildlife Management 48:464–473.

Snyder, W. A., A. J. Pernas, and J. N. Burch. 2004. Nonnative melaleuca under control at Big Cypress National Preserve. Page 15 *in* Natural Resource Year in Review 2003. Natural Resource Program Division, Natural Resource Information Division, Denver, Colorado, and National Park Service, US Department of the Interior, Washington, DC.

Stohlgren, T. J., L. L. Loope, and L. J. Makarick. 2013. Invasive plants in the United States national parks. Pages 267–283 *in* L. C. Foxcroft, P. Pyšek, D. M. Richardson, and P. Genovesi, eds. Plant invasions in protected areas. Springer, Dordrecht, Netherlands.

Stone, C. P. 1981. Research focuses on HAVO exotics. Pacific Park Science 1(3):11.

Strauss, S. Y., J. A. Lau, and S. P. Carroll. 2006. Evolutionary responses of natives to introduced species: what do introductions tell us about natural communities? Ecology Letters 9:354–371.

Sturm, M. 2008. Assessing the effects of ungulates on natural resources at Assateague Island National Seashore. Park Science 25(1):44–49.

Syphax, S. W., and R. S. Hammerschlag. 1995. The reconstruction of Kenilworth Marsh. Park Science 15(1):1,16–20.

Thompson, K. 2014. Where do camels belong? Profile Books, London, United Kingdom.

Travaglini, M. 2006. Successful partnership with The Nature Conservancy fosters corps of volunteers to tackle nonnative plants in Potomac Gorge. Pages 104–105 *in* Natural Resource Year in Review 2005. Natural Resource Program Division, Natural Resource Information Division, Denver, Colorado, and National Park Service, US Department of the Interior, Washington, DC.

Tuler, S., and C. Janda. 1991. Olympic NP mountain goat removal project subject of risk assessment report. Park Science 11(1):14–15.

Tunison, J. T., C. P. Stone, and L. W. Cuddihy. 1986. SEAs provide ecosystem focus for management and research. Park Science 6(3):10–13.

Urbach, E., K. L. Vergin, G. L. Larson, and S. J. Giovannoni. 2007. Bacterioplankton communities of Crater Lake, OR: dynamic changes with euphotic zone food web structure and stable deep water populations. Hydrobiologia 574:161–177.

van Riper, C., III, S. G. van Riper, M. L. Goff, and M. Laird 1986. The epizootiology and ecological significance of malaria in Hawaiian land birds. Ecological Monographs 56:327–344.

Vaughan, M. R., and J. Karish. 1991. Gypsy moths may alter black bear population dynamics in Shenandoah National Park. Park Science 11(4):7–8.

Vince, G. 2011. Embracing invasives. Science 331:1383–1384.

Vitousek, P. 1986. Biological invasions and ecosystem properties: can species make a difference? Pages 163–176 in H. A. Mooney and J. A. Drake, eds. Ecology of biological invasions of North America and Hawaii. Springer, New York, New York.

Vitousek, P. M., L. R. Walker, L. D. Whiteaker, D. Mueller-Dombois, and P. A. Matson. 1987. Biological invasion by Myrica faya alters ecosystem development in Hawaii. Science 238:802–804.

Von Holle, B., K. A Joseph, E. F. Largay, and R. G. Lohnes. 2006. Facilitations between the introduced nitrogen-fixing tree, Robinia pseudoacacia, and nonnative plant species in the glacial outwash upland ecosystem of Cape Cod, MA. Biodiversity and Conservation 15:2197–2215.

Von Holle, B., E. F. Largay, B. Ozimec, S. A. Clark, J. Oset, K. Lee, and C. M. Neill. 2013. Ecosystem legacy of the introduced N-fixing tree, Robinia pseudoacacia, in a coastal forest. Oecologia 172:915–924.

Watson, J. K. 1992. Hemlock woolly adelgid threatens eastern hemlock in Shenandoah National Park. Park Science 12(4):9–12.

Wheeler, J. S., R. K. Thiet, and S. M. Smith. 2013. Enhancing native plant habitat in a restored salt march on Cape Cod, Massachusetts. Park Science 29(2):42–48.

Whitworth, D. L., H. R. Carter, and J. Koepke. 2005. Recovering Xantus's murrelets on Anacapa Island. Park Science 23(2):9–10.

Wright, G. M., J. S. Dixon, and B. H. Thompson. 1933. Fauna of the national parks of the United States: a preliminary survey of faunal relations in national parks. Contribution of Wild Life Survey. Fauna Series No. 1—May 1932. US Government Printing Office, Washington, DC.

NINE

The Science and Challenges of Conserving Large Wild Mammals in 21st-Century American Protected Areas

JOEL BERGER

Introduction

Five centuries ago—when Italian Cristoforo Colombo and his three Spanish ships touched the shores of the New World—bison (*Bison bison*), grizzly bears (*Ursus arctos*), and wolves (*Canis lupus*) were found from Mexico to Alaska. Cougars (*Felis concolor*) occurred throughout what would become the contiguous United States, and wolverines (*Gulo gulo*) inhabited Michigan, California, Colorado, and New Mexico. Such wildlife grandeur occurs no more. As an ecological player, bison are absent. Wolves and grizzly bears are so geographically restricted south of Canada that they are seen most frequently only within the confines of three American parks, although they do roam beyond park boundaries. Still, the days when these species commanded awe across unbridled lands are gone. The causes are obvious.

Today more than 320 million people are within the contiguous United States. Lands are crowded. Species inimical to people or to economies or requiring large spaces are not well tolerated. As protected areas become increasingly isolated, and habitats within and beyond them change, future conservation of large mammals will become progressively difficult. Enhancing knowledge and putting it into practice will require not only understanding science but understanding and then changing human behavior. On the science front, there are many unknowns, including how climate modulates population dynamics and species persistence. Coupled with such uncertainty is the reality that animals move and park boundaries do not, an onerous combination that creates conflict when lands are crowded with people. Notwithstanding the depth of ecological knowledge about systems or species, human choices determine conservation outcomes. It's unlikely that effective conservation structures can ever be placed without a focus on people.

In this chapter, I address three contemporary conservation challenges confronting wild mammals in US national parks—insularization, long-distance migration, and climate change. I use large mammals to underscore evolving opportunities and difficulties. Such species attract disproportionate interest by park visitors, they play large ecological roles, and they have an uncanny ability to inspire while serving as ambassadors for conservation and biodiversity.

Specifically, I ask how conservation can be achieved given what we know empirically and what we do not know. I focus on parks within a mosaic of lands differing in public and private ownership, human population densities, and remoteness. Because of the inevitable expanding human population, I begin in the contiguous United States where a plethora of scientific studies reveals much about effects of isolation on animal population structure. Parks in more crowded environs are increasingly insular. Consequently, we find that many large mammals experience difficulties to disperse, which causes increased levels of inbreeding, reduced migration, and exacerbated conflicts with humans at or beyond park boundaries. Where immediate conservation goals are to enhance prospects for near-term population viability, changes in land use and the loss of open space are more likely to outstrip climate issues in urgency. I then shift to what is known about the reality of reconnecting populations, and use a case study about long-distance migration to illustrate building bridges across fragmented lands that vary in statutory jurisdictions and stakeholder input. Finally, I concentrate on climate challenges in protected areas of Arctic Alaska where uncertainties are great and, in contrast to the contiguous 48 states, where human populations are extraordinarily low.

Consequences of Insularization of Parks on Large Mammals

Backdrop

The US National Park Service (NPS) manages a total of more than 360,000 km², an area in size just smaller than Montana. More than half the area is in Alaska, with the remaining aggregate dispersed primarily across the conterminous United States. Together, this remaining land is approximately the combined size of Missouri and Florida.

From the perspective of large mammals, large spaces are unavailable because most parks are small. An inverse relationship exists between the

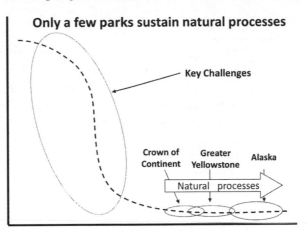

Park Size

9.1. Schematic of relationships between the number of parks and the size of parks. Only parks of the largest size embedded in mosaics of other public lands appear capable of sustaining natural processes. The key management challenges will be for ungulates and carnivores in smaller parks.

number and size of parks, with few parks sufficiently large to sustain landscape-level natural processes (fig. 9.1), as noted over the past 80 years (e.g., Wright, Dixon, and Thompson 1933; Leopold et al. 1963; Colwell et al. 2012). The mosaic of surrounding land uses has resulted in habitat degradation, loss, and fragmentation (Hilty, Lidicker, and Merenlender 2006). As a consequence, calls have been repeated for management of external events beyond protected area boundaries because these events can have dramatic effects on processes and species within parks (Keiter 2010; Austen 2011).

Concerns about park size and animal movements have been expressed since the establishment of Yellowstone National Park, even in the absence of a large number of people living nearby. In 1893, Arthur Hague commented, "Let Congress adjust the boundaries in the best interests of the Park . . . clearly defining them in accordance with the present knowledge of the country, and then forever keep this grand national reservation intact." Two decades later, William Hornaday (1913) said, "The 35,000 elk that summer in the Park are compelled in the winter to migrate to lower altitudes in order to find grass that is not under two feet of snow. In the winter of 1911–12, possibly 5,000 went south into Jackson Hole and 3,000 north into Montana." Today's concerns still focus on conflicts around park borders, but they have broadened to include the dynamics of ecological change and population viability.

Conservation evolves, and the issues of last century—overharvesting, poaching, and predator control—will not be the most pressing issues of the future. For instance, while there are 12 native ungulates that reside in NPS units, there are more than twice as many nonnative ungulates on NPS lands (Plumb et al. 2013). Invasive species, shifting communities of animals and plants, emerging diseases, and unforeseen changes will arise, just as global climate change did toward the end of the 20th century. Nevertheless, the twin threats of habitat loss due to expanding human land uses and climate change will likely be two key drivers affecting large mammals into the foreseeable future.

Management issues will always persist for parks embedded in a mosaic of private, state, and public lands. When a population becomes disconnected from other populations, its individuals often tend to suffer from the effects of isolation. Two case studies are illustrative; the first involves cougars in a dense array of human-dominated environs, and the second examines the situation facing the largest land mammal of the Western Hemisphere, the bison.

Short- to Long-Term Effects of Impermeable Landscapes

Cougars have the widest distribution of any land mammal in the New World, having once occurred across all of the contiguous United States. Populations have become isolated in different ways, but two are notable for the lessons they connote about the consequences of past persecution and modern congestion, both of which result in reduced gene flow.

About 100 years ago, the Florida panther (also called cougar) was reduced to ~30 individuals in southern and central Florida, including Everglades National Park; the nearest neighboring population was situated in the Louisiana-Texas region (Roelke, Martenson, and O'Brien 1993). Because of high levels of mating between closely related animals, inbreeding in the Florida population resulted in spermatozoan defects, cryptorchidism, and enhanced susceptibility to infectious diseases (Roelke, Martenson, and O'Brien 1993; Culver et al. 2008). Elsewhere, cougars have similarly suffered reduced gene flow as a result of increased urbanization and the inability to cross major roadways. Populations from California's Santa Ana and Santa Monica Mountains are relatively more isolated than elsewhere (Ernest et al. 2014). The former was characterized by a genetic

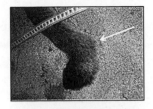

9.2. Examples of morphological deformities in populations with reduced gene flow: kinked tails in cougars from Santa Ana Mountains, California (*top*, photos courtesy of T. Winston Vickers, from Ernest et al. [2014]), and leg anomalies in bison from Badlands National Park, South Dakota (*bottom*, photos by J. Berger, from Berger and Cunningham [1994]).

bottleneck 40–80 years ago, and now—in common with Florida panthers —each of these semi-isolated California subgroups has low genetic diversity. Kinked tails (fig. 9.2), thought to be a manifestation of inbreeding depression, have been found in both Florida and Santa Ana pumas (Roelke et al. 2003; Ernest et al. 2014).

Bison, however, are probably the best example of challenges to conserving large, wide-roaming species. Today, they occupy less than 1% of their historic range, an area stretching from northern Mexico to boreal Canada and from the Atlantic seaboard to Oregon and Washington (Sanderson et al. 2008). In Badlands and Wind Cave National Parks, bison are confined by fencing. In places like Yellowstone National Park, the fencing is virtual. When animals move beyond park boundaries for very long, they are often rounded up or shot (Plumb et al. 2009). The effect is identical to being entirely fenced.

Bison are managed as closed herds (Berger and Cunningham 1994; Halbert et al. 2007), and reproductive isolation will continue until migration is induced. Nowhere other than the contiguously situated Yellowstone

and Teton National Parks is it possible for interpopulation bison movements. Bison face the near impossibility of a reconstituted metapopulation. However, management plans to move individuals across more than a
dozen federal reserves were suggested more than 20 years ago (Berger and
Cunningham 1994) and are now being designed to achieve gene flow by
exchange, or supplementation, of individuals (G. E. Plumb, unpublished
data), as is done in zoos.

With respect to isolated populations, both cougar and bison show
broadly similar responses when disconnected for generations. Cougars
in California and Florida had morphological anomalies manifested by
kinked tails or undescended testicles (Ernest et al. 2014), whereas bison
in highly inbred lineages and in the absence of new mating partners for
at least 75 years had striking limb deformities (Berger and Cunningham
1994) (see fig. 9.2), a situation that would carry strong fitness costs had
predators been present. The bison condition is further complicated since
cattle DNA is evident in most NPS bison populations, with the exceptions
of the Yellowstone and Wind Cave herds (Halbert and Derr 2008). The
body of evidence is robust—when metapopulation structure is fractured,
populations increase in demographic risk (Crooks and Sanjayan 2006).

Conservation Challenges in Impermeable Landscapes and Beyond

Implementation of conservation is onerous because the human dimension is complicated and often independent of science. Experience involving wild animals—digitally or on the ground—greatly affects perceptions
and tolerance. Cougars, for example, are often tolerated locally despite real
dangers to people, livestock, and pets. Bison are also considered dangerous,
yet they are less endured. They have potential to harm people and property
(e.g., fences) and to transmit disease to livestock. The disease issue is local, as only bison in the Greater Yellowstone ecosystem carry brucellosis
(Berger and Cain 1999; Plumb et al. 2009). The other issue—danger—is
serious, as people have been killed by bison. Moose (*Alces alces*) are also
dangerous, have killed more people (via attacks and collisions with cars),
and are far more abundant, yet they are tolerated. In comparison, human
deaths by horses and cattle in the United States average about 40 per year
(Forrester, Holstege, and Forrester 2012). Now, of course, if there were
more bison free roaming, perhaps there might be more frequent deaths.
Nevertheless, while science dictates connectivity as a means to thwart the
growing impermeability of crowded landscapes, the reality is that percep-

tions, not necessarily the facts, about species dictate what is acceptable to society.

Opinions about animal movements across both soft and hard park boundaries into porous landscapes will be further influenced by the imminent threat of danger and the size of a species, as well as its life history and status (i.e., abundant, rare, or endangered). Large carnivores like cougars or black bears (*U. americana*), or smaller carnivores such as coyotes (*C. latrans*), navigate arrays of congested private lands, roads, and other impediments including cities like Los Angeles and Chicago. Once landscapes become pervious to dispersers, the biological problems described above disappear.

As is the case with bison, large mammal movements beyond protected areas will push the limits of tolerance in some circles and will remain an issue for human dimensions, but not one lacking in ecological dimensions. While fortunately no one has died in the United States because of wolf reintroduction, livestock are killed, big game populations reduced, and some individuals feel their liberties have been abrogated. As in the bison case, perception and reality create issues when landscapes become crowded.

When little tolerance remains for ecological challenges, such as connectivity, two additional consequent challenges will grow from the insularization of large mammals. First, ungulates will attain relatively high density, especially in small NPS units where large carnivores are absent. When this occurs, vegetation structure, composition, and density are strongly affected, which can have important secondary and tertiary effects on a multitude of organisms including insects and birds (Ray et al. 2005; Ripple et al. 2015). Second, where populations remain small, vulnerability to stochastic events will increase proneness to extinction, a process exacerbated by climate change (Epps et al. 2006). Constraints associated with park size will continue to force consideration of management alternatives (Colwell et al. 2012; Plumb et al. 2013).

Corridor development continues to be suggested as a way to increase connectivity, but appreciable knowledge deficits remain and corridors will never be the panacea to enhance passage. Migratory species, like all species, can carry disease, and these in turn may increase disease risks to park resources or export them beyond park boundaries (Hess 1996). On the other hand, creating or increasing the efficiency of corridors can be a useful strategy to combat climate change by enhancing accessibility to habitats that may become suitable in the future (Beier and Gregory 2012; Hilty, Chester, and Cross 2012).

Future Prospects

Continuing pressures have further capacity to isolate populations. Energy exploration is one such pressure. On average, 50,000 new energy wells per year have been built across central North America since 2000, a pattern likely to remain (Allred et al. 2015). Another pressure emanates from expansion of human populations. While cities and towns are distributed heterogeneously and mean densities are less in the intermountain region of the United States (~10/km²) than elsewhere (22/km² for the Pacific region, 90/km² in New England) (US Census Bureau 2013), lands are increasingly occupied and less permeable. The conflation of roads, infrastructure, and habitat loss will continue to jeopardize abilities to ensure metapopulation structure.

A Disappearing Phenomenon—Long-Distance Migration—Requires Solutions That Meld People and Engage Stakeholders

Backdrop

Among ecological processes collapsing at a global scale is long-distance migration (Harris et al. 2009). Areas the size of the Arctic National Wildlife Refuge (78,051 km²) and Serengeti National Park (14,763 km²) are insufficient to capture the full range of movements of caribou (*Rangifer tarandus*) and wildebeest (*Connochaetes taurinus*). Smaller protected regions, including many of the national parks within the contiguous United States, fail to encompass the seasonal ranges for migrants. Pronghorn (*Antilocapra americana*) are a striking model. Pronghorn occur in more than 14 NPS units, yet not one is large enough to contain their normal movements throughout an annual cycle.

The largest protected area network in the contiguous United States, the 100,000 km² Greater Yellowstone Ecosystem, is composed of two national parks, four national wildlife refuges, and seven national forests. Yet the migrations of elk, mule deer (*Odocoileus heminous*), pronghorn, and bison have been either compromised or totally lost (Berger 2004). While migrations are still being discovered and refined in this comparatively wild region (Copeland et al. 2014; Sawyer et al. 2014), the scale of collapse of these ungulate migrations across most landscapes beyond the Greater Yellowstone Ecosystem is unprecedented.

The Public Face and the Park Face

Given economic realities, serious obstacles exist to protecting ample space to ensure migration and to connect seasonal ranges or populations, including competing and growing demands on public lands and the juxtaposition of private lands in and around NPS units. If parks are to function ecologically in a coupled natural-human system, collaborative networks have to be placed across broad landscapes that are already human dominated (Machlis, Force, and Burch 1997; Colwell et al. 2012). A cadre of stakeholders readily exists when parks are embedded in crowded landscapes (Hamin 2001), and among them may be varied sectors of public and park patrons, industries, homeowners, hunters, and recreation associations, as well as nongovernmental organizations (NGOs) and state and federal managers.

There are staggering impediments to conserving broad-scale migrations, some resulting from internal NPS forces and others from externalities. No parks have inventoried the bulk of their migratory species, although much is known about ungulate migrations. Nevertheless, even on a park-by-park basis, let alone under a broader NPS umbrella, numerous pragmatic questions will need to be asked, and answered, if serious attempts will be undertaken to conserve migrations.

There are many questions about migrations relevant to the NPS (Berger et al. 2014). What should be conserved—the phenomenon of migration itself, or perhaps abundant migrations only, or maybe just the rare ones? Can lost migrations be restored? Should they? Are some NPS units more important than others to focus efforts to retain migrations? From a social perspective, how should partners be identified? Do they need to be adjacent landowners or agencies? Can they be geographically distant? How will they be involved? What role should they play?

At a smaller scale and in an area of low human population density, colleagues and I coordinated many stakeholders to facilitate the creation of the Path of the Pronghorn, the popular moniker for America's first federally protected migration corridor, established in 2008 (Berger and Cain 2014) (fig. 9.3). Rather than focus on the science, we strategically addressed conservation needs, some of which first came forth by building partnerships and trust between government and private interests, and by enhancing interest in migratory phenomena across landscapes differing in political interests and economic bases (Berger and Cain 2014).

The creation of the Path of the Pronghorn ensured safe passage along an invariant route used by pronghorn for at least 6,000 years and through

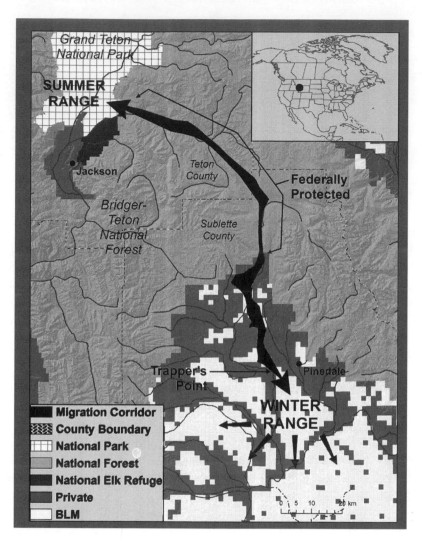

9.3. The Path of the Pronghorn in the western United States. The federally protected portion of the corridor is on US Forest Service (Bridger-Teton) lands between Grand Teton National Park and private and Bureau of Land Management properties. Map courtesy of Steve Cain, from Berger and Cain (2014).

three narrow topographical bottlenecks between summer ranges in Grand Teton National Park and less snowy wintering areas far south of NPS statutory authority. Impediments to the migration include fencing and energy development on crucial winter habitat (Beckmann et al. 2012), which also occurs for other ungulates reliant on portions of the same route (Sawyer et al. 2013). The entire round-trip distance for pronghorn migrating from the park and back exceeds 700 km, although most animals move shorter distances (Berger, Cain, and Berger 2006).

The Path of the Pronghorn resulted from public meetings and formal and informal collaborations involving industry, cattlemen associations, and NGOs, as well as discussions with county commissioners, the business community, and transportation departments, along with quiet support from state and some federal agencies. Ultimately, a 70 km long by 2 km wide pathway was protected by amendment of the US Forest Service Land Management Plan (Hamilton 2008), for which nearly 20,000 public comments were received by the federal government (Berger and Cain 2014). Related approaches have also been successful, including conservation easements where private lands may be disassociated from federal ones (Pocewicz et al. 2011).

Future Prospects

Among the central issues facing the future conservation of large mammals will be how to ensure adequate population sizes given their large spatial needs. The above subsection used one particular case study in which the human milieu and migration phenomena were juxtaposed and the conservation outcome was positive. Part of the success may have derived from Wyoming's low population density ($<2.5/km^2$), but other contributing factors include the availability of public land and the willingness of stakeholders to focus on common goals.

If migrations are to be conserved, whether in settings with an admixture of public acreage of relatively low human density or in more human-dominated areas with hard boundaries, lands will fall under diverse ownership and management, and successes will only derive from collaboration and bottom-up approaches. Failing this, however, other options remain. Animals can be shot when troublesome and beyond park borders. They can be trucked between areas where connections to suitable habitats have been severed. Migratory phenotypes can be selected against, and animals can be artificially sustained with food enhancements to reduce free-roaming behavior. Many would argue that these solutions lack creativity or imagination. They might be correct. Conservation means creating participation and investment, building consensus, and adopting an ideology that biodiversity matters.

Can a Cold-Adapted Mammal Persist in
Arctic Parks Given Climate Change?

Backdrop

Neither producers of musk nor members of the ox family, the misnamed muskoxen's closest North American relatives are mountain goats (*Oreamnos americanus*). Muskoxen (*Ovibos moschatus*) are the largest extant ungulate whose modern distribution is exclusively Arctic (Lent 1999) (fig. 9.4). Caribou, moose, and Dall sheep (*Ovis dalli*) also occur regionally in some sectors of the lower Arctic, but their distributions also transition into sub-Arctic. In the 19th and 20th centuries, moose and caribou were widespread, occurring from temperate Canada to parts of the contiguous United States from Maine to Idaho. By contrast, muskoxen are limited to permafrost, a restriction that points to a limiting role of abiotic factors in their modern

9.4. Muskoxen defensive formation with adult males (pictured *left* and *right*, with thicker horns) and adult female (*middle*); the young are not visible (*center*). Photo by J. Berger.

distribution. This is relevant for understanding possible limits to the maintenance of biodiversity in Arctic parks.

Key changes associated with polar environments include temperatures warming at two to three times the rates found elsewhere on Earth, which has changed ice and snow regimes, phenology of plant flowering and animal migrations, ecological community structure, species ranges, species life histories, and vital rates (Brodie, Post, and Doak 2012). Specific effects of abiotic factors on muskoxen are not well known; yet in both Greenland and northern Canada, population stability is more likely to occur when climate is cold and dry, in contrast to lower survival and population decline when climate is wet and warm (Vibe 1967). There has been an increase in rain-on-snow events, which encase vegetation in ice and cause population declines in wild reindeer (*R. tarandus*) (Tyler 2010). Biotic factors, such as disease predation or competition, may also play prominent, but as of yet undetermined, roles on population dynamics.

Muskoxen are probably the least studied ungulate of North America in part because research in remote, cold, and roadless areas is logistically complex and expensive. The species was extirpated from Alaska by the late 19th century owing to harvest, and population restoration commenced with reintroductions into the 1970s (Lent 1999).

Unlike some of the issues confronting large mammals in the contiguous United States, those in the Arctic differ in both kind and scale. Human population density is 0.5/km², 20 times less than the intermountain region of the United States with its relatively large national parks. Neither fenced boundaries nor insularization are issues likely to affect large mammals in Arctic parks, but climate change is, especially for species like polar bears (*U. maritimus*) and other ice-dependent obligates such as seals and walrus. Other increasing threats outside and within NPS statutory boundaries include roads and energy infrastructure. Conflicts between federal and states' rights perspectives will likely continue to have impacts on biological diversity in Alaskan parks. For the largest Arctic ungulate, only now are we beginning to understand the direct and indirect challenges.

Maintaining a Species as Ecological Conditions Deteriorate When the Science Is Uncertain

Like in most species, muskoxen demographic patterns frequently vary; in xeric climes, hot or cold events, like drought or icing, can severely affect Arctic wildlife population growth (Hansen et al. 2013). Alaskan muskoxen populations differ in their population dynamics (Schmidt and Gorn 2013), and a central question is why. Understanding the relative role of humans versus that of a warming Arctic with its suite of climatic-associated factors—increased growing season length, more rain-on-snow events, and enhanced warm temperatures—will be important in designing conservation strategies. A starting point is documenting when and where population trajectories differ, and then asking what is known of possible drivers of these differences.

Muskoxen were reintroduced in the 1970s and 1980s to three sites—the Arctic National Wildlife Refuge, the Seward Peninsula, and Cape Thompson. Muskoxen numbers increased rapidly at the first two sites, and after a few years apparently did so at Cape Thompson. Trajectories diverged widely thereafter. Across a 15-year span, the Arctic National Wildlife Refuge population dropped from about 425 animals to less than 5. The massive decline apparently occurred because of dispersal beyond the boundaries of the vast 78,000 km² refuge and from predation by grizzly bears (Reynolds, Reynolds, and Shideler 2002). The extent to which weather and/or food limitation played roles in this decline was unclear.

The other two sites were established as NPS units in 1980. Neither Cape Krusenstern National Monument (CAKR) nor Bering Land Bridge National Preserve (BELA) were locales of original muskoxen reintroduction, but were colonized on their own. The CAKR population is contiguous to the north to Cape Thompson and has increased very slowly over several decades. On the Seward Peninsula, muskoxen increased for over three decades, averaging 15% per year, and the population approached ~3,000 animals, of which a portion are within the 10,916 km² BELA. This positive growth has been reversed locally, and the population has declined 4%– 12% annually for a decade (Schmidt and Gorn 2013).

Given that muskoxen occur within a mosaic of state, borough, and federal lands with different management statutes, conservation efforts will require understanding (1) likely causes of population change and whether they stem from threats within or beyond NPS boundaries, (2) the extent to which potential drivers of change are locally manageable (e.g., mining or harvest versus climate), and (3) which, if any, NPS actions can facilitate persistence of this iconic cold-adapted representative of biodiversity.

In 2008, I initiated a project with Layne Adams on causes of variation in population growth trajectories in two broad locales: the Cape Thompson to CAKR region and the BELA region on the Seward Peninsula. The population from the former region had not grown rapidly and has been stagnant to declining. My present efforts with those of colleagues from 2008 to 2015 are intended to provide a basis for dialogue that crosses the bridge from science to conservation by understanding why populations differ in vital rates.

Sources of Variation in Muskoxen Population Dynamics: From Climate to Biological Interactions

Several interrelated drivers might explain why demographic rates at BELA and CAKR differ, including nutrition, stress, extreme climate events, parasites and disease, and predation. For instance, if food is limiting, the stagnating population (CAKR) should be characterized by individuals who are smaller, lighter, and less fecund, with other factors equal. Moreover, this population might be characterized by higher levels of glucocorticoid concentrations which signal chronic physiological stress (Sapolsky 1992; Wingfield and Romero 2001). Here the focus is on testing a food hypothesis, and I examine predictions about resource limitation based on the strong relationship between nutrition and individual growth rates in juvenile muskoxen (Peltier and Barboza 2003). An absence of differences between the BELA and CAKR animals would suggest either intersite variation in weather drivers, or perhaps biological interactions involving other community members.

Specifically, I assess muskoxen head size as a response variable and its change across different juvenile and subadult groups, pregnancy rates, and body mass because such traits are mediated by nutrition (Stewart et al. 2005). Data were derived primarily from three approaches: (1) tagging or radio-collaring more than 215 juvenile and adult females with associated measures of body mass, concentrating on areas in and adjacent to CAKR and BELA (L. Adams, unpublished data); (2) noninvasive techniques including photogrammetry (Berger 2012), from which I generated more than 700 measures of head size of one-, two-, and three-year-olds, and pregnancy and stress levels based on fecal metabolites to assess glucocorticoids (Cain et al. 2012; J. Berger, unpublished data); and (3) potential weather-related effects explored through vegetation greenness[1] and other climatic variables. Density estimates of potential carnivore predators were unavailable.

Despite lacking pertinent information on predators and muskoxen density, the data do allow assessment of the potential role of food and

---

1. Geographic Information Network of Alaska, MODIS-derived NDVI metrics, accessed 8 March 2016, http://www.gina.alaska.edu/projects/modis-derived-ndvi-metrics.

weather—both anticipated either individually or jointly—to account for population-level variation and vital rates. A metric related to nutrition, head size, was not statistically different between the CAKR and BELA sites. Additionally, had food quality varied substantially between sites, differences in adult pregnancy rates should have occurred. Furthermore, if chronic stress induced by nutritional inadequacies or other factors (e.g., predators, inclement weather) affected one population more than the other, fecal cortisol levels should have consistently differed. None of these measures differed between populations, nor were temperature, precipitation, and NDVI (normalized difference vegetation index) associated with head sizes. These findings suggest that both populations responded similarly to weather, or that weather effects were minimal (table 9.1).

Juvenile recruitment can have strong effects on population growth, especially for species in which adult survival varies little (Mills 2012). In the western Arctic, recruitment of juvenile muskoxen was inversely related to skewed adult sex ratios, and ratios decreased 4%–12% per year across 10 years (2002–2012), as subsistence and trophy hunters harvested more males (Schmidt and Gorn 2013). While hunting by humans is legally permitted in and around both CAKR and BELA, harvest is more heavily concentrated on the Seward Peninsula including in BELA. Young animals are not taken, however, so hunting can be excluded as a direct source of the variation in juvenile survival. So, too, can the differential production of offspring, since pregnancy rates were similar in CAKR and BELA. If predation pressures, especially by grizzly bears, have changed and affect juvenile survival, they may arise as an indirect consequence of the removal of adult males (Schmidt and Gorn 2013), a hypothesis in need of testing.

What is the evidence that biological interactions might play a greater proximate role than weather in affecting growth in the CAKR and BELA populations? While weather and climate have dramatic effects on northern ungulates (Post et al. 2008; Hansen et al. 2013), including localized persistence of muskoxen (Vibe 1967; Darwent and Darwent 2004), during the period for which our data exist, population trends reversed across just a few years. Alteration of sex ratios by harvest of adult males was negatively correlated with juvenile survival. For several ungulates and primates living in mixed-sex groups, adult males are associated with defense and deterrence of predatory approaches (van Schaik and Hörstermann 1994; Fischhoff et al. 2007). Whether this is the case for muskoxen is unknown. Investigation of this hypothesis using playback models in Arctic NPS units continues, and it will enable clarification of the extent to which biologi-

Table 9.1   Summary of population change in muskoxen at two NPS sites (CAKR and BELA) and at Arctic National Wildlife Refuge, and five response variables to test a food limitation hypothesis

	Cape Krusenstern (CAKR)	Bering Land Bridge (BELA)	Arctic National Wildlife Refuge	Years sampled	Comment: CAKR-BELA contrasts
Trajectory					
1st three decades[a]	~8% increase/yr	~15% increase/yr	Increase, then stable		
Last decade[a,b,c]	Stable to decline	Decline	Harsh decline ~15 yrs		
Response Variable					
Adult female mass[d]	Similar	Similar	NA	4 (2009–2012)	Only 2009 differs $p < 0.05$
Juvenile head size[e]	Similar	Similar	NA	7 (2008–2014)	No statistical differences
Subadult head size[e]	Similar	Similar	NA	7 (2008–2014)	No statistical differences
Stress level[e]	Similar	Similar	NA	5 (2008–2012)	No statistical differences
Pregnancy rates[e]	Similar	Similar	NA	5 (2008–2012)	No statistical differences

Note: All populations stem from 31 founders established on Nunivak Island in 1935–1936. Descendants reintroduced to the three mainland sites between 1969 and 1981. NA = not available.
[a]Schmidt and Gorn (2013) and references therein; Reynolds (1998)
[b]USFWS, unpublished data
[c]J. Berger, unpublished data; NPS, unpublished data
[d]USGS, unpublished data
[e]Methods described in Berger (2012); Cain et al. (2012); J. Berger and C. Hartway, unpublished data

cal interactions involving bears may be affecting muskoxen population dynamics independent of weather.

Future Prospects

Does a cold-adapted Arctic-obligate mammal have a strong possibility to persist as climate changes? Evidence so far suggests that, despite low human densities, harvest regimes may be playing an indirect role in muskoxen population declines, primarily through offtake of adult males concentrated outside NPS units. The extent to which warming temperatures and variable precipitation, as mediated by rain-on-snow events, may govern long-term survival of muskoxen is unclear. Past evidence from northern Canada and Greenland suggests warm, wet periods are challenging (Vibe 1967). If currently changing temperature and precipitation regimes are strong determinants of persistence, then immediate management may matter little.

The broader issue here is not about muskoxen per se, though their persistence as an icon of Arctic biodiversity in NPS units is of unquestionable relevance. The key matter concerns science and what we don't know, and how one might configure a plan for long-term conservation given uncertainty about biological interactions and other factors that affect species.

Despite difficulties in predicting long-term population and climatic trends, conservation of large Arctic ungulates and carnivores, including wolves, brown bears, and wolverines, requires consideration of time frames longer than half a century and suitable habitats far beyond the boundaries of existing protected areas (Klein 1982, 1992), especially given changing fire regimes, time for vegetation recovery, and broad alteration of habitat productivity (Ferguson and Messier 2000). Where not harassed, muskoxen, caribou, and other species can flourish in areas with human infrastructure including oil pipelines and wind turbines, though these areas are perhaps less appealing from aesthetic perspectives. The challenges that large mammals face in these lightly human-populated lands in the Arctic differ from those in the contiguous United States, a place where research has added amply to understanding and resolving some of the challenges associated with large mammals in parks.

Science and Conservation Challenges as Human Populations Grow

Three decades ago key tenets of conservation biology were set forth (Soulé 1986). They included protecting multiple large areas, maintaining them with buffer zones, and connecting them when and where additional land cannot be acquired. Science gives us unassailable evidence about the often negative consequences of isolation, both through experimental and field studies. Much is known about genetics and demography. We are less confident about possible effects of climate change, although knowledge accumulates rapidly.

In the case of Arctic species, there is much uncertainty on how and where cold-adapted species may persist. Polar bears are an obvious example of an ice-dependent species in serious trouble, and where currently existing protecting areas have little to do with sustaining them at contemporary levels far into the future. Climate change here is the issue for which we as individuals may have little immediate control, a situation that differs markedly from wolves. Wolf persistence beyond the boundaries of protected areas is about human dimensions and not science per se.

The polar muskoxen case study differs substantially from our other examples in the contiguous United States, and is illustrative of how and why knowledge of biological interactions is pertinent for prudent management. Despite unfettered NPS landscapes along the Chukchi Sea and associated low human densities, human subsistence and trophy hunters may be having an important indirect effect on juvenile muskoxen survival, as mediated by the loss of large males for herd protection against predators. Conjecture, however, far outstrips empiricism in this system.

On the other hand, we know that real-world complexities—many of which involve our consumptive lifestyles and our growing human populations—prevent uniform approaches to conservation. Within the more crowded landscapes of the contiguous United States, biological corridors offer effective ways to connect populations and facilitate gene flow. Science and science communication are important, and they serve as a first step in formulating conservation planning. The critical questions need not be about our resolve, the importance of biodiversity, or human dimensions, but what we want of our future landscapes. Parks have diverse missions and one is about enjoyment for future generations.

Science is, of course, relevant to scientists, but in a complex world with more than seven billion people, it is but a single currency, and rarely is it the final arbiter in decision making. When science is fused with policy, conservation practices can be furthered. In the end, however, it is education and experience that will shape and inevitably change human values. Conservation means people. If we as scientists want conservation, we need to have parks that are relevant to people.

Acknowledgments

For grant, logistic, and other help, I most gratefully thank the Alaska Department of Fish and Game, the National Park Service (Badlands, Western Arctic Parks, Grand Teton, Yellowstone; NPS offices in Fort Collins [Biological Resources Division], Nome, Anchorage, and Fairbanks; and the Beringia Heritage Program), the Liz Claiborne and Art Ortenberg Foundation, the US Geological Service, the University of Alaska, the University of Montana, the Wildlife Conservation Society, Wyoming Fish and Game, L. Adams, J. Beckmann, P. Barboza, T. Bowyer, P. Brussard, S. Cain, F. Camenzind, E. Cheng, C. Cunningham, S. Ekernas, H. Ernst, B. Frost, F. Hayes,

K. Hamilton, C. Hartway, J. Koelsch, J. Lawler, E. Leslie, B. Lowry, M. Johnson, G. Machlis, S. Mills, W. Peters, G. Plumb, R. Reading, G. Roffler, B. Schults, M. G. Scott, D. Smith, A. Vedder, W. Vickers, W. H. Weber, G. Wingard, and S. Zack.

Literature Cited

Allred, B. W., W. K. Smith, D. Twidwell, J. H. Haggerty, S. W. Running, D. E. Naugle, and S. D. Fuhlendorf. 2015. Ecosystem services lost to oil and gas in North America. Science 348:401–402.

Austen, D. J. 2011. Landscape conservation cooperatives: a science based network in support of conservation. Wildlife Professional 5:32–37.

Beckmann, J., K. Murray, R. Seidler, and J. Berger. 2012. Human-mediated shifts in animal habitat use: sequential changes in pronghorn use of a natural gas field in Greater Yellowstone. Biological Conservation 147:222–233.

Beier, P., and A. Gregory. 2012. Desperately seeking stable 50-year-old landscapes with patches and long, wide corridors. PLoS Biology 10:e1001253.

Berger, J. 2004. The longest mile: how to sustain long distance migration in mammals. Conservation Biology 18:320–332.

———. 2012. Estimation of body-size traits by photogrammetry in large mammals to inform conservation. Conservation Biology 26:769–777.

Berger, J., and S. L. Cain. 1999. Reproductive synchrony in brucellosis-exposed bison in the southern greater Yellowstone ecosystem and in non-infected populations. Conservation Biology 13:357–366.

———. 2014. Moving beyond science to protect a mammalian migration corridor. Conservation Biology 28:1142–1150.

Berger, J., S. L. Cain, and K. Berger. 2006. Connecting the dots: an invariant migration corridor links the Holocene to the present. Biology Letters 2:528–531.

Berger, J., S. L. Cain, E. Cheng, P. Dratch, K. Ellison, J. Francis, H. C. Frost, et al. 2014. Optimism and challenge for science-based conservation of migratory species in and out of US national parks. Conservation Biology 28:4–12.

Berger, J., and C. Cunningham. 1994. Bison: mating and conservation in small populations. Columbia University Press, New York, New York.

Brodie, J., E. Post, and D. Doak, eds. 2012. Climate change and wildlife conservation. University Chicago Press, Chicago, Illinois.

Cain, S. L., M. D. Higgs, T. J. Roffe, S. L. Monfort, and J. Berger. 2012. Using fecal progestagens and logistic regression to enhance pregnancy detection in wild ungulates: a case study with bison. Wildlife Society Bulletin 36:631–640.

Colwell, R., S. Avery, J. Berger, G. E. Davis, H. Hamilton, T. Lovejoy, S. Malcolm, et al. 2012. Revisiting Leopold: resource stewardship in the national parks. A report of the National Park System Advisory Board Science Committee. http://www.nps.gov/calltoaction/PDF/LeopoldReport_2012.pdf.

Copeland, H. E., H. Sawyer, K. L. Monteith, D. E. Naugle, A. Pocewicz, N. Graf, and M. J. Kauffman. 2014. Conserving migratory mule deer through the umbrella of sage-grouse. Ecosphere 5:1–16.

Crooks, K. R., and M. E. Sanjayan, eds. 2006. Connectivity conservation. Cambridge University Press, Cambridge, United Kingdom.

Culver, M, P. W. Hedrick, K. Murphy, S. O'Brien, and M. G. Hornocker. 2008. Estimation of the bottleneck size in Florida panthers. Animal Conservation 11:104–110.

Darwent, C. M., and J. Darwent. 2004. Where the muskox roamed: biogeographic distribution of tundra muskox (*Ovibos moschatus*) in the eastern Arctic. Pages 61–87 *in* R. L. Lyman and K. P. Cannon, eds. Zooarchaeology and conservation biology. University of Utah Press, Salt Lake City, Utah.

Epps, C. W., P. J. Palsbøll, J. D. Wehausen, G. K. Roderick, and D. R. McCullough. 2006. Elevation and connectivity define genetic refugia for mountain sheep as climate warms. Molecular Ecology 15:4295–4302.

Ernest, H. B., T. W. Vickers, S. A. Morrison, M. R. Buchalski, and W. M. Boyce. 2014. Fractured genetic connectivity threatens a southern California puma (*Puma concolor*) population. PLoS ONE 9:e107985.

Ferguson, M. A., and F. Messier. 2000. Mass emigration of arctic tundra caribou from a traditional winter range: population dynamics and physical condition. Journal of Wildlife Management 64:168–178.

Fischhoff, I. R., S. R. Sundaresan, J. Cordingley, and D. Rubenstein. 2007. Habitat use and movements of plains zebra (*Equus burchelli*) in response to predation danger from lions. Behavioral Ecology 18:725–729.

Forrester, J. A., C. P. Holstege, and J. D. Forrester. 2012. Fatalities from venomous and nonvenomous animals in the United States (1999–2007). Wilderness & Environmental Medicine 23:146–152.

Hague, A. 1893. The Yellowstone Park as a game reservation. Forest and Stream Publishing Company, New York, New York.

Halbert, N. D., and J. N. Derr. 2008. Patterns of genetic variation in US federal bison herds. Molecular Ecology 17:4963–4977.

Halbert, N. D., P. J. Gogan, R. Hiebert, and J. N. Derr. 2007. Where the buffalo roam: the role of history and genetics in the conservation of bison on US federal lands. Parks Science 24(2):22–29.

Hamilton, K. 2008. Decision notice & finding of no significant impact: pronghorn migration corridor forest plan amendment. US Department of Agriculture, 31 May. Accessed July 2013. http://www.fs.fed.us/outernet/r4/btnf/projects/2008/pronghorn/PronghornDN.pdf.

Hamin, E. M. 2001. The US National Park Service's partnership parks: collaborative responses to middle landscapes. Land Use Policy 18:123–135.

Hansen, B. B., V. Grøtan, R. Aanes, B.-E. Sæther, A. Stien, E. Fuglei, R. A. Ims, N. G. Yoccoz, and Å. Ø. Pedersen. 2013. Climate events synchronize the dynamics of a resident vertebrate community in the high Arctic. Science 339:313–315.

Harris, G., S. Thirgood, J. G. C. Hopcraft, J. P. G. M. Cromsigt, and J. Berger. 2009. Global decline in aggregated migrations of large terrestrial mammals. Endangered Species Research 7:55–76.

Hess, G. 1996. Disease in metapopulation models: implications for conservation. Ecology 77:1617–1632.

Hilty, J. A, C. C. Chester, and M. S. Cross, eds. 2012. Climate and conservation: landscape and seascape science, planning, and action. Island Press, Washington, DC.

Hilty, J. A., W. Z. Lidicker, and A. M. Merenlender. 2006. Corridor ecology: the science and practice of linking landscapes for biodiversity conservation. Island Press, Washington, DC.

Hornaday, W. T. 1913. Our vanishing wildlife. New York Zoological Society, Bronx, New York.

Keiter, R. E. 2010. The national park system: visions for tomorrow. Natural Resource Journal 50:71–110.

Klein, D. R. 1982. Fire, lichens, and caribou. Journal of Range Management 35:390–395.

———. 1992. Comparative ecological and behavioral adaptations of *Ovibos moschatus* and *Rangifer tarandus*. Rangifer 12:47–55.

Lent, P. C. 1999. Muskoxen and their hunters: a history. University of Oklahoma Press, Norman, Oklahoma.

Leopold, A. S., S. A. Cain, C. M. Cottam, I. N. Gabrielson, and T. L. Kimball. 1963. Wildlife management in the national parks: the Leopold Report. Unpublished report to the US Secretary of the Interior.

Machlis, G. E., J. E. Force, and W. R. Burch Jr. 1997. The human ecosystem Part I: the human ecosystem as an organizing concept in ecosystem management. Society and Natural Resources 10:347–367.

Mills, L. S. 2012. Conservation of wildlife populations: demography, genetics, and management. John Wiley & Sons, Chichester, United Kingdom.

Peltier, T. C., and P. S. Barboza. 2003. Growth in an arctic grazer: effects of sex and dietary nitrogen on yearling muskoxen. Journal of Mammalogy 84:915–925.

Plumb, G. E., R. Monello, J. Resnick, R. Kahn, K. Leong, D. Decker, and M. Clarke. 2013. A comprehensive review of ungulate management by the National Park Service: second century challenges, opportunities, and coherence. Natural Resource Report NPS/NRSS/BRMD/NRR—2013/898. National Park Service, Fort Collins, Colorado.

Plumb, G. E., P. J. White, M. B. Coughenour, and R. L. Wallen. 2009. Carrying capacity, migration, and dispersal in Yellowstone bison. Biological Conservation 142:2377–2387.

Pocewicz, A., J. M. Kiesecker, G. P. Jones, H. E. Copeland, J. Daline, and B. A. Mealor. 2011. Effectiveness of conservation easements for reducing development and maintaining biodiversity in sagebrush ecosystems. Biological Conservation 144:567–574.

Post, E., C. Pedersen, C. C. Wilmers, and M. C. Forchhammer. 2008. Warming, plant phenology, and the spatial dimension of trophic mismatch for large herbivores. Proceedings of the Royal Society B 275:2005–2013.

Ray, J., K. H. Redford, R. Steneck, and J. Berger, eds. 2005. Large carnivores and conservation of biodiversity. Island Press, Covello, California.

Reynolds, P. E. 1998. Dynamics and range expansion of a reestablished muskox population. Journal of Wildlife Management 62:734–744.

Reynolds, P. E., H. V. Reynolds, and R. T. Shideler. 2002. Predation and multiple kills of muskoxen by grizzly bears. Ursus 13:79–84.

Ripple, W. J., T. M. Newsome, C. Wolf, R. Dirzo, K. T. Everatt, M. Galetti, M. W. Hayward, et al. 2015. Collapse of the world's largest herbivores. Science Advances 1:e1400103.

Roelke, M. E., J. S. Martenson, and S. J. O'Brien. 1993. The consequences of demographic reduction and genetic depletion in the endangered Florida panther. Current Biology 3:340–350.

Sanderson, E. W., K. H. Redford, B. Weber, K. Aune, D. Baldes, J. Berger, D. Carter, et al. 2008. The ecological future of the North American bison: conceiving long-term, large-scale conservation of wildlife. Conservation Biology 22:252–266.

Sapolsky, R. M. 1992. Neuroendocrinology of the stress-response. Pages 287–324 *in* J. B. Becker, S. M. Breedlove, and D. Crews, eds. Behavioural endocrinology. MIT Press, Cambridge, Massachusetts.

Sawyer, H., M. Hayes, B. Rudd, and M. J. Kauffman. 2014. The Red Desert to Hoback mule deer migration assessment. Wyoming Migration Initiative, University of Wyoming, Laramie, Wyoming.

Sawyer, H., M. J. Kauffman, A. D. Middleton, T. A. Morrison, R. M. Nielson, and T. B. Wyckoff. 2013. A framework for understanding semi-permeable barrier effects on migratory ungulates. Journal of Applied Ecology 50:68–78.

Schmidt, J. H., and T. S. Gorn. 2013. Possible secondary population-level effects of selective harvest of adult male muskoxen. PLoS ONE 8(6):e67493.

Soulé, M. E. 1986. Conservation biology: science of scarcity and diversity. Sinauer, Sunderland, Massachusetts.

Stewart, K. M., R. T. Bowyer, B. L. Dick, B. K. Johnson, and J. G. Kie. 2005. Density-dependent effects on physical condition and reproduction in North American elk: an experimental test. Oecologia 143:85–93.

Tyler, N. J. 2010. Climate, snow, ice, crashes, and declines in populations of reindeer and caribou (Rangifer tarandus L.). Ecological Monographs 80:197–219.

US Census Bureau. 2013. Annual Estimates of the Population for the United States, Regions, States, and Puerto Rico: 1 April 2010 to 1 July 2013. http://www.census.gov/popest/data/state/totals/2013/.

van Schaik, C. P., and M. Hörstermann. 1994. Predation risk and the number of adult males in a primate group: a comparative test. Behavioral Ecology and Sociobiology 35:261–272.

Vibe, C. 1967. Arctic animals in relation to climatic fluctuations: the Danish zoogeographical investigations in Greenland. Meddelelser om Gronland 170:1–227.

Wingfield, J. C., and L. M. Romero. 2001. Adrenocortical responses to stress and their modulation in free-living vertebrates. Pages 211–236 in B. S. McEwen, ed. Handbook of physiology. Section 7, The endocrine system. Volume 4, Coping with the environment: neural and endocrine mechanisms. Oxford University Press, Oxford, United Kingdom.

Wright, G. M., J. S. Dixon, and B. H. Thompson. 1933. Fauna of the national parks of the United States: a preliminary survey of faunal relations in national parks. Contribution of Wild Life Survey. Fauna Series No. 1—May 1932. US Government Printing Office, Washington, DC.

TEN

Strategic Conversation: Stewardship of Parks in a Changing World

EDITED BY MEAGAN F. OLDFATHER,
KELLY J. EASTERDAY, MAGGIE J. RABOIN,
AND KELSEY J. SCHECKEL

In a world with rapidly changing climate, rising sea levels, invasive species, and shifting disturbance regimes, the challenges of stewardship in the national parks have never been greater. Parks are challenged with reconciling management in the face of these changes while sustaining the preservationist values embedded in history, law, and policy. To maintain and restore ecosystem functions and combat climate change, should national parks embrace species once considered nonnative to a region, organisms produced by de-extinction, or populations introduced through rewilding? Should managers use historical baselines as goals for restoration in the face of shifting climate and disturbance regimes? Is active management appropriate to resist novel ecosystems, or should trajectories of disturbance and succession be allowed to proceed unimpaired?

This strategic discussion, which transpired at the Berkeley summit "Science for Parks, Parks for Science" on 26 March 2015, focuses on the role of stewardship and science in national parks and in confronting these looming challenges. The discussion panel includes Stephanie Carlson, evolutionary ecologist and associate professor in the Department of Environmental Science, Policy, and Management at the University of California, Berkeley; Josh Donlan, founder and director of Advanced Conservation Strategies; Laurel Larsen, hydroecologist and assistant professor in the Department of Geography at the University of California, Berkeley; and Raymond Sauvajot, ecologist and associate director of natural resource stewardship and science at the National Park Service, where he has worked for over 25 years. The conversation was moderated by David Ackerly, professor in the Department of Integrative Biology at the University of California, Berkeley.

DAVID ACKERLY: I don't think it's an accident that so many of the previous speakers spoke about parks in a landscape context, embedded in a larger landscape. In some ways this goes against what may be the American ideal that parks are set aside with a fence around them. But in practice, our parks have never operated that way. I wanted to start with that theme and, Dr. Donlan, with you. We spoke this morning about what you called the coexistence model and the separation model for parks, and also about perspectives from parks in other parts of the world. *Where do you see the US parks, and the way we view the US parks, in the next century? What lessons can we learn from the rest of the world in that context?*

JOSH DONLAN: In my limited experience in the United States and more experience internationally, let's take two extremes: a preservation or separation model between humans and nature, and a coexistence model akin to a working landscape. In my view, I think that it is largely a false dichotomy. The real innovation will come from efforts that attempt to integrate these different flavors and approaches of coexistence and preservation. There's good evidence that both models can deliver biodiversity benefits in the right context. I think there's also consensus that both of those models are underperforming in general in terms of biodiversity protection. It's not one or the other. What is needed are new, innovative approaches for how to integrate these models and to find the right incentives that maximize biodiversity benefits, or in some cases biodiversity co-benefits.

ACKERLY: Dr. Carlson, you've worked a lot with migratory fish. This is an example of a species that respects no boundaries. *So where do you see the management challenges and the opportunities for the Park Service in dealing with biodiversity resources that cross boundaries and are on the move?*

STEPHANIE CARLSON: That's a great question. I think today we've heard a lot about the fact that parks aren't isolated entities, and that we need to be thinking about the larger landscape that parks are embedded in. This is really made very clear when thinking about migratory organisms that move across the boundaries of parks. I work quite a bit with anadromous fishes that are breeding and rearing in parks, particularly the Point Reyes National Seashore in this region, and then migrating to the ocean and back again to complete their life cycle. For organisms that have migrations that take them out of the park for most of their life cycle, we need to be thinking about the landscape outside the park, and how activities outside the park potentially influence these organisms' dynamics. This was the main theme of Ruth DeFries's talk earlier, that we need to be thinking about conservation on private lands, and how we can minimize impacts on organisms that are moving beyond park boundaries, which likely include many organisms found within parks.

ACKERLY: *Where are the opportunities for the Park Service in seeing the park as embedded in the larger landscape? What are the opportunities that could really enhance that mission and that vision of the parks?*

RAYMOND SAUVAJOT: Over the years, there's been an increasing realization that the mission of the National Park System depends on a perspective of looking at our parks as part of a broader conservation network. The Park Service is embedded in this broader community of effects, and some of those effects are threats. For example, we heard about air quality issues from Jill Baron. We heard about the Organic Act and keeping resources unimpaired. The National Park Service has a policy and a legal obligation to protect those resources and those values. For the Park Service to succeed in that mission, it has to recognize that those values will not persist if it isn't thinking about that broader scale. That recognition is going to force the National Park Service to look for opportunities to work outside its boundaries, and to see that as part of what natural resource conservation and natural resource management is about. It's not just worrying about the issues and the concerns that may be confined within a particular park unit. The values in a park depend on looking beyond and working with partners, and developing those relationships to ensure persistence over time.

ACKERLY: I want to transition a little bit to some of the science. Some science in the parks will be a combination of observations. We're trying, as Joel Berger suggested, to understand the causes of change, for example, in animal populations. In other cases, we might want to test whether a management intervention or restoration project will work. As scientists, what we would ideally want is three to five replicates of the restored landscape, and three to five replicates that are the control. That approach sometimes begins to make people uncomfortable. They might think, "But if it's such a good idea, how can you leave those areas untreated?" Dr. Larsen, I know this has come up in your work. *How do we balance what we might really see as the ideal design to test whether these interventions work with the desire to restore and manage for the benefit of landscapes?*

LAUREL LARSEN: This type of statistical design that you talk about works well within the context of an adaptive management framework. In reality, it's difficult to realize because oftentimes experimental manipulations are quite expensive. They affect large areas, and so we're limited by space. But there's a lot that we can learn from experiments even if we don't have full replication, particularly if it informs our knowledge of processes. Experiments that inform our knowledge of processes enable us to construct simulation models that allow us to extrapolate effects over space and time. Even if we don't achieve ideal statistical replication, it's important to do these experiments, because they do provide a data point. They enable us to implement policy on a small scale before spending huge amounts of money to implement it on much larger scales.

CARLSON: A lot of the efforts surrounding fish conservation focus on local-scale habitat restoration, and there are numerous opportunities to be learning from these many small and often unreplicated efforts. There have been many lost opportunities to learn from such studies. They could guide future experimental design and future studies. I absolutely agree that we should be trying to do this in a way that we can learn something, even if it's just a single data point. Through syntheses of multiple studies, we can begin to accumulate knowledge that can help guide restoration projects in the future.

LARSEN: There are a lot of opportunities to take advantage of natural experiments—for instance, natural disasters that perturb a system. Right now there's a big effort focused on studying the Wax Lake Delta, which is the only part of the greater Mississippi River Delta complex where we're actually gaining land because of a levee breach that happened in the 1960s. It was an engineering accident that enabled us to learn a lot about how coastal wetlands grow and build land. I think there are many of these opportunities, and they are quite effective and efficient to take advantage of.

DONLAN: I'll just add that this is recently becoming a big issue with payment from environmental services programs, where the science hasn't really kept up well with the programs compared with, say, the social sector, where randomized control trials have been used to evaluate large programs and approaches, such as conditional cash transfers, et cetera. Thus, we have a much better understanding of the performance of some of the social programs compared with payment for environmental services programs. There have been recent calls for more scientific rigor with respect to the design and evaluation of payment for environmental services programs.

ACKERLY: *Dr. Sauvajot, have you ever seen an experiment rejected because the design was just too incompatible with a park's goals?*

SAUVAJOT: I think that rarely are they rejected because of statistical design. We are doing things on the ground. There are restoration activities, and there are lots of various management actions that have been taken. But we need to look at those as opportunities to learn. We need to reach out to universities and other scientists to make sure that when those actions are taken, the information that's collected from them is obtained in a way that provides statistically robust information about how we make our decisions to inform future decisions. We need that information.

ACKERLY: Some of the experiments that are being done, intentionally or not, are species reintroductions or, as we heard about this afternoon, sometimes eradications. Dr. Donlan, I wonder if you could start. Many people may know that you've been involved in rewilding dialogues in the past. *What are the criteria for determining whether something is native or not?* Some of the things that have been proposed for reintroduction may have been gone from the system for a long time, and yet they're certainly not an alien species in the sense that they were brought from Asia or South Africa.

DONLAN: If you look a little bit back, in the 1980s Dan Jansen started to get ecologists thinking about how "history matters." Over the past 10 years, the conservation field has started to talk about and realize that "history matters"—ecological history, that is. And that perhaps, from a North American perspective, ecological history doesn't begin at 1492. It goes way further back. Thus, maybe our conservation goals should take this into account. I think there are some great national park examples that serve as case studies for this premise—examples that, if you think about them long enough, make your head hurt. Rats on Anacapa Island are a great example, where the eradication of rats, which I participated in, was justified. The biodiversity benefits were huge, and rats were clearly an invasive species. Other situations aren't so cut and dry. The Bolson tortoise, for example, is a large tortoise with a fossil record all over the southwestern United States. But it was not considered a candidate for reintroduction into Big Bend National Park because it was an "invasive species." Whether it's an invasive species really depends on your view of ecological history. The California condor, a huge conservation success story, is probably one of the best examples. They have been introduced to the Grand Canyon, but most evidence suggests that the last time condors soared there was 13,000 years ago. So whether it is a native or nonnative species really depends on your view of ecological history.

ACKERLY: To the extent that we can hold on to the idea that we don't want to introduce nonnative species into a system, Dr. Carlson, a question for you. *Does it apply equally to introducing nonnative genes?* In the face of climate change, a lot of the movement that may occur may not be just moving species but the intentional or natural movement of different genotypes across the landscape.

CARLSON: It's a fascinating question, and I think that people are perhaps a bit more comfortable thinking about moving genes to increase population resiliency than moving species to places that might become suitable in the future. You can imagine trying to move a few organisms from a warm-adapted location to a more northerly location in order to inject genes into the northerly population to help increase resiliency. This hasn't been done widely. But it's certainly something that people are discussing, actually pretty actively for corals in Australia right now. There's quite a bit of research on this possibility. I do think that this is going to become a larger part of the discussion in the future—whether we should be actively trying to introduce genes into systems to help make the recipient populations more resilient. Some people are, of course, very uncomfortable with this idea. The idea that we could be playing "God" in these systems as opposed to standing back and taking more of a hands-off approach. There are some societal decisions there to be made.

SAUVAJOT: The National Park Service, as a public agency, is responsive to a lot of those societal perceptions and values. There are policy issues, and the policies neatly define things, like what's an invasive and what's not. But when you superimpose all those nice policies on top of dynamic ecosystems, the answers are not always clear-cut. I think that we're definitely seeing now, during a period of more rapid environmental change, that the Park Service is going to have to grapple with these questions and grapple with them more frequently. It's going to open up a scientific discourse, but also a policy and a values discourse that will influence those kinds of decisions. We saw in the earlier presentations this sort of evolution, from parks being thought of as vignettes to a recognition that change occurs. In this more dynamic perspective, not only is change occurring, but it's occurring in trajectories that are very difficult to predict. So that makes this whole question—what is impairment, and how does one define that, and how does one intervene in ways to try to manage against things like resource impairment?—all that more difficult to answer.

ACKERLY: *Does the existing legislation that we have cover some of these scenarios? Are these issues legislative problems or science problems? Do we just need more science to support policy decisions? Is it within the policy mandate of the Park Service to take these problems on? Do you have everything you need to set up new policies that could tackle some of these issues?*

SAUVAJOT: We've heard several times today about the *Revisiting Leopold* report. One of the essential messages from it is that there is a dynamic environment within which parks are embedded. How does the National Park Service grapple with that? Are the policies sufficient for this kind of new world? There is a group right now in the National Park Service addressing that very question. Are current policies sufficient for addressing these challenges that the Park Service is facing as an agency? I think that, in some ways, the jury is out. I will say, though, that, as I mentioned earlier, the agency knows the threats and the challenges that resources face, and it knows that the resources themselves exist in this broader network. The mission right now, as it currently is written, says that the National Park Service has an obligation to protect those values, and the only way it can do that is to be thinking in this broader scale. So the Park Service knows that, and it can do that now, and its mission, in a sense, dictates that. It becomes more nuanced with questions and details about, what does that mean about control of invasives? And what does that mean about intervention? And what can or can't be introduced? Those are the sorts of issues that the Park Service needs to deal with and that it is beginning to deal with. I'm hoping that the next couple of days at this summit are going to provide some really interesting insights.

ACKERLY: I thought you might say "answers."

SAUVAJOT: Insights, insights!

ACKERLY: There's been a lot of talk about biological values today and the role of native and invasive species. *Dr. Larsen, as someone who thinks a lot about physical processes, is there a parallel perspective about what it means to protect and restore the physical processes that are ongoing in these ecosystems?*

LARSEN: Yeah. It's a great question. I think one of the big challenges that the Park Service and other environmental resource managers are facing in the present day is the trade-off between preserving landscape function and preserving landscape form. One of the challenges that we face in the Everglades is the fact that currently the wetland is quite compartmentalized by levees and canals that disrupt the sheet flow patterns that occurred prior to 20th-century human intervention in South Florida. There's a big effort right now to restore flow to the Everglades. But if we simply remove all those barriers to flow, and let the system flow freely in an unmanaged sense, it's likely that water levels will be much lower than they were historically, which will promote vegetation community shifts and inhibit the preservation of habitat for fish and wildlife. We're looking at the question of whether it's better to have an intensively managed system with pulsed flow releases, or to remove these barriers to flow and have a much more unmanaged system but one that is very different from what it was historically. I think that tension is representative of similar trade-offs in other parks.

ACKERLY: *Do you find that there is a public appreciation for those physical processes?* Certainly Half Dome is a physical feature in Yosemite National Park that's greatly appreciated. A lot of the history of the parks was about protecting physical features. In many ways our discussion has become much more biologically focused, maybe because biologists helped organize this meeting. *Do you think that, in communicating to the public, there is a set of values around the physical processes that are parallel to the values around native biodiversity and ecological components?*

LARSEN: There is a set of values that the different stakeholder groups have associated with those physical properties. In the Everglades, for instance, one of the worries is that by restoring flow to the system, portions of land now used by Miccosukee tribes might become flooded. They have an interest in ensuring that we understand what the system is going to do when we institute pulsed flows and remove some of these barriers to flow. Other stakeholder groups, such as the bass fishermen, value the presence of canals within the landscape. There is a diverse appreciation for the various physical features of these landscapes.

ACKERLY: Dr. Donlan, I know in your work in other parts of the world, one of your foci has been how to incentivize private landowners to engage in actions that also serve conservation. You talked about payment for ecosystem services, so money is an incentive. *What are the other values that you find really motivate people on their own lands to engage in conservation actions, especially if it's in concert with a nearby park or part of a broader landscape view?*

DONLAN: What we've generally tried to do is engage the local stakeholders—whether they're small-scale fisherman, ranchers in Tierra del Fuego, or small communities living in forests—and actually ask them what their values and needs are. One consistent thing that often comes out of that process is that the actual biodiversity benefits, while they might be important, are not at the top on their list in terms of participating in some type of incentive biodiversity conservation program. Rather, these groups are interested in how the program aligns with the way they view the world and the way they live. There's a lot of value, at least in my view, in learning and mapping the perceptions and sentiments of your target stakeholders. Then you can design a program that they are actually going to sign up for, as opposed to designing a program that maximizes the biodiversity benefits but then no one signs up for. Because these programs are largely voluntary, if no one signs up, you're not going to get the landscape-level benefits that you're targeting.

ACKERLY: There are a lot of voices that might suggest that we've had a fairly tame discussion relative to the magnitude of projected impacts that could occur in response to 21st-century climate changes, as currently projected by global models. *Are we even having the right conversation? Are we in the right ballpark of the kinds of challenges that we really might face in the next 100 years for the parks?*

CARLSON: So I'm thinking about salmon here in California, where we're basically at the southern edge of the range for several salmonids. You asked me earlier about assisted evolution. Trying to inject genes in the populations to help them become more resilient is on the minds of many people. I think an even more pressing question is that we have several salmonid species that are on the brink of extinction. Take coho salmon here in Marin County as an example—these are the southernmost wild coho salmon in the world. This past summer, which was the third year of a multiyear drought and the second driest year in California's recorded history, a decision was made to remove juvenile coho from one of the streams there and to take these organisms into a hatchery setting because there were concerns about the stream drying up. The decision to remove organisms from the wild and bring them into a captive-rearing facility is a major one. This kind of decision is made when people believe that the risk of leaving organisms in the wild is higher than the risks to them from putting them into an artificial setting. We're going to be confronted with these kinds of challenges more often, and they are going to take more and more of our time and resources. I agree that we need to be thinking of new strategies for increasing resiliency, like introducing genes into different environments, as we discussed earlier. But I think a lot of our time is actually going be eaten up through efforts to stave off extinction of species, particularly at the southern end of the range or range boundaries.

DONLAN: In my view, most of the challenges and biodiversity gains, if you look at a global scale, are going to be realized in places outside of parks and are going to be happening in places like India, as we heard from Ruth DeFries, where there's real opportunity costs for biodiversity gains. In my view, most of the strategies that we've been focused on over the past 40 years have largely been incremental innovations and probably won't serve us particularly well with respect to the challenges we have now. In addition to incremental innovation, we need to be thinking more about transformative innovation, and bringing more entrepreneurial spirits and approaches into biodiversity conservation.

LARSEN: One of the challenges is that we're not just dealing with simple latitudinal or altitudinal shifts of species. The connectivity of the landscape is very important as the ranges of species shift as a result of climate change. A lot of times species will shift into areas that might be unsuitable habitat for other reasons besides climate. For instance, in coastal regions with sea-level rise, a lot of marshes won't simply shift to higher elevations because there are hard engineering structures and urban landscapes present. We need innovation to figure out how to maintain connectivity between ecosystems and the surrounding landscape in a way that maximizes the resilience of the system.

ACKERLY: And last word to the Park Service.

SAUVAJOT: Stepping back a little bit, I think that in the National Park System and the National Park Service we have two big categories of need for science. I think everyone touched on them in different ways. There are the here-and-now challenges, the things that are eating our lunch each day: the imperiled species and the invasive plants. Some of these challenges affect our ability to manage the resources in the parks right now, and have direct implications for the experience that visitors have when they come to parks. In addition, and often they are part of a continuum, are the longer-term, broader challenges that have system-wide effects, such as climate change. You could scale up invasive species as a broader ecosystem challenge. These things will position the agency for the future. We need science to help us in both of those categories, and we need to recognize that both of those things are important. I am optimistic that, in part, the conversations that are occurring here, and the conversations that we will continue to have over the next several months and years, will provide an opportunity for parks to be places where people can learn about some of these challenges, these big-picture things like the effects of a 2°C increase in average temperature and stuff like that.

You can, as a visitor, come to a park now and see these changes happening. Why not use that as an opportunity to help raise awareness of science, to help educate people about what science is, and to impress upon them the importance of that information in helping us manage resources? One nice thing that the National Park Service does have, frankly unlike many other federal agencies, is that we provide venues for people who come willingly, with interest, to experience natural and cultural heritage. It's a window of opportunity to help raise the consciousness of the public about these challenges and to get them involved. There's an opportunity there! Even though some of these big, big challenges may not have been touched on, I think that the parks provide a really important window and stage for addressing them.

Engaging People in Parks

Engaging people in parks is essential to maintain the status of parks. Hence, engagement is a central quality of park management and core to any sustainable vision of national parks in the 21st century. Science, and in particular the social sciences, has an important role to play in understanding engagement, identifying best practices, evaluating programs and policies, and advancing our knowledge of how persons and institutions interact with parks.

Engagement is multiscaled. It can be as singular as an individual park visitor who is immersed in nature or history and deeply involved in the experience. It can be engagement within social groups—families, households, extended families, organizations (from the local church group to the national Sierra Club). And engagement can reflect institutional engagement—from robust civic action to politicized decision making.

With all these forms of engagement, the authors of the chapters in this section describe the rich interplay of knowledge and values. Ruth DeFries, describing the "tangled web" of people, landscapes, and protected areas, provides a historical perspective on the role of parks in the broader ecological and cultural landscape. Her chapter is a demonstration of the contribution history makes to understanding engagement. The shifting focus of park professionals—along with issues of scale and obvious mismatches of policy and practice—enlivens the chapter as she shows how parks function as coupled natural-human systems. She also demonstrates that sometimes past conservation strategies may be unlikely as effective tools in the future.

Thomas Dietz, in his chapter on science, values, and conflict, continues the examination of engagement via knowledge and values. But here the focus is how political struggles have shaped and will shape the US National Park System, and that park decision making can be seen as a form of adaptive risk management. Drawing from the decision-making research literature, Dietz offers recommendations for improved "public deliberation" and the use of "bridging organizations" along with other key tools to improve policy and practice. Again, this engagement is founded on and guided by values.

John Francis and colleagues focus on a specific form of engagement—the broad set of activities described as "citizen science." They provide an overview of the history and recent expansion of citizen science, and describe the BioBlitz events that have become increasingly popular in national parks. The chapter extends the BioBlitz concept beyond traditional parks to schoolyards, cities, and nations. And while they identify the trends that have led to the growth of citizen science—an available and motivated labor pool, expanding scientific interests, and new technology—the authors return to values as the foundation for "empowering an engaged and contributing community."

Edwin Bernbaum reminds us in his chapter that engagement with parks is achieved not only through science or Western forms of knowledge. His focus is on the spiritual and cultural significance of nature, and the process he examines is one of *inspiration*. He poses several significant questions on the future of visitors to parks, and makes clear that the spiritual and the sacred surely have a place in our understanding of engagement with park features and history. Using examples from indigenous cultures in Hawai'i, Alaska, and the continental United States, he demonstrates how park interpretation and education can respond and, as a result, broaden the engagement of Americans and others in the next century of America's national parks.

This theme of engagement, and at times disengagement, extends to the strategic conversation that concludes this section. It features four very different individuals with diverse experiences and much expertise on issues of parks, conservation, and science: Justin Brashares, professor of wildlife ecology at the University of California, Berkeley; Cyril Kormos, vice president for policy at The WILD Foundation; Christine Lehnertz, Pacific West regional director of the National Park Service; and Nina Roberts, professor of recreation studies at San Francisco State University. Engagement is described at different scales. Engaging people in parks, as Roberts notes, "that's where the magic happens." Discussion also centers on engaging institutions, including local and national political institutions. The conversation converges around how to ensure engagement of differing values, and the panel members speak of "dialogue," diverse voices being "at the table," communicating via a "two-way street," and making the US National Park

Service values relevant. But an undercurrent persists in the conversation that considers when it might be necessary to disengage people from sensitive areas in parks to conserve park resources. Underneath the conversation is the difficult issue of power—*who has the power to decide about the future of our national parks?* In a democracy, and at the centennial of the extraordinary US National Park System, it's a healthy and essential question.

ELEVEN

The Tangled Web of People, Landscapes, and Protected Areas

RUTH DEFRIES

Introduction

For millennia, people have set aside land to protect flora and fauna. Utilitarian purposes, such as protecting forests for hunting, harvesting products, and grazing elephants used in battle, motivated protection in some places (Kautilya 1992). Cultural, spiritual, and religious sensibilities motivated protection as well—for example, sacred groves, which still persist in many parts of the world (Bhagwat and Rutte 2006).

Modern-day conservation has a similar variety of motivations for protecting land from human exploitation. Yellowstone National Park, the first officially recognized protected area in the world, was designated in 1872 and "set apart as a public park or pleasuring-ground for the benefit and enjoyment of people."[1] Subsequently, protected areas have been justified on the basis of protecting watersheds, providing habitat for iconic species, conserving natural resources, promoting tourism, and safeguarding the intrinsic value of nature, as well as scenic beauty and recreation (Watson et al. 2014).

Globally, terrestrial protected areas currently are a substantial land use. Approximately 14% of the land surface was under some form of protection in 2014 (Deguignet et al. 2014). Levels of protection range from strict restrictions barring human use to provisions for sustainable use of natural resources (see Dudley [2008] for definitions of the International Union for the Conservation of Nature [IUCN] categories). The extent and number of

---

1. Forty-Second Congress of the United States of America, Transcript of Act Establishing Yellowstone National Park (1872), National Archives and Records Administration, Washington, DC.

protected areas increased rapidly since 1970, particularly in the period between 1970 and 1995, with globalization and priorities of conservation
organizations driven by rapid land-use change and habitat fragmentation (Zimmerer, Galt, and Buck 2004; West, Igoe, and Brockington 2006)
(fig. 11.1). Coverage is unevenly distributed, with relatively high proportions of land under protection in the Americas and eastern and southern
Africa and low proportions in South Asia and North Eurasia (see table 1
in West, Igoe, and Brockington [2006]). The average size of newly declared
protected areas has decreased markedly over the last few decades, which
has expanded the interface between protected and nonprotected areas even
faster than the expansion of the area under protection (Naughton-Treves,
Holland, and Brandon 2005; Palomo et al. 2014).

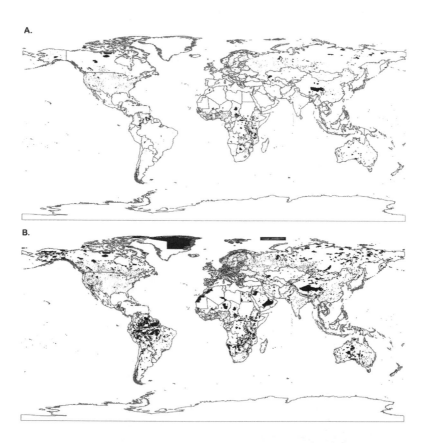

11.1. The coverage of terrestrial protected areas (*A*) before 1970 and (*B*) in 2014.
Data from the World Database on Protected Areas (Deguignet et al. 2014).

Although the area under protection has increased, the current network of protected areas is inadequate to conserve biodiversity in the face of continuing habitat fragmentation and climate change. Gap analyses of ranges of threatened species indicate that regions with high levels of endemism are particularly in need of additional protection to preserve habitat for threatened species (Rodrigues et al. 2004; Pouzols et al. 2014; Jenkins et al. 2015). Moreover, downgrading, downsizing, and degazettement (removal from protected status) of protected areas is eroding the area under protected status (Mascia et al. 2014).

Despite inadequate coverage to protect threatened species, protected areas remain the primary tool for conservation of biodiversity. Other, more recent instruments for conserving biodiversity include payments for ecosystem services, decentralized management, and forest certification schemes. The evidence base for assessing the effectiveness of these recent instruments is weak (Miteva, Pattanayak, and Ferraro 2012; Naeem et al. 2015). Many studies point toward the general effectiveness of protected areas for reducing deforestation, even accounting for remoteness and other covariates and potential spillover effects that could displace deforestation outside protected areas (e.g., Andam et al. 2008; Joppa, Loarie, and Pimm 2008; Gaveau et al. 2009; Sims 2010; Ferraro and Hanauer 2011). Fires within protected areas are also generally fewer than in nonprotected areas (Nelson and Chomwitz 2011). On the other hand, there is a major shortfall in the effectiveness of the management of protected areas (Leverington et al. 2010; Watson et al. 2014), and there is little evidence to assess the effectiveness of protected areas in improving socioeconomic conditions (Andam et al. 2010; Miteva, Pattanayak, and Ferraro 2012).

The extensive area of land currently under protection is remarkable considering the intense demands to produce food and fiber for the world's growing and increasingly affluent population. Agriculture, the land use with the most direct relevance for civilization's survival, covers almost 50% of the land surface (Foley et al. 2005), and protected areas are the second-most extensive land use. As is the case with any land use, particularly one that has expanded as rapidly as protected areas, competing objectives from different stakeholders lead to conflicts and involve trade-offs (DeFries, Foley, and Asner 2004). For example, local populations living in and around protected areas understandably prioritize their livelihood needs for land and forest products over the conservation agenda enforced by local managers and promoted by scientists based in faraway places.

Like other places where people use land—whether croplands, pastures, or cities—protected areas are inherently social-ecological systems (also known as coupled human-natural systems, human ecological systems, or human-environment systems) (Turner et al. 2003). Social-ecological systems are complex, dynamic, integrated systems in which humans and nature interact, and are characterized by feedbacks and nonlinearities (Berkes, Folke, and Colding 2000). For example, people living in protected areas are coupled with ecological systems through reliance on biological resources such as wood, medicines, and wild foods. Feedbacks occur when human use alters the ecological conditions that provide the resource, which in turn alters the availability of the resource and affects social systems. Managers, political leaders, local communities, flora, fauna, nutrients, water, and soil are all parts of a holistic whole in the conceptualization of protected areas as social-ecological systems (Cumming et al. 2015).

This chapter traces the evolution of approaches toward studying and managing protected areas as social-ecological systems, followed by examination of social-ecological processes operating at multiple spatial scales: inside protected areas, surrounding protected areas, and in the larger landscape encompassing protected area networks. Processes at all these scales are influenced by national- and global-scale dynamics that set priorities and allocate financial resources. The chapter concludes with next steps in the evolution of managing protected areas to account for the reality that protected areas are embedded within larger socio-ecological settings. The focus is on protected areas in the Global South where high biodiversity and rapid land-use change currently converge, creating a priority for conservation.

The Evolution of Managing Protected Areas as Social-Ecological Systems

Most protected areas today have people residing within their administrative boundaries; for example, 85% of protected areas in Latin America are inhabited (Colchester 2004). Millions more live on the fringes of protected areas. High-biodiversity areas with conservation priority generally overlap in space with rural, poor, and often indigenous populations in the tropics whose livelihoods depend on local ecosystems. This intersection exacerbates the complexities and ethical dimensions of establishing and managing protected areas.

Protected areas have not always been recognized as social-ecological systems. The original conception of Yellowstone in the 1830s by the painter George Catlin was as a "nation's park" set aside to preserve wilderness including Native Americans. When the establishment of Yellowstone was put into law about 40 years later, Native Americans were excluded (Nash 1970; Colchester 2004). The exclusionary "fortress" model of conservation, based on the premise that nature can only be preserved if devoid of people, spread to other places around the world. The number of people who have subsequently been displaced by protected areas, or "conservation refugees," is unknown. Estimates include 600,000 tribal people displaced by protected areas in India (Nash 1970) and between 1 and 16 million on the continent of Africa (Geisler and De Sousa 2001). Even less is known about the impacts of displacement on their well-being, although many historical examples exist about denial of rights to land and natural resources and criminalization of traditional land-use practices (see West, Igoe, and Brockington [2006] and Brockington and Igoe [2006] for a summary of this literature).

By the 1970s, as the area under protection began its upward trend, the view of protected areas as scenic treasures had evolved to encompass their value for conserving biodiversity. The rights of indigenous and other people living in parks were not yet high on the agenda (Watson et al. 2014). By the 1980s, with increasing contact between protected areas and local people, the international conservation community recognized that conservation needed to encompass the realities of people in and around protected areas. Two rationales justified this view: ethical considerations that hardly need an explanation and the realization that conservation cannot be effective without local resource-users whose actions affect biodiversity on a daily basis. In other words, protected areas were increasingly recognized as socio-ecological systems, although this terminology may not have been used explicitly. In 1982, consensus at the World Parks Congress in Bali was that "protected areas in developing countries will survive only insofar as they address human concerns." (Naughton-Treves, Holland, and Brandon 2005).

Management to reconcile the well-being of local populations and conservation met with mixed success. With the trend toward decentralized rather than top-down management (Ostrom 2008), considerable investments from conservation organizations and international development agencies were directed toward projects under various terms including integrated conservation and development projects (ICDPs), community-based management, and eco-development. While generalizations are difficult based on anecdotal case studies, a body of studies indicates widespread underachievement from ICDPs (Wells and McShane 2004; Palomo et al. 2014). Reported problems with ICDPs relate more to the implementation than the principle of managing protected areas to benefit both local populations and conservation. Contributing factors for the disappointing results include naive assumptions that local communities share the same values as the conservation agenda, rapid implementation that cannot sufficiently address the deep complexities of socio-ecological systems, and unrealistic expectations that significant benefits could be accrued from protected areas and equitably shared. Lack of clarity of objectives and real conflicts between aspirations of local people and conservation plagued the laudable push to account for local people's needs in conservation (Brown 2002; Adams et al. 2004). Moreover, external, powerful interests such as mines, dams, and roads, which were out of the control of protected area managers, could have impacts on biodiversity at least as great as local communities.

By the turn of the millennium, myths of widespread win-win solutions fell by the wayside, with the possible exception of ecotourism that potentially benefits local people if opportunities are available for them to participate. From the conservation perspective, attention shifted toward corridors and networks to connect protected areas and foster movements of organisms between them (Palomo et al. 2014). From the social perspective, rights for indigenous peoples were codified into international law (Colchester 2004).

Recent attention has turned toward the role of protected areas in maintaining ecosystem services such as food provisioning for people living in local proximity, watershed protection for people downstream, and carbon sequestration with global benefits (Watson et al. 2014). With questions unresolved about how to balance the often-competing goals of local populations and conservation, some researchers have examined the role of protected areas in poverty alleviation (see section below on socio-ecological interactions surrounding protected areas).

This brief recount of the evolution of trends in conservation reveals the unavoidable realities that protected areas are embedded within social-ecological systems, involve multiple stakeholders, and bring to the fore differing values about which land uses are in the best interest of society. Future management will continue to grapple with these difficult problems for which there is no single or "right" answer.

Inside to Outside: Socio-ecological Dynamics of Protected Areas at Different Scales

As protected area management incorporates socio-ecological systems, it is useful to consider these dynamics according to varying spatial scales. These dynamics and their management implications differ across scales: inside

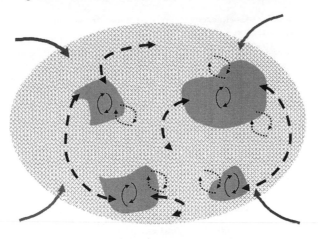

11.2. Schematic of socio-ecological processes and decision makers that operate at different scales: (a) within protected areas (*thin solid lines*), (b) between protected areas and the surroundings (*dotted lines*), (c) at the landscape level in networks of protected areas (*dashed lines*), and (d) at the national and global scale (*thick solid lines*). The landscape matrix that includes settlements, agriculture, and other nonprotected land uses is represented by cross-hatching, and protected areas are in gray.

the boundaries of protected areas, surroundings in proximity to protected areas, and the larger landscape that encompasses networks of protected areas (fig. 11.2, table 11.1). All of these dynamics are influenced by national- and global-scale processes that trickle through to finer scales.

Socio-ecological Dynamics within Protected Areas

People are part of ecosystems in a variety of ways. Between 50% and 100% of stricter protected areas in South America and Asia have people using or living within them (Brockington et al. 2006). Traditional lifestyles of people living in protected areas include hunting, foraging, collecting plants for medicines and other uses, and setting fire. While an overly romantic view considers these uses to be part of "nature" and undamaging to biodiversity, an equally unsupported view is that these uses necessarily cause damage. In some cases, such as in the Amazon, indigenous reserves play a major role in preserving forest and halting deforestation (Nepstad et al. 2006). In other cases, unsustainable human use undoubtedly has a negative impact on biodiversity. Mines and timber extraction within protected areas, either legal or illegal, can also have a major impact for conservation (Rangarajan and Shahabuddin 2006).

As noted above, an undetermined but large number of people have

Table 11.1   Examples of socio-ecological processes and decision makers operating at
the different scales illustrated in figure 11.2

Scale	Examples of processes	Relevant decision makers
Inside protected areas	Use of resources for liveli-hoods of people living inside protected area; relocation	Protected area managers
Between protected areas and local surroundings	Human-wildlife conflict; collection of fuelwood and NTFPs by local communities; livestock grazing; impacts from ICDPs	Local communities; local NGOs
Landscapes encompassing multiple protected areas	Commercial extraction of timber and minerals; tourism demand; infrastructure development in corridors	Administrative units in landscape; state and national governments; private sector

Note: NTFP = nontimber forest product. ICDP = integrated conservation and development project.
Outcomes at all three scales are influenced by national and global processes (such as climate
change), national policies for conservation and other sectors that affect habitats (such as highway
development), and shifting priorities of international NGOs.

been relocated from protected areas in the presumed interest of conservation. Enforcement and new legislation suggest that millions more face relocation in the future (Brockington and Igoe 2006; Brockington, Igoe, and Schmidt-Soltau 2006). Relocation of people living within protected areas is among the most sensitive topics for conservationists. The effectiveness of relocation for conservation in any particular place depends on whether people are damaging, beneficial, or neutral for promoting biodiversity. The answer to this question is likely to be context specific, that is, dependent on population density, the extent to which local people use resources from the protected area, and ecological conditions that affect regeneration.

Evidence is scanty to assess either the benefit to biodiversity or the well-being of people following relocation (Brockington and Igoe 2006; Brockington, Igoe, and Schmidt-Soltau 2006). Some evidence suggests that relocation can improve access to health care, transportation, electricity, jobs, and overall quality of life (e.g., Karanth 2007). Other evidence indicates loss of culture, poor nutrition, and violation of human rights (e.g., Colchester 2004). The outcomes are likely to depend on the context-dependent details of how managers implement relocation schemes, conditions where people relocate, and myriad other details that are difficult to unravel to decipher generalizations.

With the blatant injustices of the past now recognized, research on the effectiveness of relocation schemes for biodiversity and the well-being of relocated people is urgently needed to guide future decisions and approaches that balance the trade-offs.

Socio-ecological Dynamics Surrounding Protected Areas

A number of socio-ecological processes affect biodiversity and resources for people in and around protected areas. Such processes include migrations of organisms beyond protected area boundaries, hydrological flows, transport of air and water pollution, disease, and fire (Hansen and DeFries 2007). In the 1970s, recognition of these processes led to the concept of biosphere reserves that establish gradients of decreasing human use to buffer protected areas (Palomo et al. 2014). More recently, zones of interaction or park-centered ecosystems have been defined to delineate those areas with strongest interaction with protected areas based on ecological principles (DeFries, Karanth, and Pareeth 2010; Hansen et al. 2011).

People living in surroundings of protected areas influence dynamics inside protected areas. Surrounding communities may graze livestock, hunt bushmeat, and collect timber and other products from within the boundaries of many protected areas. These uses contribute to food security and income (Food and Agriculture Organization 2014). For example, greater access to bushmeat is associated with higher protein in children's diets in Madagascar (Golden et al. 2011) and reduced stunting in central Africa (Fa et al. 2015). People in the surroundings can also encroach into protected areas to expand cropland, pasture, or tree plantations (e.g., Curran et al. 2004).

Conversely, dynamics within protected areas influence people living in their surroundings. The ability of local people to gain food, income, and livelihood needs from functional ecosystems in protected areas is positive for people. On the negative side, livestock predation and crop raiding by wildlife roaming beyond protected area boundaries is a major hardship for farmers and herders nearby and can lead to retaliatory killing of wildlife (Barua, Bhagwat, and Jadhav 2013). Tourism, which becomes more prevalent as people have discretionary income, is a double-edged sword. On one hand, it can create economic opportunities for local people and contribute to support for conservation as visitors have opportunities to appreciate nature. On the other hand, extensive land-use changes and infrastructure associated with tourism can sever connectivity and usurp land from agriculture (Karanth and DeFries 2011; Karanth et al. 2012).

The impacts of protected areas on poverty alleviation for local populations on the fringes of protected areas are particularly relevant as economic aspirations and the need to conserve biodiversity grow in tandem through-out the developing world. The outcome is unclear and the evidence base is weak to draw general conclusions. Protected areas could exacerbate poverty traps by forgoing options for agricultural development and exploitation of natural resources. Or protected areas could alleviate poverty by provid-ing employment opportunities from tourism and improved connectiv-ity through roads and other infrastructure that provide access to markets, health care, and education.

The dual goals of poverty alleviation and conservation can create a mis-match of objectives, the former to alter the system to a new state away from poverty traps and the latter to maintain the system to conserve biodiver-sity (Barrett, Travis, and Dasgupta 2011). In Costa Rica and Thailand, two relatively economically advanced countries, people living around parks are poorer than the national average, but the net impact of protected areas has been alleviation of poverty (Andam et al. 2010). Benefits to poverty allevia-tion did not overlap in space with benefits for conservation at a fine scale (Ferraro, Hanauer, and Sims 2011), suggesting that win-win options are not in play. Two-thirds of the poverty alleviation in Costa Rica is attributable to tourism (Ferraro and Hanauer 2014). Other empirical studies indicate that the impact of protected areas is context specific (see the papers in a special feature on biodiversity conservation and poverty traps in Barrett, Travis, and Dasgupta [2011]). Clearly, considerable research is required to develop general principles about conditions that lead to poverty alleviation or exac-erbation for people living around protected areas.

Socio-ecological dynamics surrounding protected areas imply responsi-bilities for management beyond those pertaining to the ecological system within protected area boundaries. Managers need to address damages of human-wildlife conflict through compensation and other measures, estab-lish communication with sometimes hostile communities, control tour-ism, and consider the repercussion of their actions for poverty alleviation. These responsibilities often are not congruent with the training and back-ground of managers of protected areas.

Socio-ecological Dynamics at the Landscape Level

As early as the First World Conference on National Parks in Seattle in 1962, it was clear that protected areas cannot be big enough to capture large-scale ecological dynamics such as the flows of water, air, and nutrients; migrations of large-ranging species; and large-scale atmospheric processes affecting climate. The influential Leopold Report to the US Secretary of the Interior noted that "few of the world's parks are large enough to be in fact self-regulatory ecological units; rather, most are ecological islands subject to direct and indirect modification by activities and conditions in the surrounding area" (Leopold et al. 1963). Similarly, protected areas are not isolated from social and economic forces shaped by processes in the larger landscape. Such processes include economic activity in urban areas that drives tourism, downstream demand for ecosystem services such as watershed protection, and demographic changes in the structure of human populations.

The landscape level encompasses multiple protected areas that form networks, with processes in protected areas dependent on other protected areas in the network and on the landscapes between them. Networks and corridors are well-established approaches to address conservation needs for connectivity across the landscape. For example, the Mesoamerican Corridor was established by central American countries and Mexico in the late 1990s to create a land bridge between North and South America (Grandia 2007). Corridors can also have negative consequences—for example, by enabling disease to spread between protected areas (Altizer, Bartel, and Han 2011). From a socio-ecological perspective, corridors traverse nonprotected landscapes and interface with local populations, raising questions about how human-wildlife conflicts can be minimized along corridors, who decides land-use priorities for the landscape, and how to balance conservation and development goals (Altizer, Bartel, and Han 2011).

A landscape approach aims to balance multiple social, economic, and environmental objectives where land uses compete. Such approaches differ from management of individual protected areas by accounting for needs for biological connectivity, economic connectivity (e.g., roads), watershed integrity, and other processes that operate at the landscape level. Although a developing area of research, principles for a landscape approach include adaptive management, multifunctionality, participatory monitoring, and recognition of multiple stakeholders with clear rights and responsibilities (Sayer et al. 2013).

With acceleration of economic growth and the much-needed expansion of roads and other infrastructure (Laurance and Balmford 2013), the need to view protected areas as embedded within larger landscapes becomes increasingly pertinent. Protected area managers have little authority over decisions in the larger landscape. The mismatches in spatial and temporal scales between ecological and governance processes calls for new ways to make decisions about balancing competing interests (Cumming et al. 2015).

Socio-ecological Influences from National and Global Levels

Policies and priorities set at national and global scales influence the dynamics that occur at finer scales. From an ecological perspective, anthropogenic climate change caused by emissions of greenhouse gases far from a particular protected area or landscape can have major repercussions on the ability to maintain suitable habitats (Hannah et al. 2007). From a social perspective, national priorities that allocate resources for enforcement, management, and compensation influence effectiveness of protected areas to conserve biodiversity (Bruner et al. 2001), and distal market forces, or "teleconnections," lead to habitat conversion to produce goods consumed far from the location of production (Liu et al. 2013). At the global level, priorities of international donors, nongovernmental organizations (NGOs), and multilateral treaties determine, through funding decisions, where conservation occurs, which species receive conservation attention, and how impacts on local communities are addressed. The international Convention on Biological Diversity, for example, sets the target for protection at 17% of land area by 2020 and requires national action plans (Secretariat of the Convention on Biological Diversity 2014).

These national- and global-scale processes trickle down to influence socioeconomic dynamics at all scales: namely, within protected areas (e.g., through resources allocated to conservation of a particular species or to relocation of people), within local surroundings (e.g., by shifting attention to local communities and poverty alleviation as described above), and within larger landscapes (e.g., through other sectors such as water, energy, and transport that alter habitats and connectivity outside protected areas).

These socio-ecological dynamics operating at multiple scales shape both conservation outcomes and human well-being, as illustrated in the example of the central Indian highlands (box 11.1). Research is emerging that integrates social and ecological dimensions in conservation science—for example, by studying China's Wolong Nature Reserve as an integrated system that provides both giant panda habitat and economic benefits from ecotourism (Liu et al. 2007; He et al. 2008). Such research will become increasingly important to manage protected areas as socio-ecological systems.

Beyond the Boundaries: New Frontiers for Protected Areas

The history of protected areas shows that they cannot be isolated from their social-ecological setting. Whether people in and around protected areas are a net positive or net negative for conservation, and whether protected ar-

Box 11.1 Socio-ecological Dynamics in Conservation Landscapes of the Central Indian Highlands

India is a megadiverse country that holds remaining populations of endangered, iconic species such as tigers (*Panthera tigirs*) (fig. 11.3) and Asian elephants (*Elephas maximus*). With a human population over 1.2 billion, of which 70% is rural and dependent on local resources for their livelihoods, humans coexist at high densities with wildlife. A long history of sacred spaces and cultures that revere wildlife makes conservation a deep-rooted value. Currently, approximately 5% of the land area is under protected status.

The central Indian highlands (fig. 11.4) exemplify the linkages between social and ecological dynamics at the various scales discussed in this chapter.This landscape lies at the heart of the country and is a "global priority landscape for tiger conservation" (Sanderson et al. 2006). The landscape contains many small protected areas, ranging in size from less than 100 to slightly over 2,000 ha. The matrix between the protected areas includes rice paddies and small agricultural fields. Forest cover, which generally remains only in hillier parts of the landscape, provides connectivity between some of the protected areas (Dutta et al. 2015). The landscape is also home to several tribal and

11.3. A tigress in Tadoba Tiger Reserve. Photo courtesy of Jit Bajpai.

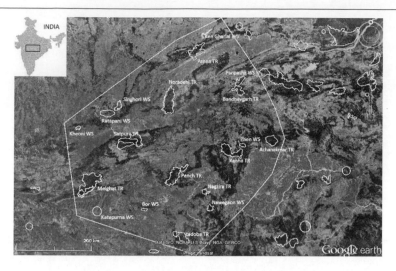

11.4. Approximate boundaries of the central Indian highlands (*polygon*) and protected areas (*outlined in white*). Note that remaining forest corridors connecting protected areas are in darker shades of gray.

indigenous groups and other rural populations that rely on forests for fuelwood, forest grazing, nontimber forest products, and other livelihood needs.

As in many protected areas, people in the central Indian highlands have lived for centuries within the boundaries of what are now designated as protected areas. Many people and villages are currently relocating outside protected areas through government schemes (Read 2015). In surroundings of protected areas, Joint Forest Management has been in place for several decades with the aim of decentralizing management of resources (Agarwala 2014). Recently, the growth of wildlife-related tourism has led to rapid expansion of resorts and associated infrastructure surrounding protected areas (Karanth and DeFries 2011). At the landscape level, rapid expansion of transport networks and other infrastructure to meet development needs threatens to sever connectivity critical for the genetic viability of large mammals as they move between small protected areas (Sharma et al. 2012).

These intensifying socio-ecological dynamics that are occurring over a range of scales illustrate the critical need to incorporate social-ecological factors in conservation and protected area management. These needs include ensuring the well-being of relocated peoples, managing tourism to benefit local communities without harming sur-

Box 11.1 (*continued*)

rounding ecosystems, resolving conflicts between humans and wild-life, and balancing the development needs for infrastructure expansion and the conservation needs for connectivity between protected areas. Recognizing these needs, the local government has developed plans for corridors (Madhya Pradesh Forest Department 2013). The central Indian highlands is one of many landscapes around the world facing similar management challenges to support wildlife conservation, local livelihoods, and economic development.

eas are a net positive or net negative for people, depends on the specific context and stakeholder. Generalizations are not possible based on limited evidence from case studies about such questions as the impact of relocations on biodiversity and people's well-being, and under what conditions protected areas alleviate or exacerbate poverty. These questions are fruitful areas for research.

Three mismatches stand in the way of incorporating socio-ecological systems within protected area management: objectives, spatial scale, and governance. First, the objectives for protected areas have evolved over the decades from a single focus on scenery and recreation to multiple foci, including protection of ecosystems services and development benefits for local communities. The changing objectives for conservation have paralleled emerging evidence that apparently "pristine" landscapes have actually been shaped by long-term use by people, such as indigenous cultivation in the Amazon (Fairhead and Leach 1998; Posey and Balick 2006). While the ethical and practical needs to merge conservation and development agendas are frequently discussed, reality often divides these two objectives into distinct camps. Approaches to conservation need to recognize that human aspirations and economic advancement are critical to society. Conservation will have limited success if the objective is to trump broader societal goals. Conversely, development at all costs too often overlooks possibilities to incorporate conservation goals.

For the second mismatch, protected areas are managed according to the spatial footprint of their administrative boundaries. In reality, socio-ecological processes link protected areas to their surroundings and to the larger landscape. Management of protected areas needs to foresee the processes operating at these larger scales and to integrate them into management. For example, growth in domestic tourism with urbanization and economic growth is an outcome of forces beyond the boundaries of reserves but requires management attention in protected areas.

Governance is a third and related mismatch. Socio-ecological processes in the larger landscape affect protected areas, but protected area managers have no control over land-use decisions in the larger landscape. New ways to govern at the landscape level are needed to account for competing objectives among multiple stakeholders.

As with all human endeavors, conservation and protected area management are evolving processes. Approaches that account for socio-ecological processes within protected areas, in their surroundings, and in larger networks of protected areas will only become more relevant for conservation success as aspirations for economic growth and rights of all peoples are realized around the world.

Acknowledgments

Thanks to Steve Beissinger and David Ackerly for providing helpful guidance on the manuscript, and to them and others for organizing the conference "Science for Parks, Parks for Science."

Literature Cited

Adams, W. M., R. Aveling, D. Brockington, B. Dickson, J. Elliot, J. Hutton, D. Roe, B. Vira, and W. Wolmer. 2004. Biodiversity conservation and the eradication of poverty. Science 306:1146–1149.

Agarwala, M. 2014. Forest degradation and governance in Central India: evidence from ecology, remote sensing and political ecology. Columbia University, New York, New York.

Altizer, S., R. Bartel, and B. Han. 2011. Animal migration and infectious disease risk. Science 331:296–302.

Andam, K., P. Ferraro, A. Pfaff, G. Sanchez-Azofeifa, and J. Robalino. 2008. Measuring the effectiveness of protected area networks in reducing deforestation. Proceedings of the National Academy of Sciences USA 105:16089–16094.

Andam, K., K. Sims, A. Healy, and M. B. Holland. 2010. Protected areas reduced poverty in Costa Rica and Thailand. Proceedings of the National Academy of Sciences USA 107:9996–10001.

Barrett, C., A. Travis, and P. Dasgupta. 2011. On biodiversity conservation and poverty traps. Proceedings of the National Academy of Sciences USA 108:13907–13912.

Barua, M., S. Bhagwat, and S. Jadhav. 2013. The hidden dimensions of human-wildlife conflict: health impacts. opportunity and transaction costs. Biological Conservation 157:309–316.

Berkes, F., C. Folke, and J. Colding, eds. 2000. Linking social and ecological systems: management practices and social mechanisms for building resilience. Cambridge University Press, Cambridge, United Kingdom.

Bhagwat, S., and C. Rutte. 2006. Sacred groves: potential for biodiversity management. Frontiers in Ecology and the Environment 4:519–524.

Brockington, D., and J. Igoe. 2006. Eviction for conservation: a global overview. Conservation and Society 4:424–470.

Brockington, D., J. Igoe, and K. Schmidt-Soltau. 2006. Conservation, human rights and poverty reduction. Conservation Biology 20:250–252.

Brown, K. 2002. Innovations for conservation and development. The Geographical Journal 168:6–17.

Bruner, A. G., R. E. Gullison, R. E. Rice, and G. A. B. da Fonseca. 2001. Effectiveness of parks in protecting tropical biodiversity. Science 291:125–128.

Colchester, M. 2004. Conservation policy and indigenous peoples. Environmental Science and Policy 7:145–153.

Cumming, G., C. R. Allen, N. Ban, D. Biggs, H. Biggs, D. Cumming, A. DeVos, et al. 2015. Understanding protected area resilience: a multi-scale, social-ecological approach. Ecological Applications 25:299–319.

Curran, L. M., S. N. Trigg, A. K. McDonald, D. Astiani, U. M. Hardiono, P. Siregar, I. Caniago, and E. Kasischke. 2004. Lowland forest loss in protected areas of Indonesian Borneo. Science 303:1000–1003.

DeFries, R., J. Foley, and G. P. Asner. 2004. Land use choices: balancing human needs and ecosystem function. Frontiers in Ecology and the Environment 2:249–257.

DeFries, R., K. Karanth, and S. Pareeth. 2010. Interactions between protected areas and their surroundings in human-dominated tropical landscapes. Biological Conservation 143:2870–2880.

Deguignet, M., D. Juffe-Bignoli, J. Harrison, B. MacSharry, N. Burgess, and N. Kingston. 2014. 2014 United Nations List of Protected Areas. UNEP-WCMC, Cambridge, United Kingdom.

Dudley, N. E. 2008. Guidelines for Applying Protected Area Management Categories. IUCN, Gland, Switzerland.

Dutta, T., S. Sharma, B. H. McRae, P. S. Roy, and R. DeFries. 2015. Connecting the dots: mapping habitat connectivity for tigers in central India. Regional Environmental Change, 1–15.

Fa, J., J. Olivero, R. Real, M. Faraf, A. Marquez, J. Vargas, S. Ziegler, et al. 2015. Disentangling the relative effects of bushmeat availability on human nutrition in central Africa. Scientific Reports 5:8168.

Fairhead, J., and M. Leach. 1998. Reframing deforestation. Global analysis and local realities: studies in West Africa. Routledge, London, United Kingdom.

Ferraro, P., and M. Hanauer. 2011. Protecting ecosystems and alleviating poverty with parks and reserves: "win-win" or tradeoff? Environmental Resources and Economics 48:2.

———. 2014. Quantifying causal mechanisms to determine how protected areas affect poverty through changes in ecosystem services and infrastructure. Proceedings of the National Academy of Sciences USA 111:4332–4337.

Ferraro, P., M. Hanauer, and K. Sims. 2011. Conditions associated with protected area success in conservation and povery reduction. Proceedings of the National Academy of Sciences USA 108:13913–13918.

Foley, J., R. DeFries, G. P. Asner, C. G. Barford, G. B. Bonan, S. R. Carpenter, F. S. I. Chapin, et al. 2005. Global consequences of land use. Science 309:570–574.

Food and Agriculture Organization. 2014. Protected areas, people and food security: an FAO contribution to the World Parks Congress, Sydney, 12–19 November 2014. FAO, Rome, Italy.

Gaveau, D., J. Epting, O. Lyne, M. Linkie, I. Kumara, M. Kanninen, and N. Leader-Williams. 2009. Evaluating whether protected areas reduce tropical deforestation in Sumatra. Journal of Biogeography 36:2165–2175.

Geisler, C., and R. De Sousa. 2001. From refuge to refugee: the African case. Public Administration and Development 21:159–170.

Golden, C., L. Fernald, J. S. Brashares, B. Rasolofoniaina, and C. Kremen. 2011. Benefits of wildlife consumption to child nutrition in a biodiversity hotspot. Proceedings of the National Academy of Sciences USA 108:19653–19656.

Grandia, L. 2007. Between Bolivar and bureaucracy: the Mesoamerican biological corridor. Conservation and Society 5:478–503.

Hannah, L., G. F. Midgley, S. Andelman, M. Araujo, G. Hughes, E. Martinez-Meyer, R. Pearson, and P. H. Williams. 2007. Protected area needs in a changing climate. Frontiers in Ecology and the Environment 5:131–138.

Hansen, A., C. Davis, N. Pikielek, J. Gross, D. Theobald, S. Goetz, F. Melton, and R. DeFries. 2011. Delineating the ecosystems containing protected areas for monitoring and management. BioScience 61:363–373.

Hansen, A. J., and R. DeFries. 2007. Ecological mechanisms linking nature reserves to surrounding lands. Ecological Applications 17:974–988.

He, G., X. Chen, W. Liu, S. Bearer, S. Zhou, L. Cheng, H. Zhang, Z. Ouyang, and J. Liu. 2008. Distribution of economic benefits for ecotourism: a case study of Wolong Nature Reserve for giant pandas in China. Environmental Management 42:1017–1025.

Jenkins, N., K. Van Houtan, S. L. Pimm, and J. Sexton. 2015. US protected lands mismatch biodiversity priorities. Proceedings of the National Academy of Sciences USA 112:5081–5086.

Joppa, L. N., S. R. Loarie, and S. L. Pimm. 2008. On the protection of "Protected Areas." Proceedings of the National Academy of Sciences USA 105:6673–6678.

Karanth, K. 2007. Making resettlement work: the case of India's Bhadra Wildlife Sanctuary. Biological Conservation 139:315–324.

Karanth, K., and R. DeFries. 2011. Nature-based tourism in Indian protected areas: new challenges for park management. Conservation Letters 4:137–149.

Karanth, K., R. DeFries, A. Srivathsa, and V. Sankaraman. 2012. Wildlife tourists in India's emerging economy: potential for a conservation constituency? Oryx 46:382–390.

Kautilya. 1992. The Arthashastra. Edited, rearranged, and translated by L. N. Rangarajan. Penguin Books India, New Delhi, India.

Laurance, W., and A. Balmford. 2013. Land use: a global map for road building. Nature 495:308–309.

Leopold, A. S., S. A Cain, C. M. Cottam, I. N. Gabrielson, and T. L. Kimball. 1963. Wildlife management in the national parks: the Leopold Report. Unpublished report to the US Secretary of the Interior.

Leverington, F., K. Costa, H. Pavese, A. Lisle, and M. Hockings. 2010. A global analysis of protected area management effectiveness. Environmental Management 46:685–698.

Liu, J., T. Dietz, S. R. Carpenter, M. Alberti, C. Folke, E. F. Moran, A. N. Pell, et al. 2007. Complexity of coupled human and natural systems. Science 317:1513–1516.

Liu, J., V. Hull, M. Batistella, R. DeFries, T. Dietz, F. Fu, T. Hertel, et al. 2013. Framing sustainability in a telecoupled world. Ecology and Society 18:art26.

Madhya Pradesh Forest Department. 2013. Management Plan of Kanha Pench Corridor from 2012–13 to 2021–22. Unpublished report. Madhya Pradesh Forest Department, Madhya Pradesh, India.

Mascia, M., S. Pailler, R. Krithivasan, V. Roshchanka, D. Burns, M. Mlotha, D. Murray, and N. Peng. 2014. Protected area downgrading, downsizing, and degazettement (PADDD) in Africa, Asia, and Latin America and Caribbean, 1900–2010. Biological Conservation 169:355–361.

Miteva, D., S. Pattanayak, and P. Ferraro. 2012. Evaluation of biodiversity policy instruments: what works and what doesn't? Oxford Review of Economic Policy 28:69–92.

Naeem, S., J. Ingram, A. Varga, T. Agardy, P. Barten, G. Bennett, E. Bloomgarden, et al. 2015. Get the science right when paying for nature's services. Science 1403:1206–1207.

Nash, R. 1970. The American invention of national parks. American Quarterly 22: 726–735.

Naughton-Treves, L., M. B. Holland, and K. Brandon. 2005. The role of protected areas in conserving biodiversity and sustaining local livelihoods. Annual Review of Environment and Resources 30:219–252.

Nelson, A., and K. Chomwitz. 2011. Effectiveness of strict vs multiple use protected areas in reducing tropical forest fires: a global analysis using matching methods. PLoS ONE 6:e22722.

Nepstad, D., S. Schwartzman, B. Bamberger, M. Santilli, D. Ray, P. Schlesinger, P. Lefebvre, et al. 2006. Inhibition of Amazon deforestation and fire by parks and indigenous lands. Conservation Biology 20:65–73.

Ostrom, E. 2008. The challenge of common-pool resources. Environment 50:8–21.

Palomo, I., C. Montes, I. Martin-Lopez, M. Garcia-LLorente, P. Alcorlo, and M. Mora. 2014. Incorporating the social-ecological approach in protected areas in the Anthropocene. BioScience 64:181–191.

Posey, D., and M. Balick, eds. 2006. Human impacts on Amazonia: the role of traditional knowledge in conservation and development. Columbia University Press, New York, New York.

Pouzols, F., T. Toivonen, E. Do Minin, A. Kukkala, P. Kullberg, J. Kurrstera, J. Lehtomaki, et al. 2014. Global protected area expansion is comrpomised by projected land-use and parochialism. Nature 516:383–386.

Rangarajan, M., and G. Shahabuddin. 2006. Displacement and relocation from protected areas: towards a biological and historical synthesis. Conservation and Society 4:359–378.

Read, D. 2015. Legitimacy, access, and the gridlock of tiger conservation: lessons from Melghat and the history of central India. Regional Environmental Change, April. doi:10.1007/s10113-015-0780-7.

Rodrigues, A. S. L., S. Andelman, M. I. Bakarr, L. Boitani, T. M. Brooks, R. M. Cowling, L. Fishpool, et al. 2004. Effectiveness of the global area network in representing species diversity. Nature 428:640–643.

Sanderson, E. W., J. Forrest, C. Loucks, J. R. Ginsberg, S. Dinerstein, J. Seidensticker, P. Leimburger, et al. 2006. Setting priorities for the conservation and recovery of wild tigers: 2005–2015. The Technical Assessment. WCS, WWF, Smithosnian, and NFWF-STF, New York, New York, and Washington, DC.

Sayer, J., T. Sunderland, J. Ghazoul, J.-L. Pfund, D. Sheil, E. Mijaard, M. Venter, et al. 2013. Ten principles for a landscape approach to reconciling agriculture, conservation, and other competing land uses. Proceedings of the National Academy of Sciences USA 110:8349–8356.

Secretariat of the Convention on Biological Diversity. 2014. Global Biodiversity Outlook 4. Montreal, Canada.

Sharma, S., T. Dutta, J. Maldonado, T. Wood, H. Panwar, and J. Seidensticker. 2012. Forest corridors maintain historical gene flow in a tiger metapopulation in the highlands of central India. Proceedings of the Royal Society B 280:20131506.

Sims, K. 2010. Conservation and development: evidence from Thai protected areas. Journal of Environmental Economics and Management 60:94–114.

Turner, B. L., II, P. A. Matson, J. McCarthy, R. W. Corell, L. Christensen, N. Eckley, G. K. Hoverlsrud-Broda, et al. 2003. Illustrating the coupled human-environment system for vulnerability analysis: three case studies. Proceedings of the National Academy of Sciences USA 100:8080–8085.

Watson, J., N. Dudley, D. Segan, and M. Hockings. 2014. The performance and potential of protected areas. Nature 515:67–73.

Wells, M., and T. McShane. 2004. Integrating protected area management with local needs and aspirations. Ambio 33:513–519.

West, P., J. Igoe, and D. Brockington. 2006. Parks and people: the social impact of protected areas. Annual Review of Anthropology 35:251–277.

Zimmerer, K. S., R. E. Galt, and M. V. Buck. 2004. Globalization and multi-spatial trends in the coverage of protected-area conservation (1980–2000). Ambio 33:520–529.

TWELVE

Science, Values, and Conflict in the National Parks

THOMAS DIETZ

Introduction

The US national parks and the National Park Service (NPS) have dealt with conflict from their inception; a history of US national parks is a history of conflict. Some sources of conflict that began in the 19th century persist in the 21st century. Global environmental change will add another layer of conflict. These conflicts make the job of the NPS much more difficult. Conflicts about the right course of action will mean that most decisions will have to balance multiple and often antagonistic points of view. Thus, park decision making will usually involve conflict management. Once a decision is made, it becomes part of a legacy that will either inflame or reduce conflict around the next effort to make a decision.

In this chapter, I examine the bases of conflicts about US national parks, noting that our multiple expectations of these parks make conflicts inevitable and persistent. Uncertain facts and differing and often uncertain values challenge our ability to make sound decisions. Research on environmental decision making provides insights into these conflicts and suggests some paths forward. The science of environmental decision making can be a "science for the parks." In turn, research on these conflicts can help advance our understanding of environmental decision making, in the spirit of "parks for science."

Insights from environmental decision-making research can aid in diagnosing the conflicts that often underlie national park management decisions. Science can suggest ways to resolve conflicts and make good decisions in the face of disagreement about facts and values. It can help us learn from experience. But like Tolstoy's unhappy families, each conflict is unique: "All happy families resemble one another, each unhappy family is unhappy in its own way" (Tolstoy 1995). Science can help diagnose and prescribe, but the prescription must be carefully designed around the unique aspects of particular parks and their challenges. To put it differently, there are no panaceas, no overarching solutions for coping with the conflicts that often arise around national parks. Rather each decision and the conflicts it entrains will require its own diagnosis, drawing on both the science of environmental decision making and an understanding of the specific issues in play (Dietz, Ostrom, and Stern 2003; Ostrom 2007).

Since decision making is the focus of this discussion, it is useful to define what I mean by a good decision. I emphasize three criteria: factual competence, value competence, and social learning (Dietz 2003, 2013a). First, a good decision should take account of what is known and the degree of uncertainty in that knowledge. In the context of national park management, it is important to scientifically assess the ecological dynamics of parks, the ways people rely on or connect with parks, and the beliefs and values of interested and affected parties. Second, a good decision should be value competent, taking account of both value differences and value uncertainty. Most decisions involving national parks will affect multiple outcomes that people care about (e.g., ease of visitor access, local jobs, and protection of biodiversity), and people will differ in the importance they assign to these outcomes—people's values will differ. Given the complexity of many decisions about managing national parks, most people will have mixed feelings about most options being considered, and their views may change as new information emerges and new perspectives are articulated. So, in that sense, values are uncertain. Third, given uncertainty in our understanding of facts and values, and the likelihood that both may change over time, a good decision provides opportunities for social learning and corrections in the face of new information.

Why Conflicts about National Park Management Are Inevitable

The US national parks are often seen as crown jewels among American institutions, "the best idea we ever had" (Stegner 1948).[1] Yet they have always been the site of fierce conflicts. From the decision to dam Hetch Hetchy

1. A. MacEachern, Who had "America's Best Idea"?, NiCHE, 23 October 2011, accessed 24 March 2016, http://niche-canada.org/2011/10/23/who-had-americas-best-idea/.

Valley in Yosemite National Park in 1913 to dozens of current issues, the history of US national parks can easily be viewed as a series of conflicts that helped define the modern environmental movement (Runte 1997; Spence 2000; Worster 2008; Bidwell 2009; Sellars 2009; Orr and Humphreys 2012; Foresta 2013). Each conflict has its own particular features. But there are two aspects of US national parks that make conflict inevitable.

The first source of conflict is embedded in American culture and values. A distinction between self-interest and altruism, between private property and the commons, is fundamental to American values and environmental decision making (Dietz, Fitzgerald, and Shwom 2005; Steg and de Groot 2012; Dietz 2015). The national parks are a quintessential commons, held in trust by the federal government for the US population. In contrast, a strong theme in American values emphasizes private property rights, a value enshrined in the Bill of Rights of the Constitution. The idea that private property owners have rights that cannot be abrogated by the government has always been part of US politics. National parks crystallize conflicts between altruism and self-interest in at least three ways.

First, the creation and expansion of national parks requires either the government acquisition of privately held land or the transfer to the NPS of public land managed by other agencies for other purposes. In the case of management transfer, it is likely that the NPS will be much more restrictive in what uses it allows on the land. For example, on land transferred to the NPS from the US Forest Service, logging and grazing will likely be prohibited, and that will concern those who had benefited from logging or grazing there in the past. In such conflicts, traditional users often feel that their private interests are being sacrificed for the common good.

Second, the impacts of park management decisions do not stop at the park borders but can have effects on nearby private property. Since its inception and especially as a result of George Wright's efforts, the NPS has paid special attention to what we now call biodiversity (Wright, Dixon, and Thompson 1932). But management of animals, and especially of charismatic megafauna, is often controversial (Gore et al. 2011; National Research Council 2013). For example, grizzly bear (*Ursus arctos*), bison (*Bison bison*), cougars (*Felis concolor*), elk (*Cervus canadensis*), wolves (*Canus lupus*), and many other animals move across park boundaries, and are often viewed as having a damaging effect on local game and domestic animals through disease transmission and predation (Bidwell 2009). In the future, the "spillover" of animals across national park boundaries could intensify. Many species will be in parks where ecosystems are being altered by global environmental change. They may move outside the parks, seeking better habitat. In most cases, there are no migration corridors that would allow vagile species to find more suitable environments, and establishing such corridors generates additional conflicts around private property (Shafer 2014, 2015).

Third, a thread of American politics argues that private property is superior to government control. This has led to serious proposals to privatize the US national parks (Schwartz 2005). Thus, the national parks crystallize conflict between those who view them as the common property of the nation and those who feel that decisions made about the parks impinge on private property rights.

A second set of conflicts comes from the 1916 Organic Act, which directs the NPS "to provide for the enjoyment of [the scenery and the natural and historic objects and the wild life] in such manner and by such means as will leave them unimpaired for the enjoyment of future generations."[2] By calling for a balance between current use and benefits to future generations, the Organic Act may be one of the first US laws advocating what we now call sustainability (Grober 2012; Caradona 2014). It thus entrains all the complexities and uncertainties that arise in trying to make decisions that take account of the needs of future generations, including determining what will be important to them, whether to discount benefits and costs that occur in the future, how to incorporate risk and uncertainty into decisions, and so forth (National Research Council 1999a; Norton 2005; Rockwood, Stewart, and Dietz 2008; Dietz, Rosa, and York 2009; Ostrom 2009; Neumayer 2010).

At root, these conflicts reflect differences in the kinds of services Americans want from national park ecosystems. The Millennium Ecosystem Assessment divides ecosystem services into several categories, based on the ways they contribute to human well-being (Reid et al. 2005). *Supporting services* include provision of pollinators, recharge of groundwater, nutrient cycling, and soil formation. *Provisioning services* include extraction of food, fuel, fiber, and other materials. *Regulating services* include control of climate and biogeochemical cycles. *Cultural services* include recreation, aesthetic appreciation, education, and spiritual satisfaction.

Most national parks provide multiple services and their value can be substantial. Richardson et al. (2014) estimate that improved water quality in the Everglades would yield benefits of $1.8 billion across all categories

2. 16 USC § 1.

of services. Banasiak, Bilmes, and Loomis (2015) suggest that the vegetation in US national parks sequesters about 17.5 million metric tons of carbon dioxide (CO_2) per year, providing an estimated economic value for this regulating service of some $700 million per year based on the social cost of carbon emissions.

But the fundamental reason for the existence of US national parks is the provision of cultural services. This makes national parks unique: all other federally managed lands balance the provision of cultural services with other services. When some people want provisioning services from an ecosystem and others want other types of services, conflict is certain (Berger et al. 2014; DeFries, this volume, ch. 11). Ranchers near Yellowstone National Park are concerned that wolves crossing the park borders will prey on their cattle and that elk will transmit disease to them, yet these large mammals are greatly valued by park visitors. Loggers who can harvest timber from national forest lands around Olympic National Park will oppose transfer of parcels of that land to the NPS, even though cultutral, supporting, and regulating services to the public may be enhanced under NPS management. Conflicts about different forms of cultural services can also be intense. In particular, the desire to use national parks for recreation often conflicts with aesthetic concerns, with the park as a sacred site, or with the park as a preserve of biodiversity (Bernbaum, this volume, ch. 14; Wilson, this volume, ch. 1).

Global environmental change may overshadow all these other conflicts as a massive disruptive force (see Grimm and Jacobs [2013] and the rest of this special issue; Fisichelli et al. 2014; Monahan and Fisichelli 2014; Estenoz and Bush 2015; Gonzalez, this volume, ch. 6; Grorud-Colvert, Lubchenco, and Barner, this volume, ch. 2). Climate change includes changes in temperature, precipitation regimes, hydrology, fire regimes, ocean acidity, and sea level. At the same time, changes in biogeochemical cycles, long-distance transport of pollutants, and the movement of invasive species are affecting national parks (Baron et al., this volume, ch. 7; Simberloff, this volume, ch. 8). The stresses of global environmental change may be more challenging for national parks than for any other US institution or sector. The mission of US national parks, to "conserve unimpaired" a heritage of landscapes, seascapes, and biodiversity, is fundamentally threatened by global environmental change. No course of action can prevent most national parks from being very different ecosystems in 2115 than they were in 2015. As a result, global environmental change impedes the core mission of national parks.

Uncertainty of Facts and Values Complicates Environmental Decision Making

Good decisions must handle facts and values competently and allow for learning. The facts that underpin decisions about national parks will always be uncertain, and there will be value differences among those interested in or affected by decisions about parks. Indeed, because of the complexity of what is at stake in decisions about national parks, values are also frequently uncertain. Here I consider the implications of facts and values that are both in conflict and uncertain. The problem of social learning will be discussed later in the chapter.

Uncertain Facts

Scientific facts are always uncertain. Our confidence in science is based on the accumulation of ostensible (easy to observe) and repeatable observations (Rosa 1998; Dietz 2013b; York 2013). Those areas of science that are the most ostensible and repeatable are seen as the most "solid"; they have the least uncertainty. An example of such highly ostensible and repeatable science is our understanding of gravity developed by rolling balls down an inclined plane. The results of this experiment are easy to see—they are ostensible. And they are highly repeatable—I estimate that this experiment has been performed over 40 million times in high schools and universities around the world (T. Dietz, unpublished data). Highly ostensible and repeatable research helps create trust in science among decision makers and members of the public.

However, most of the science on which national park management is based has neither high ostensibility nor high repeatability. The life history parameters of a species, the toxicity of a compound, or the principles of island biogeography may have been well established in general. But applying those results to a specific park ecosystem, and especially an ecosystem evolving under global environmental change, inevitably involves substantial uncertainty.

To further complicate matters, our understanding of global environmental change involves a cascade of uncertainty. We cannot be certain of the future trajectory of greenhouse gas emissions. Even if we could, the results of climate models are uncertain. Even if they were certain at the large scale, downscaling to the spatial scale needed to make decisions for a particular park introduces uncertainty. Even if the downscaled projections were certain, the response of ecosystems to climate change is not. And even if the ecosystem response was known with certainty, the ultimate effects of a policy once it is implemented will always be uncertain.

The uncertainty of the relevant science means that expert judgment about the facts and their uncertainties will be especially important in decision making. As a result, trust in the individuals and organizations doing research and assessing the state of knowledge becomes central to conflicts. Unfortunately, when local decisions need to take account of global environmental change, as nearly all will, science for national parks may become embroiled in larger political dynamics. Trust in science has become increasingly politically polarized. Over the last 40 years, overall trust in science has been roughly constant among moderates and liberals but has declined substantially among conservatives (Gauchat 2011, 2012; Hamilton 2014). Active campaigns to emphasize scientific uncertainty about climate change have also led to a decline in belief in climate change among conservatives (McCright 2000; McCright and Dunlap 2003; Michaels 2008; McCright and Dunlap 2011a, 2011b; McCright et al. 2013; McCright, Dunlap, and Xiao 2013). When the NPS has to rely on uncertain science about climate change in decision making, these trends may influence trust in those decisions.

While the NPS cannot be expected to fully counter these national trends, clearly a premium must be placed on processes that build trust in the science used for decisions about the parks. A key step is to be explicit about uncertainty and how it is estimated. For example, the NPS Climate Change Response Program conducts research on how climate change will affect national parks, and works with specific parks to develop an understanding of the local implications of climate change (e.g., Fisichelli et al. 2013, Fisichelli et al. 2014; Fisichelli et al. 2015). In doing so, the evaluation is careful to take into consideration multiple possible trajectories of climate change and thus multiple suites of impacts. In some cases this is done with formal quantitative models; in other cases multiple scenarios capture the uncertainty without estimating probabilities. Tools from the environmental decision sciences can aid in assessing and communicating the uncertainty in these analyses. There are well-developed procedures for eliciting expert views about uncertainty when formal modeling is not feasible (Morgan, Pitelka, and Shevliakova 2001; Martin et al. 2012; McDaniels et al. 2012). Careful guidance on how best to express uncertainty in assessments intended to inform decision making has emerged in several domains, including climate change (Moss and Schneider 2000; Moss 2004; Pidgeon and Fischhoff 2011) and ecosystem dynamics (Schindler and Hilborn 2015). Later in the chapter, I discuss processes that can enhance trust in science by linking scientific analysis undertaken for national parks with public deliberation.

Conflicting and Uncertain Values

We are all familiar with uncertain facts and the difficulties they cause in making decisions. And we are familiar with value conflicts that come from different interests and preferences regarding the future of national parks. But what do I mean by uncertain values? Even if we knew with certainty the facts of what will happen in the future, each of us faces substantial challenges applying our values to the complex choices that must be made.

Global environmental change makes clear why most of us will face uncertainty as we try to use our values to decide the best course of action for national parks. Because global environmental change will almost certainly change national parks no matter what is done, and because preserving the legacy of parks is a core goal of the NPS, there will seldom be "win-win" options. It is plausible that the features that motivated the creation of some national parks could be lost or at least greatly diminished, such as the glaciers of Glacier National Park, the Joshua trees of Joshua Tree National Park, and the saguaros of Saguaro National Park. Box 12.1 discusses the problem of the wolf population in Isle Royale National Park and illustrates the ethical complexity of managing parks in the face of global change—there are no obvious "best" choices. As in Isle Royale, in many national parks there will be hard decisions about whether to make heroic interventions to save an iconic feature. In other parks, management decisions will have to be made with the understanding that the iconic features may be lost no matter what actions are taken. In most parks, strenuous efforts to minimize local stressors on the ecosystem will be required to reduce risks from global environmental change (Scheffer et al. 2015). The prospect of losing iconic park features and the uncertainty involved will drive conflict. Nearly everyone has trouble making decisions in the face of risk and uncertainty, and most people place more value on losses than on equivalent gains (Kahneman and Tversky 1979; Fischhoff and Kadvany 2011; Kahneman 2011). These characteristics make it even more difficult to reach agreement about the best course of action.

How Can We Make Better Decisions?

Since our understanding of the outcomes of management decisions will always be uncertain, especially under the influence of global environmental

Box 12.1 Global Environmental Change and Isle Royale National Park

Isle Royale National Park provides an example of how difficult it can be to determine the appropriate course of action in the face of global environmental change. The dynamics of the wolf-moose population on Isle Royale is a classic of ecological research, and perhaps the longest field study we have of predator-prey dynamics (Allen and Mech 1963; Peterson 1977; Hedrick et al. 2014; Peterson et al. 2014; Mlot 2015). While moose (*Alces alces*) have been there since at least the early 20th century, wolves (*Canus lupus*) seem to have colonized the island in the late 1940s using a winter ice bridge from the Canadian mainland. Since the population of wolves has always been small, it has been subject to the vagaries of small population dynamics, including inbreeding, founder effects, and genetic drift. Until recently, new wolves would occasionally enter the population via the ice bridge and counter those effects. It seems likely that the ice bridge will form much less often in the future than it has in the past due to climate warming. Over the last 16 years the bridge has formed only twice, whereas the historical average was roughly two years out of three. So the chance of new wolves migrating to the island has almost certainly decreased as a result of anthropogenic climate change. If there are no introductions of wolves via either the ice bridge or direct human intervention, Isle Royale's wolves will be much more vulnerable to local extinction. Indeed, by spring 2016 the wolf population has apparently dropped to just two individuals (Mlot 2015; Peterson and Vucetich 2016). The absence of a viable wolf population will affect the moose population and, as a result, the ecosystems throughout the entire island. This is just one climate-related change facing Isle Royale (Fisichelli et al. 2013).

Humans have altered the environment globally. How, if at all, should the NPS respond to those alterations at Isle Royale National Park? I submit that, for most of us, the value basis for deciding what to do is hard to determine with any great certainty. Even among people who agree on the importance of protecting the biosphere from human stresses, there will be conflicts about the best course of action for Isle Royale. Reasonable arguments can be made for and against the introduction of new wolves (Vucetich, Nelson, and Peterson 2012;

Box 12.1 (*continued*)

Cochrane 2013; Gostomski 2013; Mech 2013; Vucetich, Peterson, and Nelson 2013; Cochrane 2014). Some hold that human intervention is warranted because human actions altered the climate in a way that makes natural colonization much less likely. But others argue that the best course of action in a national park is not to intervene, even if this makes it likely that the wolves will go locally extinct and the Isle Royale ecosystem will be substantially altered as a result.

Part of the complexity in deciding what to do is that, unlike geological formations that are the iconic features of many US national parks, the wolf-moose system on Isle Royale is less than 100 years old. Is that sufficiently old for the wolves to be considered an aspect of Isle Royale worthy of preserving for future generations? The problem may be further complicated by the recent arrival of three wolves in Michipicoten Island Provincial Park in eastern Lake Superior. There are no moose on that island, but the ecosystem is dominated by woodland caribou (*Rangifer tarandus*), another large herbivore that will be prey for the wolves. One might argue that this new predator-prey system in some sense compensates for the changed dynamic on Isle Royale, but one could also argue that this development is not relevant to deciding what to do on Isle Royale.

change, decision making for US national parks is risk management. We need to learn as we proceed; ideally a risk management approach should adapt as circumstances change and as science advances (National Research Council 2010a, 2010b, 2010c, 2010d, 2011; Rosa, Renn, and McCright 2013). Adaptive risk management (ARM) is defined as "an ongoing decision-making process that takes both known and potential risks and uncertainties into account, and periodically updates and improves plans and strategies as new information becomes available" (National Research Council 2010b).

Ideas of ARM are already in play in NPS decision making and in natural resource and biodiversity management in general, but more needs to be done (Cross et al. 2012; Cross et al. 2013; Stein et al. 2013; Rannow et al. 2014; Sharp et al. 2014). Approaches to ARM developed to deal with decision making for national parks will have to be shaped to the overall challenges faced by the NPS and modified to respond to the special challenges of each local context. Consider the Isle Royale example. The NPS has developed scenarios for how the island's ecosystem might change in the face of climate change and is continuing its assessment and planning process (Fisichelli et al. 2013). An ARM strategy would opt for a tentative course of action (e.g., do or do not bring new wolves to the island) based on the best available assessment of the science and of the value implications of the alternatives. But an ARM approach would also regularly assess the factors that contribute to uncertainty in the outcome—whether the ice bridge forms, the dynamics of the wolf population, the dynamics of the moose population, subsequent changes to the island's ecosystems, public views about the ecosystem changes, and so on. Decisions about introducing new wolves and other management strategies could be updated year by year, or at least until the ecosystem undergoes irreversible changes. Isle Royale, like every US national park, is unique, and the strategies for ARM there might not be suitable elsewhere. However, there are a number of general lessons that have emerged from the larger literature on environmental decision making that may be of help as the NPS and its allies cope with the special challenges of ARM in the 21st century.

Of course, no approach to problems as complex as those faced by US national parks is perfect. As Doremus (2007, 2010) has clearly articulated, adaptive management can add a substantial burden of complexity, and with that a need to devote considerable time and money to the process. She prudently calls for adaptive use of adaptive management. There will be situations in which the benefits of ARM do not outweigh the costs. The general logic of ARM would benefit all NPS decision making by considering, even briefly, what outcomes would be like under different assumptions about the future and by consulting with those interested in or affected by a decision. This process can be rather simple and informal when choices are limited, or when the consequences are modest or easily reversible. In other situations, when the stakes are large, the uncertainty is high, and the outcomes of a decision are hard to reverse, more complex forms of ARM are warranted. For example, deciding on the appropriate response to the decline of the Isle Royale wolf population requires consideration of what may be substantial changes to ecosystems that will be irreversible on a scale of many decades or longer and, especially in the face of climate change, entrain a high degree of uncertainty. Similarly, consideration of dam removals involves high stakes, irreversibility, and at least moderately high uncertainty. In such cases, investing in a carefully designed ARM process seems warranted.

Social Learning on Networks

Formal decision-making authority for US national parks rests with the director of the NPS and ultimately with the executive branch and Congress. But policy decisions are always shaped by a policy network that includes not just those formally designated as decision makers, but also a broader network of interested parties. Interested and affected parties include other federal agencies and communities near national parks. They also include the broad community of those who care deeply about parks and who assign great importance to the ecosystem services the parks provide, including both regular visitors and those who have never visited national parks but nonetheless value them. The policy network includes professionals working for formal organizations that engage with US national parks, members of those organizations, a large community of researchers, members of local communities, and local government officials. Because the US national parks are viewed as a part of the national heritage, interest in the long-term trajectory of park landscapes and ecosystems is probably broader and more intense than it is for nearly any other ecosystems in the United States. This creates a particular challenge—the need to engage citizens who are interested in the parks and their fate but who have no formal organizational affiliations related to the parks.

Because of this network of interested and affected individuals and organizations, ARM for US national parks must engage a broader community than would be the case for management of a resource that engages narrower interests and with simpler management directives. Organizational learning is essential to the long-term success of ARM. But in the case of US national parks, it is not just learning on the part of the NPS that is required. Rather, ARM will be most successful when interested or affected parties are aware of and change their views in response to evolving knowledge. Such social learning on networks can lead to improved decision making over time (Henry 2009; Frank et al. 2011; Frank et al. 2012; Henry and Vollan 2014; Lemieux et al. 2015). So a goal of ARM is to facilitate such learning by all involved.

The dynamics of policy networks will make social learning challenging. Substantial literature shows that policy networks evolve under the influence of homophily and biased assimilation (Sabatier and Weible 2007; Weible and Sabatier 2007; Henry 2011; Henry, Prałat, and Zhang 2011; Henry and Vollan 2014). Homophily is the tendency to engage mostly with those who are similar to us and to avoid those who are different. In policy networks, this can lead to the formation of inward-looking cliques with substantial communications within the clique and limited communication outside it. Biased assimilation is the tendency to accept assertions of fact and arguments about values that are consistent with one's own views. When coupled with homophily, biased assimilation can lead to policy networks composed of relatively isolated cliques. It can lead to beliefs that are homogenous within cliques but very different between cliques. This polarization process can substantially exacerbate conflict, degrade trust in those outside one's clique, make it harder to find consensus, and substantially retard the prospects for social learning (Henry and Vollan 2014).

In the face of these challenges, four complementary approaches can facilitate adaptive decision making and social learning. None is a panacea, and each must be designed with attention to the particular circumstances in which it will be deployed. But these approaches do offer some hope for helping the NPS and those engaged with it to make better decisions and learn as they cope with the conflicts of the 21st century.

Linking Scientific Analysis and Public Deliberation

A series of US National Research Council reports, starting with *Understanding Risk*, have called for supporting assessment and decision making by linking scientific analysis with public deliberation (National Research Council 1996, 1999b, 2008, 2010b, 2013). The idea has its origins in the arguments of Dewey (1923) and was further developed by Habermas (1970, 1996). Its potential has been articulated in a number of areas of environmental decision making, including biodiversity management (Dietz and Stern 1998; National Research Council 1999b, 2013), climate change (National Research Council 2010b), impact assessment (Dietz 1984, 1987, 1988), and risk analysis (Tuler and Webler 1995; National Research Council 1996; Renn 2008; Renn and Schweitzer 2009; Rosa, Renn, and McCright 2013).

The analytic deliberative approach is illustrated in figure 12.1. The core of the process is iterative communication between the public and those doing research in support of a decision. The communication process starts with codesign of the process itself by the public and the researchers. In this context, the public is defined as all parties interested in or affected by a decision (Dewey 1923). Thus, the public for decisions about US national parks is very diverse, including those in proximity to a park and who may have an economic stake in decisions about that park, but also generations of park visitors and the broad population who care about cultural services, even from parks they will never visit. The research agenda is shaped by the

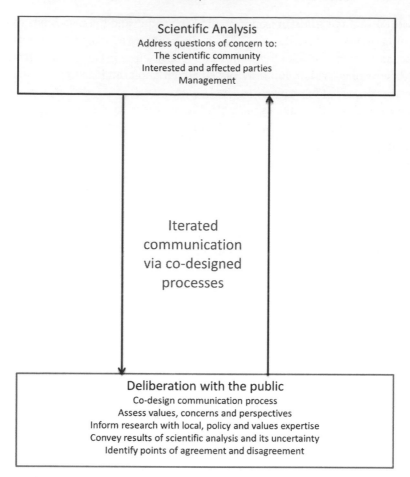

12.1. The analytic deliberative process adapted from
the US National Research Council (2008).

concerns not only of scientists but also of the public and managers. The
latter groups do not determine the design of studies or the conclusions
drawn from them. But their advice can be essential in making sure the right
questions are asked and that local context is given proper attention. The
deliberative process uses engagement with interested and affected parties
to assess values, concerns, and perspectives, to provide locally grounded
expertise, to give a forum for scientists to describe their conclusions and
associated uncertainties, and to provide a chance for all parties to attempt
to find common ground.

One question that immediately arises is whether engagement with the public, while nearly always mandated by US federal law, is beneficial. The National Research Council examined what is known about public participation by reviewing roughly 1,000 empirical studies, codifications of practitioner experience, and theoretical analyses. The report concluded: "When done well, public participation improves the quality and legitimacy of decisions and builds the capacity of all involved to engage in the policy process" (National Research Council 2008). The key phrase here is "done well." The report acknowledges that public participation is often done poorly. Because the challenges to effective public participation will vary substantially across contexts, there is no single approach that can be universally applied. Rather, the report suggests 15 design principles (described in table 12.1). The specific principles engage three aspects of the process of linking analysis and deliberation: the commitment of the agency, the nature of the process, and the way uncertainty is handled. Deciding how these principles should be implemented requires a diagnosis of the particular decision under consideration and the broader context in which it will be made. For example, when a decision has large-scale and irreversible effects, the design principles imply the NPS should make a major commitment to the process. In contrast, a smaller and more informal process might be suitable for less consequential decisions. How the process is run will also vary across contexts. In some cases a long history of entrenched conflict suggests a process that devotes considerable effort to building mutual understanding and trust, even while acknowledging that consensus may not be reached. When positions are less entrenched, less attention needs to be given to mechanisms of conflict management and resolution. Assessments of uncertainty can be relatively informal and based on carefully elicited and articulated expert judgment, or for difficult and consequential decisions, uncertainty assessments can involve formal analysis of a suite of quantitative projections. Given the diagnosis of the context, the design principles provide guidance on how to structure a process of linked scientific analysis and public deliberation.

The process of linking scientific analysis with public deliberation in national park decision making has not been extensively studied. But the National Research Council review of research on this kind of process in other natural resource management contexts yields several general conclusions that likely apply to parks as well (National Research Council 2008). Linking analysis and deliberation helps get the right science by identifying questions that matter to interested and affected parties. That is, it helps target science that will matter in decision making. Engaging with those who are most likely to care about or be affected by a decision helps ensure that

Table 12.1   Design principles for public participation

Process	Design principles
Agencies proceed with	(1) clarity of purpose, (2) a commitment to use the process to inform actions, (3) adequate funding and staff, (4) appropriate timing in relation to decisions, (5) a focus on implementation, and (6) a commitment to self-assessment and learning from experience.
Procedure must be	(7) inclusive, (8) collaborative in problem formulation and process design, (9) transparent, and (10) based on good-faith communication.
Attend to uncertainty by	(11) ensuring transparency of decision-relevant information and analysis, (12) paying explicit attention to both facts and values, (13) promoting explicitness about assumptions and uncertainties, (14) including independent review of official analysis and/or engaging in a process of collaborative inquiry with interested and affected parties, and (15) allowing for iteration to reconsider past conclusions on the basis of new information.

Source: National Research Council (2008).

the research questions asked will address their concerns. In the end, the conclusions from research may not support the prior beliefs of those who are interested in a decision. But if they feel they have had a voice in what questions are addressed, they are more likely to accept unwelcome conclusions. Linking research to public deliberation also helps get the science right by eliciting insights from individuals who may have special expertise regarding local contexts. Hearing diverse views is especially important when scientific conclusions do not have the benefit of high ostensibility and repeatability, and when multiple outcomes are at stake (Rosa and Clarke 2012; Dietz 2013b; Rosa, Renn, and McCright 2013; Harding 2015). Linking analysis and deliberation helps the public understand the science, including the inherent uncertainty in it. Thus one benefit of the process of linking analysis and deliberation is that it can help overcome the problems generated by homophily and biased assimilation in policy networks. Linked analysis and deliberation can be an important tool for building trust and understanding, and for finding compromises that take account of multiple values. Of course, given the conflicts inherent in the mission of the NPS and the value differences among interested and affected parties, many decisions will leave some parties dissatisfied. The evidence suggests that an effective process linking analysis and deliberation can increase acceptance of decisions by those who do not feel the decision fully serves their interest. However, since the research that leads to these conclusions has mostly been conducted in contexts outside US national parks, important insights could be gained by studying such processes around NPS decision making.

Engaging Multiple Forms of Expertise

At least five forms of expertise should be engaged for effective ARM. *The first is scientific expertise on coupled human and natural systems.* This kind of expertise uses the scientific method to establish what we know about the facts relevant to decisions and also assesses the degree of uncertainty about facts. In the Isle Royale National Park example, this is primarily expertise about the local ecosystem and its dynamics and the historical and likely future dynamics of climate in the Great Lakes region.

The second is scientific expertise on decision making, including appropriate ways to handle uncertainty; assess values; make trade-offs; resolve conflicts; design effective policies, programs, and institutions; and evaluate what follows from decisions so as to allow for adaptation. Because the value implications of alternative courses of action on Isle Royale are so complex, groups with the requisite expertise include social scientists who understand public values and ethicists and decision scientists who can help design processes of ARM suitable for the local context.

The third expertise is grounded in local knowledge, sometimes called traditional ecological knowledge. This can reside in Native American communities, in other local communities, and in NPS staff and professionals who know the park from long experience. In the Isle Royale National Park case, it would be valuable to consult with Native Americans as well as NPS staff, researchers, and long-term visitors. Several Lake Superior Native American communities have expressed concern about wolf hunts, and they may have important insights on the problem of the Isle Royale wolves as well (Burns 2014; ICTMN Staff 2012; VanEgeren 2012).

Fourth, it is useful to also engage political and policy expertise. This expertise resides in those active in the policy network, and it can help identify what is feasible and what is not, and where there is trust and where there is none, and can help guide discussions toward policies that will be feasible to adopt and implement and are likely to be effective. In the case of Isle Royale, this includes NPS staff and the scientific community committed to long-term research on the ecosystem, concessionaires and others whose livelihood is affected by park management, and members of the public who are actively engaged with the park.

Finally, because the US national parks are a public trust for the whole nation, assessment of concerns and values cannot be limited to those of visitors or of the communities in the vicinity of parks. The national interest must be considered. Minimally this can be done by engaging policy experts from advocacy groups representing various value positions. However, in many cases it would be useful to assess the values and concerns of the larger public as well. *One can define this as a fifth form of expertise, value expertise.* Here every member of the public interested in or affected by a decision is an expert on her or his own values. However, the unprecedented implications of the decisions to be made, the complexity of coupled human and natural systems, and the great uncertainties about the consequences of our decisions make applying values to decisions difficult—this is what I have called value uncertainty. Most of us can benefit from using the tools of environmental decision making to help think through and better understand our own values as they apply to issues facing parks (Cramer, Dietz, and Johnston 1980; Arvai, Gregory, and McDaniels 2001; Florig et al. 2001; Morgan et al. 2001; Cobb and Thompson 2012). For some decisions, rather routine processes will suffice, whereas in other cases an exceptional effort may be warranted (Chess, Dietz, and Shannon 1998; Renn and Schweitzer 2009).

Drawing again on Isle Royale as an example, the NPS, with the support of the Great Lakes Integrated Sciences and Assessments Center, conducted a workshop that identified multiple possible scenarios of how climate change will affect the park (Fisichelli et al. 2013). The literature contains a suite of arguments for how best to handle the problem of the wolf population decline. Several future steps might aid in decision making. Public values regarding the future of the park could be assessed via surveys and workshops. An assessment of risks of ecosystem changes could be developed based on estimates of probabilities about the future of key drivers such as the formation of the ice bridge, the chances of a reinvigorated wolf population via either ice bridge migration or reintroduction by the NPS, and cascading changes in the ecosystem. The scenarios already developed for Isle Royale are the critical first step in this assessment, and expert judgment, rather than formal modeling, would probably be appropriate to develop plans around more and less likely futures. Finally, an exercise among interested and affected parties to detail value trade-offs could help clarify points of agreement and disagreement regarding future management strategies for the park. Depending on the time and resources available, these analyses could be done with "off the shelf" approaches, or could constitute a research program that would both aid in decision making and contribute to fundamental knowledge.

Building Evaluation into Every Program and Policy

The need for evaluation comes from appropriate humility regarding what we know. The idea that we need to learn from policy experience was articulated by Dewey (1923) and refined by Campbell (1969). We have sophisticated methods for understanding causal influence and generalizability in both experimental and nonexperimental studies (Cook and Campbell 1979; Frank 2000; Cook, Campbell, and Shadish 2002; Frank and Min 2007; Frank et al. 2013). But the application of these methods to policies and programs intended to enhance sustainability and to protect biodiversity globally has been too limited (Miteva, Pattanayak, and Ferraro 2012). The US national parks could benefit from more systematic use of evaluation.

Each decision should be considered as an experiment whose outcome is uncertain. Then part of every decision is specifying how it will be possible to learn from the decision, thinking through how adequate records will be maintained and eventually analyzed. Of course, as Doremus (2007, 2010) has made clear, it is not feasible to do an exhaustive analysis of every decision. Some decisions have minimal impacts, are routine, and have implications that are well understood. For other decisions, it is simply not feasible to conduct a substantial post hoc analysis. In those cases, it will suffice to simply revisit the decision at some point in the future and informally assess outcomes. But decisions that are addressing novel circumstances induced by global environmental change or are otherwise innovative warrant post hoc evaluation so that the NPS can learn from them. The vast differences across the current suite of 59 national parks, 230 other natural resource units, and 118 other units managed by the NPS provides a natural laboratory for understanding how variation in policies and programs and variation in context contribute to outcomes. Analysis of this variation can be a powerful tool for social learning and ARM. But this potential will be realized if and only if a process of monitoring and evaluation is built into decisions.

Build Bridging Institutions and Networks

Asking national park scientists and managers to become experts at risk communication and managing participatory processes is of course not realistic given their already heavy workload. The use of bridging organizations is an efficient way to enhance social learning on policy networks (Frank et al. 2012; Bidwell, Dietz, and Scavia 2013). Information needed for decision making flows through networks based on trusted relationships, while homophily, biased assimilation, and clique formation can block information flow (Leonard et al. 2011; Henry and Vollan 2014). Often there are organizations within policy networks that have good relationships both with decision makers and with most of the kinds of expertise that are needed in support of good decision making. Such organizations can bridge the various forms of expertise and can link experts and decision makers, helping to make connections, translate, build trust, and enhance social learning on the network. Network analysis can help inform where existing networks are adequate to support such communication and where efforts to build new bridges are needed.

The NPS is already using this approach to address climate change. For example, as noted above, the Great Lakes Integrated Sciences and Assessments Center (GLISA) worked with the NPS Climate Change Response Program to develop and interpret scenarios to support climate adaption at Isle Royale (Fisichelli et al. 2013).[3] This was a case of a bridging unit within the NPS working with a university-based bridging center funded by the National Oceanic and Atmospheric Administration. The goal of the effort was to provide information that would be useful to decision making by Isle Royale National Park managers. GLISA provided expertise on climate dynamics in the Great Lakes region, the NPS Climate Change Response Program provided expertise on the effects of climate change on ecosystems, and the Isle Royale National Park staff provided expertise on the island ecosystem. In addition to the report and its analysis, all parties involved learned about the perspectives and expertise of the others involved, building the basis for future collaborations. GLISA and the NPS Climate Change Response Program are now working with other NPS units in the Great Lakes region to develop comparable reports. That work is more efficient because of the relationships established in the first project. In the future, these relationships can be expanded to bring in other partners. One emerging opportunity comes through GLISA's work on climate adaptation with Native American and First Nations communities in the region. This part-

---

3. See also Great Lakes Integrated Sciences and Assessments Center, Isle Royale National Park: Climate change scenario planning, accessed 10 March 2016, http://glisa.msu .edu/projects/isle-royale-national-park-climate-change-scenario-planning.

nership will be able to facilitate discussions between them and the NPS when that is appropriate for developing climate adaptation plans.

Conclusions

The challenges the US national parks face in the 21st century reflect the long-standing goals set for the parks, goals that are inspirational but that also guarantee conflicts. These conflicts will be magnified by global environmental change. By 2115, nearly every US national park ecosystem will have changed substantially from what it is today, and decisions made now will have huge impacts on the nature of those changes. Like Tolstoy's unhappy families, every decision about national parks and every conflict that arises as a result is unique. But scientific analysis is centrally about extracting lessons from across seemingly unique circumstances. ARM implies that we should respect the particular features of each national park and each decision, and design processes responsive to the context. But at the same time, ARM requires that we learn as we proceed, both by engaging multiple forms of expertise and by studying the outcomes of decisions already made. Building bridges across multiple kinds of expertise can facilitate the flow of information needed to inform decisions and build trust even in the face of uncertain facts and conflicting and uncertain values.

The US national parks can contribute a great deal to the study of environmental decision making. They are special among American institutions because of their deep commitment to both current use and long-term stewardship, and because of their profound emphasis on the importance of the spiritual, scientific, educational, and aesthetic values of ecosystems and landscapes. This makes these parks a unique testbed for examining how long-term uncertainty and long-term value commitments influence conflict and decisions.

Lessons from the environmental decision sciences cannot fully resolve the conflicts the US national parks will face since many of these conflicts are embedded in the core intent of the parks and in core American values. But the environmental decision sciences can provide insights into the sources of conflict, an understanding of the beliefs and values that interested and affected parties bring to the discussion, and, perhaps most important, tools and concepts for making the best possible decisions in the face of conflict and uncertainty. In that sense, the environmental decision sciences are truly a "science for the parks."

Acknowledgments

This work was supported in part by Michigan AgBio Research and by GLISA. The manuscript benefited greatly from comments by Linda Kalof and especially Holly Doremus.

Literature Cited

Allen, D. L., and L. D. Mech. 1963. Wolves versus moose on Isle Royale. National Geographic 123:200–219.

Arvai, J. L., R. Gregory, and T. McDaniels. 2001. Testing a structured decision approach: value-focused thinking for deliberative risk communication. Risk Analysis 21: 1065–1076.

Banasiak, A., L. Bilmes, and J. Loomis. 2015. Carbon sequestration in the US national parks: a value beyond visitation. John F. Kennedy School of Government, Harvard University, Cambridge, Massachusetts.

Berger, J., S. L. Cain, E. Cheng, P. Dratch, K. Ellison, J. Francis, H. C. Frost, et al. 2014. Optimism and challenge for science-based conservation of migratory species in and out of US national parks. Conservation Biology 28:4–12.

Bidwell, D. 2009. Bison, boundaries, and brucellosis: risk perception and political ecology at Yellowstone. Society & Natural Resources 23:14–30.

Bidwell, D., T. Dietz, and D. Scavia. 2013. Fostering knowledge networks for climate adaptation. Nature Climate Change 3:610–611.

Burns, G. 2014. Native American opponents of Michigan wolf hunt call rationale irrational. MLive, 13 July. Accessed 25 June 2015. http://www.mlive.com/news/detroit/index.ssf/2014/07/native_american_opponents_of_m.html.

Campbell, D. T. 1969. Reforms as experiments. American Psychologist 24:409–429.

Caradona, J. L. 2014. Sustainability: a history. Oxford University Press, Oxford, United Kingdom.

Chess, C., T. Dietz, and M. Shannon. 1998. Who should deliberate when? Human Ecology Review 5:45–48.

Cobb, A. N., and J. L. Thompson. 2012. Climate change scenario planning: a model for the integration of science and management in environmental decision-making. Environmental Modelling & Software 38:296–305.

Cochrane, T. 2013. Island complications: should we retain wolves on Isle Royale? The George Wright Forum 30:313–325.

———. 2014. Rejoinder to Discernment and precaution: a response to Cochrane and Mech. The George Wright Forum 31:94–95.

Cook, T. D., and D. T. Campbell. 1979. Quasi-experimentation. Rand McNally, Chicago, Illinois.

Cook, T. D., D. T. Campbell, and W. R. Shadish. 2002. Experimental and quasi experimental designs for generalized causal inference. Houghton Mifflin, New York, New York.

Cramer, J. C., T. Dietz, and R. Johnston. 1980. Social impact assessment of regional plans: a review of methods and a recommended process. Policy Sciences 12:61–82.

Cross, M. S., P. D. McCarthy, G. Garfin, D. Gori, and C. A. Enquist. 2013. Accelerating adaptation of natural resource management to address climate change. Conservation Biology 27:4–13.

Cross, M. S., E. S. Zavaleta, D. Bachelet, M. L. Brooks, C. A. Enquist, E. Fleishman, L. J. Graumlich, et al. 2012. The Adaptation for Conservation Targets (ACT) framework: a tool for incorporating climate change into natural resource management. Environmental Management 50:341–351.

Dewey, J. 1923. The public and its problems. Henry Holt, New York, New York.

Dietz, T. 1984. Social impact assessment as a tool for rangelands management. Pages 1613–1634 in National Research Council/National Academy of Sciences, ed. Developing strategies for rangelands management. A report prepared by the committee on developing strategies for rangeland management. Westview Press, Boulder, Colorado.

———. 1987. Theory and method in social impact assessment. Sociological Inquiry 57:54–69.

———. 1988. Social impact assessment as applied human ecology: integrating theory and method. Pages 220–227 in R. Borden, J. Jacobs, and G. R. Young, eds. Human ecology: research and applications. Society for Human Ecology, College Park, Maryland.

———. 2003. What is a good decision? Criteria for environmental decision making. Human Ecology Review 10:60–67.

———. 2013a. Bringing values and deliberation to science communication. Proceedings of the National Academy of Sciences USA 110:14081–14087.

——— 2013b. Epistemology, ontology, and the practice of structural human ecology. Pages 31–52 in T. Dietz and A. K. Jorgenson, eds. Structural human ecology: essays in risk, energy, and sustainability. WSU Press, Pullman, Washington.

———. 2015. Environmental value. Pages 329–349 in T. Brosch and D. Sander, eds. Handbook of value: perspectives from economics, neuroscience, philosophy, psychology and sociology. Oxford University Press, Oxford, United Kingdom.

Dietz, T., A. Fitzgerald, and R. Shwom. 2005. Environmental values. Annual Review of Environment and Resources 30:335–372.

Dietz, T., E. Ostrom, and P. C. Stern. 2003. The struggle to govern the commons. Science 301:1907–1912.

Dietz, T., E. A. Rosa, and R. York. 2009. Environmentally efficient well-being: rethinking sustainability as the relationship between human well-being and environmental impacts. Human Ecology Review 16:113–122.

Dietz, T., and P. C. Stern. 1998. Science, values and biodiversity. BioScience 48:441–444.

Doremus, H. 2007. Precaution, science, and learning while doing in natural resource management. Washington Law Review 82:547–579.

———. 2010. Adaptive management as an information problem. North Carolina Law Review 89:1455–1498.

Estenoz, S., and E. Bush. 2015. Everglades restoration science and decision-making in the face of climate change: a management perspective. Environmental Management 55:876–883.

Fischhoff, B., and J. Kadvany. 2011. Risk: a very short introduction. Oxford University Press, Oxford, United Kingdom.

Fisichelli, N., C. Hawkins Hoffman, L. Welling, L. Briley, and R. Rood. 2013. Using climate change scenarios to explore management at Isle Royale National Park: January 2013 workshop report. Natural Resource Report NPS/NRSS/CCRP/NRR—2013/714. National Park Service, Fort Collins, Colorado.

Fisichelli, N. A., S. R. Abella, M. Peters, and F. J. Krist. 2014. Climate, trees, pests, and weeds: change, uncertainty, and biotic stressors in eastern US national park forests. Forest Ecology and Management 327:31–39.

Fisichelli, N. A., G. W. Schuurman, W. B. Monahan, and P. S. Ziesler. 2015. Protected area tourism in a changing climate: will visitation at US national parks warm up or overheat? PLoS ONE 10:e0128226.

Florig, H. K., M. G. Morgan, K. E. Jenni, B. Fischoff, P. S. Fischbeck, and M. L. DeKay. 2001. A deliberative method for ranking risks (I): overview and test-bed development. Risk Analysis 21:913–921.

Foresta, R. A. 2013. America's national parks and their keepers. Routledge, New York, New York.

Frank, K., I.-C. Chen, Y. Lee, S. Kalafatis, T. Chen, Y.-J. Lo, and M. C. Lemos. 2012. Network location and policy-oriented behavior: an analysis of two-mode networks of coauthored documents concerning climate change in the Great Lakes Region. Policy Studies Journal 40:492–515.

Frank, K. A. 2000. The impact of a confounding variable on a regression coefficient. Sociological Methods and Research 29:147–194.

Frank, K. A., S. Maroulis, D. Belman, and M. D. Kaplowitz. 2011. The social embeddedness of natural resource extraction and use in small fishing communities. Pages 302–332 in W. W. Taylor, A. J. Lynch, and M. G. Schechtler, eds. Sustainable fisheries: multi-level approaches to a global problem. American Fisheries Society, Bethesda, Maryland.

Frank, K. A., S. Maroulis, M. Q. Duong, and B. Kelcey. 2013. What would it take to change an inference? Using Rubin's causal model to interpret the robustness of causal inferences. Educational Evaluation and Policy Analysis 35:437–460.

Frank, K. A., and K.-S. Min. 2007. Indices of robustness for sample representation. Sociological Methodology 37:349–392.

Gauchat, G. 2011. The cultural authority of science: public trust and acceptance of organized science. Public Understanding of Science 20:751–770.

———. 2012. Politicization of science in the public sphere: a study of public trust in the United States, 1974 to 2010. American Sociological Review 77:167–187.

Gore, M. L., M. P. Nelson, J. A. Vucetich, A. M. Smith, and M. A. Clark. 2011. Exploring the ethical basis for conservation policy: the case of inbred wolves on Isle Royale, USA. Conservation Letters 4:394–401.

Gostomski, T. 2013. Are Isle Royale wolves too big to fail? A response to Vucetich et al. The George Wright Forum 30:96–100.

Grimm, N. B., and K. L. Jacobs. 2013. Evaluating climate impacts on people and ecosystems. Frontiers in Ecology and the Environment 11:455–455.

Grober, U. 2012. Sustainability: a cultural history. Green Books, Devon, United Kingdom.

Habermas, J. 1970. Towards a rational society. Beacon Press, Boston, Massachusetts.

———. 1996. Between facts and norms: contributions to a discourse theory of law and democracy. MIT Press, Cambridge, Massachusetts.

Hamilton, L. 2014. Do you trust scientists about the environment? Carsey School of Public Policy, Durham, New Hampshire.

Harding, S. 2015. Objectivity and diversity: another logic of scientific research. University of Chicago Press, Chicago, Illinois.

Hedrick, P. W., R. O. Peterson, L. M. Vucetich, J. R. Adams, and J. A. Vucetich. 2014. Genetic rescue in Isle Royale wolves: genetic analysis and the collapse of the population. Conservation Genetics 15:1111–1121.

Henry, A. D. 2009. The challenge of learning for sustainability: a prolegomenon to theory. Human Ecology Review 16:131–140.

———. 2011. Ideology, power, and the structure of policy networks. Policy Studies Journal 39:361–383.

Henry, A. D., P. Prałat, and C.-Q. Zhang. 2011. Emergence of segregation in evolving social networks. Proceedings of the National Academy of Science USA 108:8605–8610.

Henry, A. D., and B. Vollan. 2014. Networks and the challenge of sustainable development. Annual Review of Environment and Resources 39:583–610.

ICTMN Staff. 2012. Minnesota ignores Indians, allows wolf hunting. Indian Country Today Media Network, 5 July. Accessed 25 June 2015. http://indiancountrytoday medianetwork.com/2012/07/05/minnesota-ignores-indians-allows-wolf-hunting -121922.

Kahneman, D. 2011. Thinking fast and slow. Farrar, Straus & Giroux, New York, New York.

Kahneman, D., and A. Tversky. 1979. Prospect theory: an analysis of decision making under risk. Econometrica 47:263–291.

Lemieux, C. J., J. Thompson, D. S. Slocombe, and R. Schuster. 2015. Climate change collaboration among natural resource management agencies: lessons learned from two US regions. Journal of Environmental Planning and Management 58:654–677.

Leonard, N. J., W. W. Taylor, C. I. Goddard, K. A. Frank, A. E. Krause, and M. G. Schechter. 2011. Information flow within the social network structure of a Joint Strategic Plan for Management of Great Lakes Fisheries. North American Journal of Fisheries Management 31:629–655.

Martin, T. G., M. A. Burgman, F. Fidler, P. M. Kuhnert, S. Low-Choy, M. McBride, and K. Mengersen. 2012. Eliciting expert knowledge in conservation science. Conservation Biology 26:29–38.

McCright, A. M. 2000. Challenging global warming as a social problem: an analysis of the conservative movement's counter-claims. Social Problems 47:499–522.

McCright, A. M., K. Dentzman, M. Charters, and T. Dietz. 2013. The influence of political ideology on trust in science. Environmental Research Letters 8:044029.

McCright, A. M., and R. E. Dunlap. 2003. Defeating Kyoto: the conservative movement's impact on US climate change policy. Social Problems 50:348–373.

———. 2011a. Cool dudes: the denial of climate change among conservative white males in the United States. Global Environmental Change 21:1163–1172.

———. 2011b. The politicization of climate change and polarization in the American public's views of global warming, 2001–2010. Sociological Quarterly 52:155–194.

McCright, A. M., R. E. Dunlap, and C. Xiao. 2013. Perceived scientific agreement and support for government action on climate change in the USA. Climatic Change 119:511–518.

McDaniels, T., T. Mills, R. Gregory, and D. Ohlson. 2012. Using expert judgments to explore robust alternatives for forest management under climate change. Risk Analysis 32:2098–2112.

Mech, L. D. 2013. The case for watchful waiting with Isle Royale's wolf population. The George Wright Forum 30:326–332.

Michaels, D. 2008. Doubt is their product: how industry's assault on science threatens your health. Oxford University Press, New York, New York.

Miteva, D. A., S. K. Pattanayak, and P. J. Ferraro. 2012. Evaluation of biodiversity policy instruments: what works and what doesn't? Oxford Review of Economic Policy 28:69–92.

Mlot, C. 2015. Inbred wolf population on Isle Royale collapses, but other wolves adopt a new Lake Superior island. Science **348**:383.

Monahan, W. B., and N. A. Fisichelli. 2014. Climate exposure of US national parks in a new era of change. PloS ONE **9**:e101302.

Morgan, K. M., M. L. DeKay, P. S. Fischbeck, M. G. Morgan, B. Fischoff, and H. K. Florig. 2001. A deliberative method for ranking risks (II): evaluation of validity and agreement among risk managers. Risk Analysis **21**:923–937.

Morgan, M. G., L. F. Pitelka, and E. Shevliakova. 2001. Estimates of climate impacts on forest ecosystems. Climatic Change **51**:251–257.

Moss, R., and S. H. Schneider. 2000. Uncertainties in the IPCC TAR: recommendations to lead authors for more consistent assessment and reporting. Pages 33–51 *in* R. Pachauri, T. Taniguchi, and K. Tanaka, eds. Guidance papers on the cross-cutting issues of the Third Assessment Report of the IPCC. World Meteorological Organization, Geneva, Switzerland.

Moss, R. H. 2004. Improving information for managing an uncertain future climate. Global Environmental Change **17**:4–7.

National Research Council. 1996. Understanding risk: informing decisions in a democratic society. National Academies Press, Washington, DC.

———. 1999a. Our common journey: a transition toward sustainability. National Academies Press, Washington, DC.

———. 1999b. Perspectives on biodiversity: valuing its role in an ever changing world. National Academies Press, Washington, DC.

———. 2008. Public participation in environmental assessment and decision making. National Academies Press, Washington, DC.

———. 2010a. Adapting to the impacts of climate change. National Academies Press, Washington, DC.

———. 2010b. Advancing the science of climate change. National Academies Press, Washington, DC.

———. 2010c. Informing an effective response to climate change. National Academies Press, Washington, DC.

———. 2010d. Limiting the magnitude of climate change. National Academies Press, Washington, DC.

———. 2011. America's climate choices. National Academies Press, Washington, DC.

———. 2013. Using science to improve the BLM Wild Horse and Burro Program: a way forward. National Academies Press, Washington, DC.

Neumayer, E. 2010. Weak versus strong sustainability: exploring the limits of two opposing paradigms. 3rd ed. Edward Elgar, Cheltenham, Gloucester, United Kingdom.

Norton, B. G. 2005. Sustainability: a philosophy of adaptive ecosystem management. Oxford University Press, New York, New York.

Orr, S. K., and R. L. Humphreys. 2012. Mission rivalry: use and preservation conflicts in national parks policy. Public Organization Review **12**:85–98.

Ostrom, E. 2007. A diagnostic approach for going beyond panaceas. Proceedings of the National Academy of Sciences USA **104**:15181–15187.

———. 2009. A general framework for analyzing sustainability of social-ecological systems. Science **325**:419–422.

Peterson, R. O. 1977. Wolf ecology and prey relationships on Isle Royale. National Park Service, Washington, DC.

Peterson, R. O., and J. A. Vucetich. 2016. Ecological studies of wolves on Isle Royale, Annual Report 2016. Michigan Technological University, Houghton, Michigan.

Peterson, R. O., J. A. Vucetich, J. M. Bump, and D. W. Smith. 2014. Trophic cascades in a multicausal world: Isle Royale and Yellowstone. Annual Review of Ecology, Evolution, and Systematics 45:325–345.

Pidgeon, N., and B. Fischhoff. 2011. The role of social and decision sciences in communicating uncertain climate risks. Nature Climate Change 1:35–41.

Rannow, S., N. A. Macgregor, J. Albrecht, H. Q. Crick, M. Förster, S. Heiland, G. Janauer, et al. 2014. Managing protected areas under climate change: challenges and priorities. Environmental Management 54:732–743.

Reid, W. V., H. A. Mooney, A. Cropper, D. Capistrano, S. R. Carpenter, K. Chopra, P. Dasgupta, et al. 2005. Ecosystems and human well-being: synthesis. Island Press, Washington, DC.

Renn, O. 2008. Risk governance: coping with uncertainty in a complex world. Earthscan, London, United Kingdom.

Renn, O., and P.-J. Schweitzer. 2009. Inclusive risk governance: concepts and application to environmental policy making. Environmental Policy and Governance 19:174–185.

Richardson, L., K. Keefe, C. Huber, L. Racevskis, G. Reynolds, S. Thourot, and I. Miller. 2014. Assessing the value of the Central Everglades Planning Project (CEPP) in Everglades restoration: an ecosystem service approach. Ecological Economics 107:366–377.

Rockwood, L. L., R. E. Stewart, and T. Dietz, eds. 2008. Foundations of environmental sustainability: the coevolution of science and policy. Oxford University Press, Oxford, United Kingdom.

Rosa, E. A. 1998. Metatheoretical foundations for post-normal risk. Journal of Risk Research 1:15–44.

Rosa, E. A., and L. Clarke. 2012. Collective hunch? Risk as the real and the elusive. Journal of Environmental Studies and Science 2:39–52.

Rosa, E. A., O. Renn, and A. M. McCright. 2013. The risk society revisited: social theory and governance. Temple University Press, Philadelphia, Pennsylvania.

Runte, A. 1997. National parks: the American experience. University of Nebraska Press, Lincoln, Nebraska.

Sabatier, P. A., and C. M. Weible. 2007. The advocacy coalition framework: innovation and clarification. Pages 189–222 in P. A. Sabatier, ed. Theories of the policy process. Westview Press, Boulder, Colorado.

Schwartz, M. H. 2005. Privatization: an overview—introduction and summary. The George Wright Forum 22:6–11.

Scheffer, M., S. Barrett, S. Carpenter, C. Folke, A. J. Green, M. Holmgren, T. Hughes, et al. 2015. Creating a safe operating space for iconic ecosystems. Science 347:1317–1319.

Schindler, D. E., and R. Hilborn. 2015. Prediction, precaution, and policy under global change. Science 347:953–954.

Sellars, R. W. 2009. Preserving nature in the national parks: a history: with a new preface and epilogue. Yale University Press, New Haven, Connecticut.

Shafer, C. L. 2014. From non-static vignettes to unprecedented change: the US National Park System, climate impacts and animal dispersal. Environmental Science & Policy 40:26–35.

———. 2015. Land use planning: a potential force for retaining habitat connectivity in the greater Yellowstone ecosystem and beyond. Global Ecology and Conservation 3:256–278.

Sharp, R. L., C. J. Lemieux, J. L. Thompson, and J. Dawson. 2014. Enhancing parks and protected area management in North America in an era of rapid climate change through integrated social science. Journal of Park and Recreation Administration 32:1–18.

Spence, M. D. 2000. Dispossessing the wilderness: Indian removal and the making of the national parks. Oxford University Press, Oxford, United Kingdom.

Steg, L., and J. I. M. de Groot. 2012. Environmental values. Pages 81–92 *in* S. Clayton, ed. The Oxford handbook of environmental and conservation psychology. Oxford University Press, New York, New York.

Stegner, W. 1948. Marking the sparrow's fall: the making of the American west. Henry Holt and Company, New York, New York.

Stein, B. A., A. Staudt, M. S. Cross, N. S. Dubois, C. Enquist, R. Griffis, L. J. Hansen, et al. 2013. Preparing for and managing change: climate adaptation for biodiversity and ecosystems. Frontiers in Ecology and the Environment 11:502–510.

Tolstoy, L. 1995. Anna Karenina. Translated by Louise and Aylmer Maude. Oxford University Press, Oxford, United Kingdom.

Tuler, S., and T. Webler. 1995. Process evaluation for discursive decision making in environmental and risk policy. Human Ecology Review 2:62–71.

VanEgeren, J. 2012. Brother wolf: Native Americans say upcoming wolf hunt is premature and disrespectful. The Cap Times, 25 June. Accessed 25 June 2015. http://host.madison.com/news/local/govt-and-politics/capitol-report/brother-wolf-native-americans-say-upcoming-wolf-hunt-is-premature/article_8e8f8bf6-fc61-11e1-94aa-001a4bcf887a.html.

Vucetich, J., R. O. Peterson, and M. P. Nelson. 2013. Response to Gostomski. The George Wright Forum 30:101–102.

Vucetich, J. A., M. P. Nelson, and R. O. Peterson. 2012. Should Isle Royale wolves be reintroduced? A case study on wilderness management in a changing world. The George Wright Forum 29:126–147.

Weible, C. M., and P. A. Sabatier. 2007. The advocacy coalition framework: innovations and clarifications. Pages 189–220 *in* P. A. Sabatier, ed. Theories of the policy process. Westview Press, Boulder, Colorado.

Worster, D. 2008. A passion for nature: the life of John Muir. Oxford University Press, New York, New York.

Wright, G. M., J. S. Dixon, and B. H. Thompson. 1933. Fauna of the national parks of the United States: a preliminary survey of faunal relations in national parks. Contribution of Wild Life Survey. Fauna Series No. 1—May 1932. US Government Printing Office, Washington, DC.

York, R. 2013. Metatheoretical foundations of post-normal prediction. Pages 19–29 *in* T. Dietz and A. K. Jorgenson, eds. Structural human ecology: new essays in risk, energy and sustainability. Washington State University Press, Pullman, Washington.

THIRTEEN

The World Is a Park: Using Citizen Science to Engage People in Parks and Build the Next Century of Global Stewards

JOHN FRANCIS, KELLY J. EASTERDAY, KELSEY J. SCHECKEL, AND STEVEN R. BEISSINGER

Introduction

Many of us live in a shrink-wrapped, online world. We purchase our groceries and gifts from Internet businesses and get our messages and news from social media. Gone are the days of hunting and gathering vegetables for our meals and being exposed to the elements that make other creatures thrive or die. Half of humanity now lives in urban areas (Population Reference Bureau 2014) but rarely connect with their neighbors, have only a vague idea of where their water and food come from, and cannot comprehend how much less biodiversity our planet will support in 2050 under the huge needs of a human population expected to exceed nine billion. Yet people treasure their pets, nurture their houseplants, and use their computer screen savers to remind them what wild landscapes look like. They watch videos of parks and may occasionally visit one. But do they understand the fabric that holds the glaciers to their peaks or that maintains corridors so pronghorn antelope (*Antilocapra americana*) can survive the passing seasons?

Thankfully some changes are afoot. Just as human technology touches every square meter of the planet—often with the toll of consumption and pollution—the same technology can connect people to nature, teaching them how to identify organisms, measure their attributes or needs, and determine how their numbers or distributions are changing on our watch. Smartphones record georeferenced observations or photos of a bird or plant; when multiplied by thousands of users, they present a visual record of migration or changes in phenology (Hurlbert and Liang 2012; Schwartz,

Betancourt, and Weltzin 2012; Supp et al. 2015). Science can now occur at the fingertips of millions of interested citizens, rather than reside solely in the hands of a relatively few, highly trained scientists. "Citizen science" offers the potential to foster emotional connections to nature while contributing to knowledge about nature (Dickinson et al. 2012). By participating in the collection of data and aspects of the scientific process, often in partnership with professional scientists, the general public can reawaken their delicate relationship with nature while empowering an engaged and contributing community.

This chapter examines the opportunities to connect people and nature through the growth of citizen science in parks. We begin with a brief discussion of the evolution of citizen science, its opportunities, and its challenges. We then examine the largest organized citizen science event taking place in parks, the BioBlitz, and particularly the collaborative effort between the US National Park Service (NPS) and the National Geographic Society (NGS) to create a new generation of global stewards. Next, we discuss the potential for connecting people to biodiversity by expanding the citizen science engagement model of biodiversity discovery used in BioBlitzes to venues beyond parks. We conclude by considering the challenges for scientists that work with citizens and with data generated by citizen scientists.

Parks are some of the best breeding grounds for enhancing the connection between science and human values. US national parks, with their goal to be conserved unimpaired for the enjoyment of future generations as specified in the Organic Act of 1916, may play a special role in catalyzing people to see themselves as one with nature. They are emblems of intact natural and cultural heritage, and act as laboratories where people can see how nature works and can visualize themselves as a part of that process. But other locations can also serve as an impetus to transfer this feeling across boundaries into unassuming landscapes closer to home. Using parks as models for scientific engagement and then following in the schoolyard and beyond, a movement can build to rediscover the planet and to rewild its people, ending with the notion that everyone has a backyard that is a park.

The Evolution of Citizen Science

More than a century before citizen science became a buzzword, citizens were involved in the collection of data for science. Of course, most science was conducted by amateurs before the professionalization of science in the late 19th century (Miller-Rushing, Primack, and Bonney 2012). With the growth of professional scientists over the past 150 years, however, contributions by nonprofessionals became marginalized. Nevertheless, disciplines such as meteorology, astronomy, and ornithology have a long history of contributions by citizens in North America. The first Secretary of the Smithsonian Institution, Joseph Henry, established the Meteorological Project in 1847 as a dispersed network of volunteers that tracked storm and weather patterns across North America. The network of volunteers grew quickly, and within a decade after establishment had a pool of over 600 volunteers spanning six countries.[1] Citizen involvement in the frontier of biodiversity assessment at the continental scale in North America probably began with the National Audubon Society's Christmas Bird Count in the early 20th century and was accelerated by the North American Breeding Bird Survey that was initiated in 1966 (Miller-Rushing, Primack, and Bonney 2012). In US national parks, citizen scientists have long been contributing their wildlife and botanical observations, writing their sightings in logbooks maintained at visitor centers and ranger stations. Even President Theodore Roosevelt in the early 20th century recorded observations of birds in the back and front yards of the White House, now an NPS unit.

Technological advances have fundamentally changed the engagement of citizens in science over the past 30 years by increasing the ability to reach larger and more diverse pools of volunteers and to quickly and easily amass the data they collect. First and foremost, the Internet and the World Wide Web expanded the human reach beginning in the 1990s but accelerating in the early 21st century. Professional scientists no longer had to rely on their known networks of colleagues, friends, and family, and the slow pace set by publication and access to libraries, but could access a pool of contributors that transcended traditional boundaries. By tapping into the hobbies and passions of nonprofessional scientists, professional scientists could offer to citizens more ways of engaging and participating in scientific endeavors than were previously obtainable or imaginable. For the scientist, the Internet also served as a portal that greatly increased project visibility, functionality, and accessibility (Bonney et al. 2014). Nearly as important as the Internet, however, was the development of handheld devices, such as smartphones with built-in global positioning system (GPS) capabilities and high-definition cameras, and the growth of personal sensor technol-

1. Citizen science at the Smithsonian, accessed 20 March 2016, www.si.edu/content/governance/pdf/Archives_06-2011.pdf.

ogy that now allows the nonprofessional to participate in real-time data collection (Dickinson et al. 2012; Newman et al. 2012). This, in turn, has fueled the growth of mobile phone apps for nature observation that facilitate the identification and/or reporting of biodiversity observations. Thus, the combination of handheld and other small monitoring devices that could be connected to the Internet opened possibilities for citizens to collect and submit data quickly and efficiently.

The widening pool of potential engagement that has resulted from the untapped resource of labor, the expanding scale of scientific inquiry, and the increasing use of technology has led to the establishment of a unique partnership between scientists and volunteers to document, observe, and discover the world in a much richer and more expansive scale than ever before. Rising on the foundation of citizen science efforts that began over a century ago, the newly established Citizen Science Association (http://citizenscienceassociation.org/) amassed 1,000 new members within the first five days (Citizen Science Association, pers. comm.), demonstrating the appetite for a growing commitment to the potential force of citizen engagement. As a result of this expanding scale of human reach, scientific studies also began to expand in scale, tracking the distributions, migrations, and phenologies of species at continental scales (Silvertown 2009; Hochachka et al. 2012; Ries and Oberhauser 2015).

The growth of citizens as contributors to science can also be seen in the scientific literature. We documented the growth of citizen contributions to science by searching Scopus (http://www.elsevier.com/solutions/scopus), the largest abstract and citation database of peer-reviewed literature, on 17 September 2015 using the keywords "citizen science," "volunteer monitoring," "community science," "civic science," and "volunteered geographic information" in combination to obtain a diverse sample of papers with citizen involvement. We limited our search to scientific articles and reviews. Scientific papers started appearing sporadically in the 1970s (four total) and 1980s (three total), although our search missed some prior contributions that analyzed data generated by early citizen science programs like the Christmas Bird Count that were not captured by these keywords (e.g., Davis 1950; Preston 1958). Publications that involved citizen science grew rapidly over the past decade as Internet usage expanded, and accelerated greatly over the last five years as smartphones and personal sensor technology penetrated markets (fig. 13.1). Thus, the involvement of citizens in data generation is growing, as is the number of scientific products resulting from their efforts.

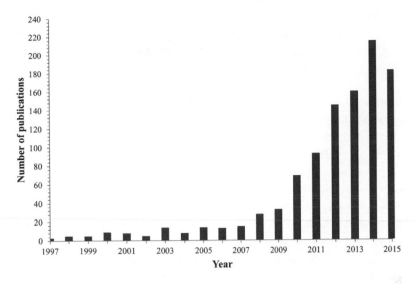

13.1. Number of scientific publications produced annually based on a search conducted on 5 September 2015 in the online database Scopus using the following combination of keywords: "citizen science," "volunteer monitoring," "community science," "civic science," and "volunteered geographic information."

The major goals of citizen science efforts in conservation are to engage people with science for stewardship in natural settings and to reinvigorate an appreciation of nature, not only in remote national parks, but often through the study of "ordinary nature" that is easily accessible and usually composed of common species and habitat types (Devictor, Whittaker, and Beltrame 2010; Wiggins and Crowston 2011). Here citizen science can make important contributions to the growing field of urban ecology and coupled natural-human systems research (DeFries, this volume, ch. 11). Moreover, citizen contributions allow access to private property, which is often inaccessible to researchers. For example, in California 45% of the forests and woodlands that cover nearly one-third of the state are privately owned (Fire and Resource Assessment Program 2010). The collection of information made possible by citizen observations during everyday activities in their private domains produces immediate input for scientific inquiry. Therefore, partnerships between scientists and citizens may expand the capture of ecological phenomena, regardless of jurisdictional boundaries (public or private). If everyone's backyard is a park, then every backyard has the potential to yield useful, scientific data.

The BioBlitz: Engaging People with Nature through Biodiversity Discovery in Parks

What Is a BioBlitz?

Citizen science has grown within the NPS over the past 20 years through intensive field studies based on biodiversity discovery often commonly called BioBlitzes (Plumb et al. 2014). A BioBlitz attempts to inventory all species in a defined area, often a park or protected area, typically within a 24-hour period. BioBlitzes are designed now to unite scientists with families, students, and the general public to explore and study natural systems in areas ranging from neighborhood parks and schoolyards to state and federal lands. The direct engagement of scientists with the public that characterizes a BioBlitz distinguishes it from many "virtual citizen science" programs in which citizen scientists report their observations through the Internet without direct interactions with scientists (Wiggins and Crowston 2011; Reed, Rodriguez, and Rickhoff 2012). A BioBlitz typically has three main goals: (1) promote greater knowledge, understanding, and appreciation of the incredible biodiversity present in parks; (2) provide park management with accurate, comprehensive, and current biodiversity data to help determine the health of the park's ecosystems and known species; and (3) inspire young people to become stewards of their region's natural resources and leaders in the care and preservation of biodiversity.

The first BioBlitz took place in 1996 at Kenilworth Park and Aquatic Gardens, an NPS unit located in Washington, DC, along the Anacostia River. It was organized by Sam Droege of the US Geological Survey and Dan Roddy of the NPS, and attended primarily by local scientists and naturalists. Few species were expected to be encountered because Kenilworth is a small park surrounded by residential and industrial development, but more than 900 species were tallied.[2] This event was followed in 1997 by an All Taxa Biodiversity Inventory in the Great Smoky Mountains National Park (Sharkey 2001; Plumb et al. 2014). It grew from the vision of Daniel Janzen to enact a similar inventory in Guanacaste National Park in Costa Rica with the Costa Rican National Institute for Biodiversity (INBio), but for financial and political reasons the effort evolved into a survey of selected taxa over five conservation areas (Gámez-Lobo et al. 1997; Kaiser

---

2. USGS Patuxent Wildlife Research Center, Species list: (May 31–June 1, 1996) Kenilworth Park and Aquatic Gardens—BioBlitz, accessed 20 March 2016, http://www.pwrc.usgs.gov/blitz/species.html.

1997). The All Taxa Biodiversity Inventory is an ongoing collaborative effort between the NPS and Discover Life in America (http://www.dlia.org/), a small nonprofit organization based in the Great Smoky Mountains, to discover every species within the park though the efforts of professional and citizen scientists. As of 19 September 2015, it has tallied 18,545 species in the park including 951 species new to science and 8,095 new to the park.

Over the past two decades there has been rapid growth of biological diversity discovery and monitoring work in US national parks that involve the public in BioBlitz-like activities (fig. 13.2). These activities often focused on understudied taxa, such as insects and other groups of invertebrates, fungi, and nonvascular plants, but have also included blitzes for birds, bats, and mammals, and even paleoblitzes to discover new fossils (Selleck 2014). Over 30 BioBlitz-like activities occur per year in national parks (fig. 13.2), and they have been conducted in over 119 park units. These range from small, one-time activities to annual coordinated surveys. Another 100 BioBlitz-like activities are likely to occur in US national parks with the centennial celebration of the NPS in 2016 (Selleck 2014).

In 2004, John Francis was appointed to the NPS Advisory Board as the Science Subcommittee chair. In 2006, Francis initiated a series of BioBlitzes in advance of the NPS centennial in 2016 that would combine the com-

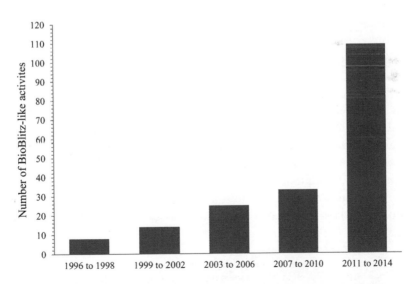

13.2. Growth in the number of BioBlitz-like activities taking place in US national parks over the past two decades. Data and figure modified from Selleck (2014).

munication and networking strengths of the NGS with the growing interest of the NPS in increasing awareness of biodiversity in the parks by engaging scientists with the public (table 13.1). The BioBlitz locations, determined in consultation with NPS associate directors Mike Soukup, Bert Frost, and Raymond Sauvajot and with the NPS Biological Resource Management Division's Elaine Leslie, were selected to represent the breadth of NPS unit types, regions, and habitats. There was also a desire to maximize participation of audiences from urban communities within a one-hour drive of national parks. Timing of these events was oriented toward spring or fall, so as to be during the school year to enable classroom engagement. At each site, the number of species identified and number of new species added to the park lists, the number of participants, and the coverage of the activities in the parks has grown considerably. These BioBlitzes were often combined with festivals that included music, arts, games, and community exhibits, where local stakeholders had the opportunity to build an audience for conservation activities that unite protected areas with the surrounding community.

The BioBlitz idea has also caught on in parks outside the United States. Countries that have implemented BioBlitzes in parks or similar sites include Australia, Canada, Ireland, Israel, Malaysia, New Zealand, Portugal, Spain, Sweden, Taiwan, Trinidad and Tobago, and the United Kingdom.[3] At the 2014 World Parks Congress in Sydney, a BioBlitz was held on the conference grounds,[4] taking advantage of the surrounding restored wetlands and the opportunity for the public and experts alike to study and celebrate protected areas, the focus of the congress.

While a major goal of a BioBlitz is to engage citizens with biodiversity discovery, species that are new to park inventories are often discovered and occasionally species that are new to science have been identified. There was a new bee discovered in Santa Monica Mountains National Recreation Area (Gibbs 2010) and a new tardigrade in Biscayne National Park (Miller, Clark, and Miller 2012). More commonly, a BioBlitz results in adding tens to hundreds of new species to park inventory lists of biodiversity (see table 13.1). An additional benefit has been to increase the engagement of scientists with national parks following the BioBlitzes. For example, new studies and a new grant-making program supporting research projects

3. *Wikipedia*, s.v. BioBlitz, accessed 20 March 2016, https://en.wikipedia.org/wiki/BioBlitz.

4. New South Wales Office of Environment and Heritage, BioBlitz: World Parks Congress BioBlitz 2014, last updated 21 July 2015, accessed 20 March 206, http://www.environment.nsw.gov.au/research/bioblitz.htm.

Table 13.1 BioBlitz activities conducted in US national parks as a partnership between the US National Park Service and the National Geographic Society

Park	Location	Date	Species observed	Student participants	Total participants	First park records for
Rock Creek Park	Washington, DC	2007	661	NA	NA	NA
Santa Monica Mountains National Recreation Area	Los Angeles, CA	2008	1,361	1,000	NA	NA
Indiana Dunes National Lakeshore	Outside Chicago, IL	2009	1,700	2,000	NA	20 types of rove beetles and a few tardigrades
Biscayne National Park	Miami, FL	2010	1,027	1,300	2,500	11 species of lichen and 22 species of ants
Saguaro National Park	Tucson, AZ	2011	859	2,000	5,500	400 species, with species new to science
Rocky Mountain National Park	Outside Denver, CO	2012	490	2,000	5,000	1 lizard, 9 insect, and 13 nonvascular plant species
Jean Lafitte National Historical Park and Preserve	New Orleans, LA	2013	570	1,500	3,000	Several new invasive insect species
Golden Gate National Recreation Area	San Francisco, CA	2014	2,505	2,700	9,000	80 species; sightings of 15 endangered species
Hawai'i Volcanoes National Park	Volcano, HI	2015	416	850	6,000	60 new species of fungi

Note: NA = not available.

were stimulated by the events in Santa Monica National Park (R. Sauvajot, pers. comm.).

BioBlitz Engagement, Education, and Outreach through Biodiversity Discovery

BioBlitzes are designed to connect youth, especially those in urban landscapes, to the wonders of nature. The events can include online lesson plans and activities,[5] and often have been paired with a full day of face-to-face professional development with NGS and NPS staff. Skills such as observing, taking field notes, identifying species, and mapping observations are included in both classroom and field sessions (fig. 13.3). The wealth of experiences has recently been enhanced with videos illustrating how to conduct such events, which can include backyard or schoolyard BioBlitzes. For many young participants, a BioBlitz marks their first visit ever to a national park.

Stories of inspiration and illumination abound with each event. One class session at Malibu Lagoon in the Santa Monica Mountains Recreation Area was punctuated by a cheer of "Thank you!" from more than a dozen students from the Los Angeles Unified School District, half of whom were said to have never set foot in the ocean! A giant desert centipede racing across a trail at Saguaro National Park on a night walk became a teachable moment, as an entomologist from a nearby light-trapping station captured the amazing six-inch long creature for a class of 20 students. He described the difference between a sting and a bite and his own encounter with this species, which could give "quite a pop." The students' eyes lit up with this expert tale and saw, with new respect, the animal scurry off into the desert darkness.

Education at BioBlitzes occurs at all levels. Scientists and naturalists volunteer to take groups of about 10–20 adults and children on predetermined taxonomic surveys (e.g., of birds, plants, insects, and fungi) for two to four hours. Having multiple eyes at work on the survey increases the detection of species and also yields interpretive moments in which the deep expertise of the scientist can be shared with those along for the experience. Scientists become aware of new study opportunities, and the public becomes energized with the excitement that can unfold in a park setting and as part of a tangible contribution to research.

5. E.g., see http://education.nationalgeographic.com/education/programs/bioblitz/?ar_a=1 (accessed 21 March 2016).

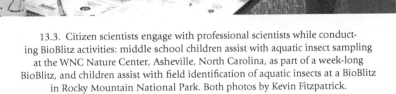

13.3. Citizen scientists engage with professional scientists while conducting BioBlitz activities: middle school children assist with aquatic insect sampling
at the WNC Nature Center, Asheville, North Carolina, as part of a week-long
BioBlitz, and children assist with field identification of aquatic insects at a BioBlitz
in Rocky Mountain National Park. Both photos by Kevin Fitzpatrick.

Outreach on biodiversity appreciation is built into each BioBlitz. Adding to the experience of participating in an inventory with as many as 150–300 scientists, Biodiversity Festivals attract 50–100 exhibits associated with BioBlitzes to offer a way for representatives of the local community, such as nongovernmental organizations and government agencies, to build constituencies and to provide engaging activities that can extend in impact beyond the event. Attendees explore interactive exhibitor booths that feature environmental organizations, wildlife groups, health organizations, and demonstrations of cutting-edge science and technology. In Biscayne National Park, the festival added a Biodiversity University that allowed children of all ages to earn stamps toward a degree by completing a number of hands-on exercises at various stations. In addition, the Youth Ambassador Program provided leadership opportunities for young people interested in inspiring their peers and communities to be stewards of the environment. Attendance during the first eight years reached a pinnacle with 9,000 participants at the Golden Gate National Recreation Area BioBlitz during a rainy weekend in 2014.

Coverage of BioBlitzes online and in print media has been substantial and has extended their influence well beyond the event attendees. On the NGS website alone over 100,000 visitors explored the pages and read blogs from seven BioBlitzes. An additional virtual field trip conducted in association with the 2011 BioBlitz at Saguaro National Park enrolled more than 100,000 students and allowed classrooms around the country to observe a BioBlitz in progress. The website enjoys continued viewership as a legacy of this one event alone. Local news coverage is a regular opportunity. For the BioBlitz in San Francisco's Golden Gate National Recreation Area, coverage on the nationally televised program *Good Morning America* drew the attention of a national audience, as should the upcoming 2016 BioBlitz headquartered in Washington, DC.

The World Is a Park: Extending the BioBlitz Engagement Model

Perhaps nowhere is there more potential for connecting people to biodiversity through citizen science than in the schoolyard, for which the NGS recently adapted the BioBlitz model of engagement through discovery.[6] Although field trips are often part of the education curricula in many schools,

6. See http://education.nationalgeographic.com/media/schoolyard-bioblitz/ (accessed 21 March 2016).

it can be expensive and challenging for school districts to send students on weeklong trips or even day visits to outdoor education centers for instruction. Instead, looking to the schoolyard as the new wild can model biodiversity discovery and natural resource awareness without needing a bus or a special venue.

Students and teachers can lead a class- or school-wide annual biodiversity inventory or BioBlitz on their school grounds and use the event to assess environmental health. Students can compare their discoveries with results from previous years, and ask questions about how and why certain insects and plants change from year to year in abundance and phenology. Camera traps as well as small-mammal traps can be used to look for other creatures. Students can participate in habitat improvement by planting bird- or butterfly-friendly plants and enriching environments to attract a greater diversity of natural flora and fauna. As a result of repeating surveys and additional training, teachers can become local experts and elevate public land into examples of natural ecological communities to appreciate, creating thousands of versions of Michele Obama's pollinator garden on the White House South Lawn.

Bringing nature to kids in this way can have lasting impact. Juan Martinez's inspiring life story is a great example.[7] Relegated to Eco Club at Dorsey High School in South Central Los Angeles, he found an alternative to gang membership, first growing chili peppers in the schoolyard and later going on a camping trip to Grand Teton National Park. After graduating high school, Martinez formed the Natural Leaders Network, became a key spokesperson for the Children in Nature Network, and was recognized as an NGS Emerging Explorer. The antidote to what Richard Louv (2005) called "Nature Deficit Disorder" in his popular book is "an easy fix," Martinez said. "All you have to do is get outside."

Once people become engaged in discovering nature in parks and other sites, it's a short step to recognizing the beauty of their own backyard and the need to nurture and protect it. Discussion is building about the idea of urban protected areas (Tyrzna et al. 2014), including the concept of turning the city of London into a national park.[8] This project, championed by NGS Emerging Explorer Daniel Raven Ellison, recently gained the support of the London Assembly (Kirk 2015). Even more noteworthy is the slow but steady process of turning the Federated States of Micronesia into a world park (Nobel 2010).

7. See http://ed.ted.com/on/UZKehzbt (accessed 21 March 2016).
8. See http://www.greaterlondonnationalpark.org.uk/about/ (accessed 21 March 2016).

International citizen science efforts devoted to biodiversity discovery are increasingly possible with web-served data collection programs. The Global Snapshot of Biodiversity, which is part of the NGS Great Nature Project, set a Guinness world record in 2014 with the largest online photo album of animals (more than 104,000 images).[9] The effort employed numerous social networks as well as the increasingly popular app iNaturalist that allowed sightings to become research-grade observations within the Global Biodiversity Information Facility, which requires confirmed identification by two individuals. The Global Snapshot of Biodiversity from 15 to 25 May 2015 engaged participants to go outside to take photos of their encounters with plants, animals, and fungi, and upload them online as part of a global snapshot of biodiversity.[10] It attracted participants from 102 countries who recorded more than 40,000 observations of 8,000+ species of plants and animals.

Challenges for Scientists Working with Citizens and Citizen Data

By developing connections between people and nature, especially by engaging broad and global audiences with professional scientists in the discovery and assessment of biodiversity, citizens become increasingly aware of the importance of biodiversity and natural ecosystems. Moreover, citizens can also produce large volumes of data collected at large geographic scales that cannot be attained by professional investigators individually or in teams. Both the large volume of data and the large scale of study create challenges for professional scientists working with citizen science.

The BioBlitz and the rapidly developing technologies that are connecting scientists directly with citizens in the quest to understand nature require scientists to be effective communicators. Encouragement from federal funding agencies for scientists to engage in both public education and outreach is a relatively new form of academic contribution that expands the traditional "publish or perish" ethos. Since 1997, the National Science Foundation (NSF) has judged grant applications in part based on their potential "broader impacts" in addition to their intellectual merit. Increasing attention and training in social media and popular press writing has allowed academics to engage with a broader audience, which has also been advanced by the NSF.[11] Books like *Don't Be Such a Scientist* (Olson 2009)

9. See http://greatnatureproject.org/photos/guinness (accessed 21 March 2016).

10. See http://greatnatureproject.org/events/global-snapshot-2015/ (accessed 21 March 2016).

11. E.g., see http://www.nsf.gov/discoveries/disc_summ.jsp?cntn_id=114406 (accessed 21 March 2016).

have prodded scientists to adopt new ways of communicating that use less jargon and resemble storytelling so that their research may engage larger audiences. Scientists that develop these skills provide good models for the growing armies of citizens engaged in large-scale collection of data and analysis.

Citizen science and big data in general are challenging the norms of what skills a scientist requires. Currently, there is recognition that the skill sets specific to traditional academic disciplines, such as geography, earth science, biology, and ecology, are inadequate to handle the large volumes of data that citizen science can produce (Lin 2013; Marx 2013). Good management and analysis of citizen science data requires a transdisciplinary approach and the development of skills outside traditional scientific disciplines (Hampton et al. 2013). There is an increasing need for researchers to expand their skill sets to include computer science and data science to facilitate synthesis, analysis, and visualization (Peters 2010; Hampton and Parker 2011). The demand for data science professionals and spatial data science professionals, both recently minted titles, is much greater than the supply available in public, private, and academic sectors. Universities and private corporations have acknowledged this deficit, developing new classroom and online courses. The push to gain inference from large, dispersed, and often messy data sets and to communicate their meaning has led to advancements in data mining, data cleaning, and analytical techniques. In addition, one of the most promising recent advancements in communicating science to the general public has been the generation of visualization tools that aim to make data more interactive and accessible to broad audiences.[12] A visualization component creates new ways of communicating science writ large and expands the potential for engaging broad audiences.

Despite the great potential for citizens to contribute large amounts of data and insights to scientific research, issues of data quality often arise. Challenges related to the use of biodiversity occurrence data collected by citizen scientists include observer variability, bias, detection probability, and misrepresentation of data, although all of these can also occur in data generated by professional scientists. Just like scientists, citizens vary in their skill sets and abilities to perform tasks such as species identification, data collection, and data analysis (Dickinson, Zuckerberg, and Bonter 2010). Such variation can lead to biased results, which could affect the integrity of a specific study, depending on its scientific goals (Newman et al. 2012).

12. E.g., see http://nsf.gov/news/special_reports/scivis/dates.jsp (accessed 21 March 2016).

Adequate training at the beginning of a study or a BioBlitz to familiarize participants with the data collection protocols and techniques can reduce variability among participants, increase the quality of data collected, and increase the scientific literacy of the public (Bonney et al. 2009; Dickinson, Zuckerberg, and Bonter 2010). Nevertheless, issues of sampling bias, including the tendency to overreport some species or phenomena and to underreport others, often remain with data collected opportunistically by citizens. Recent advances in statistical methods have developed ways to represent these types of biases yet still reap the potential of the data (Isaac et al. 2014). Additional advances are sure to emerge.

Conclusions

The growth of BioBlitz events around the world has provided an opportunity for citizens to record the everyday elements of nature they observe and wish to understand. The delight of identifying and tracking organisms is extending from parks to schoolyards and backyards, and will become more commonplace for the global citizen. Questions like "what kind of wasp is that?" will be answered with ever more ease online. The occurrence of range extensions of species, the arrival of invasive species, and even occasionally the discovery of new species will be the result of an uploaded photo identified by experts and, perhaps soon, by expert online computer systems as algorithms and reference photo libraries expand. As the inventory mounts, so too will our ability to analyze and map Earth's biodiversity and its shifts in the coming decades that will accompany climate change and other environmental threats. By monitoring the health of the natural world, global citizens often become aware of the delicate relationship they have with nature. Through citizen participation, programs like the BioBlitz help grow the vision of shared responsibility for our planet. Moreover, as threats to biological diversity mount and hazards to the integrity of parks build on their peripheries (Wilson, this volume, ch. 1; Defries, this volume, ch. 11), political processes that adjudicate their fates will depend on balancing values. People who value biodiversity and parks from an emotional connection they developed while participating in citizen science are likely to vote in support of those values. Biodiversity discovery leads to engaged stewardship.

A BioBlitz can be costly and time consuming, especially if the objective is to draw large crowds and prepare teachers and local institutions extensively in advance of an event. Nevertheless, the amount of online guidance to support BioBlitz events, such as curricula and online video tutorials, has grown greatly over the years. As approaches improve, there is an increasing need for evaluation of new approaches in inventory, public participation, and relative impact. The most important results of BioBlitzes may be measured in decades when a young explorer becomes the next leading taxonomist or forceful member of our global conservation community.

In conclusion, our ubiquitous use in the 21st century of cell phones, cameras, and sensing devices that constantly record location and other ambient variables has increased the ease of collecting data and lowered the barrier for people from all backgrounds to contribute to citizen science. The same technology that is sometimes blamed for deteriorating our connection with nature may simply be reinventing it, and if done well, a more engaged and active outdoor community should arise.

Literature Cited

Bonney, R., C. B. Cooper, J. Dickinson, S. Kelling, T. Phillips, K. V. Rosenberg, and J. Shirk. 2009. Citizen science: a developing tool for expanding science knowledge and scientific literacy. BioScience 59:977–984.

Bonney, R., J. L. Shirk, T. B. Phillips, A. Wiggins, H. L. Ballard, A. J. Miller-Rushing, and J. K. Parrish. 2014. Next steps for citizen science. Science 343:1436–1437.

Davis, D. E. 1950. The growth of starling, Sturnus vulgaris, populations. Auk 67:460–465.

Devictor, V., R. J. Whittaker, and C. Beltrame. 2010. Beyond scarcity: citizen science programmes as useful tools for conservation biogeography. Diversity and Distributions 16:354–362.

Dickinson, J. L., J. Shirk, D. Bonter, R. Bonney, R. L Crain, J. Martin, T. Phillips, and K. Purcell 2012. The current state of citizen science as a tool for ecological research and public engagement. Frontiers in Ecology and the Environment 10:291–297.

Dickinson, J. L., B. Zuckerberg, and D. N. Bonter. 2010. Citizen science as an ecological research tool: challenges and benefits. Annual Review of Ecology, Evolution, and Systematics 41:149–172.

Fire and Resource Assessment Program. 2010. California's forests and rangelands: 2010 assessment. Fire and Resource Assessment Program, California Department of Forestry and Fire Protection, Sacramento, California.

Gámez-Lobo, R., T. Lovejoy, R. Solórzano, and D. Janzen. 1997. Costa Rican all-taxa survey. Science 277:17–21.

Gibbs, J. 2010. Revision of the metallic species of Lasioglossum (Dialictus) in Canada (Hymenoptera, Halictidae, Halictini). Zootaxa 2591:1–382.

Hampton, S. E., and J. N. Parker. 2011. Collaboration and productivity in scientific synthesis. BioScience 61:900–910.

Hampton, S. E., C. A. Strasser, J. J. Tewksbury, W. K. Gram, A. E. Budden, A. L. Batcheller, C. S. Duke, and J. H. Porter. 2013. Big data and the future of ecology. Frontiers in Ecology and the Environment 11:156–162.

Hochachka, W. M., D. Fink, R. A. Hutchinson, D. Sheldon, W.-K. Wong, and S. Kelling.

2012. Data-intensive science applied to broad-scale citizen science. Trends in Ecology & Evolution 27:130–137.

Hurlbert, A. H., and Z. Liang. 2012. Spatiotemporal variation in avian migration phenology: citizen science reveals effects of climate change. PLoS ONE 7:e31662.

Isaac, N. J. B., A. J. van Strien, T. A. August, M. P. de Zeeuw, and D. B. Roy. 2014. Statistics for citizen science: extracting signals of change from noisy ecological data. Methods in Ecology and Evolution 5:1052–1060.

Kaiser, J. 1997. Unique, all-taxa survey in Costa Rica "self-destructs." Science 276:893.

Kirk, A. 2015. Plans to turn London into a national park get a boost from the London Assembly. City A.M., 4 June. Accessed 21 March 2016. http://www.cityam.com/217204/plans-turn-london-national-park-get-boost-london-assembly.

Lin, T. 2013. Imagining data without division. Quanta Magazine, 30 September. Accessed 30 November 2015. https://www.quantamagazine.org/20130930-imagining-data-without-division/.

Louv, R. 2005. Last child in the woods. Algonquin Books, Chapel Hill, North Carolina.

Marx, V. 2013. Biology: the big challenges of big data. Nature 498:255–260.

Miller, W. R., T. Clark, and C. Miller. 2012. Tardigrades of North America: *Archechiniscus biscaynei*, nov. sp. (Arthrotardigrada: Archechiniscidae), a marine tardigrade from Biscayne National Park, Florida. Southeastern Naturalist 11:279–286.

Miller-Rushing, A., R. Primack, and R. Bonney. 2012. The history of public participation in ecological research. Frontiers in Ecology and the Environment 10:285–290.

Newman, G., A. Wiggins, A. Crall, E. Graham, S. Newman, and K. Crowston. 2012. The future of citizen science: emerging technologies and shifting paradigms. Frontiers in Ecology and the Environment 10:298–304.

Nobel, J. 2010. Will Micronesia be the first nationwide park? Time Magazine, 10 February. Accessed 30 November 2015. http://content.time.com/time/world/article/0,8599,1959020,00.html.

Olson, R. 2009. Don't be such a scientist. Island Press, Washington, DC.

Peters, D. P. C. 2010. Accessible ecology: synthesis of the long, deep, and broad. Trends in Ecology & Evolution 25:592–601.

Plumb, G., E. O. Wilson, S. Plumb, and P. Ehrlich. 2014. Biodiversity and national parks: what's relevance got to do with it? Park Science 31:14–16.

Population Reference Bureau. 2014. World population data sheet. Population Reference Bureau, Washington, DC. http://www.prb.org/pdf14/2014-world-population-data-sheet_eng.pdf.

Preston, F. W. 1958. Analysis of the Audubon Christmas counts in terms of the lognormal curve. Ecology 39:620–624.

Reed, J., W. Rodriguez, and A. Rickhoff. 2012. A framework for defining and describing key design features of virtual citizen science projects. Pages 623–625 *in* Proceedings of the 2012 iConference. 7–10 February 2012, Toronto, Ontario, Canada.

Ries, L., and K. Oberhauser. 2015. A citizen army for science: quantifying the contributions of citizen scientists to our understanding of monarch butterfly biology. BioScience 65:419–430.

Schwartz, M. D., J. L. Betancourt, and J. F. Weltzin. 2012. From Caprio's lilacs to the USA National Phenology Network. Frontiers in Ecology and the Environment 10:324–327.

Selleck, J. 2014. National parks and biodiversity discovery. Park Science 31:11–13.

Sharkey, M. J. 2001. The all taxa biological inventory of the Great Smoky Mountains National Park. Florida Entomologist 84:556–564.

Silvertown, J. 2009. A new dawn for citizen science. Trends in Ecology & Evolution 24:467–471.

Supp, S. R., F. A. La Sorte, T. A. Cormier, M. C. W. Lim, D. R. Powers, S. M. Wethington, S. Goetz, and C. H. Graham. 2015. Citizen-science data provides new insight into annual and seasonal variation in migration patterns. Ecosphere 6:art15.

Trzyna, T. 2014. Urban protected areas: profiles and best practice guidelines. Best practice protected area guidelines series no. 22. IUCN, Gland, Switzerland

Wiggins, A., and K. Crowston. 2011. From conservation to crowdsourcing: a typology of citizen science. Pages 1–10 in Proceedings of the 44th Hawai'i International Conference on System Sciences. 4–7 January 2011, Koloa, Hawai'i.

FOURTEEN

The Spiritual and Cultural Significance of Nature: Inspiring Connections between People and Parks

EDWIN BERNBAUM

Introduction

To gain the lasting support of the general public as well as local communities, parks need to ground their interpretation, management, and conservation programs in not only solid scientific research and practice, but also deeply held spiritual, cultural, and aesthetic values and ideas that will engage and inspire people to care for nature and, when necessary, to make sacrifices to protect the environment. Without this kind of enthusiastic and enduring support, no matter how good the science, parks, in particular national parks, will lose the special place they hold in the public imagination, and elected officials will reduce the funding needed for their adequate operation and for their very existence. This occurred recently when the state of California proposed closing a large number of state parks when faced with a major governmental budget deficit. It was only the outcry of the general public, and the actions of organizations representing their interests, that prevented many park closures, which would have had disastrous consequences for the environments and infrastructures of the affected parks (Dolesh 2012).

A key threat to continued public support of parks is their limited visitor base. Most visitors to US national parks are middle-class white Americans and foreign tourists. Relatively few of the so-called minorities—African Americans, Hispanic Americans, and Asian Americans—visit US national parks (Floyd 2001). With demographic change occurring in the United States, these ethnic groups are rapidly gaining political and economic influence. In California, minorities taken as a whole are now the majority of the population in most metropolitan areas (Armendariz 2011). If parks are not engaging minorities, they will not develop an interest in supporting parks, and the future of the US national parks and other park systems will not be ensured. In addition, if members of these ethnic groups have no experience of nature in parks, environmental conservation in general will be threatened.

Many parks, especially US national parks, have sites that are sacred or have other special significance for Native Americans, Native Hawaiians, and Native Alaskans. Rather than interfere with traditional practices at these sites, park managers need to welcome and involve indigenous peoples in interpretation and management as key stakeholders. Having lived in and interacted with the environments of parks for centuries and millennia, indigenous peoples have knowledge and experience that can contribute greatly to conservation and scientific research. Park managers need to acknowledge and respect their values, traditions, ideas, and ancestral ties to the land. Park managers need to work with indigenous cultures to develop their support—for example, through programs of comanagement that benefit them as well as the parks (Leaman 2013).

The task of engaging people in parks, in particular national parks, faces, therefore, three major challenges, which can be framed in the form of three questions: (1) How to attract people and build deep-seated, long-lasting support for parks and conservation? (2) How to broaden the limited visitor base of parks to ensure the future of the US National Park Service? And (3) how to respect and engage Native Americans, Native Hawaiians, and Native Alaskans for whom parks contain sacred and other cultural sites of special significance?

The spiritual and cultural significance of nature has a key role to play in helping address the first challenge by inspiring connections between people and parks. By "spiritual and cultural significance of nature," I mean the inspirational, spiritual, cultural, aesthetic, historic, social, and other meanings, values, ideas, and associations that natural features, ranging from mountains and rivers to forests and wildlife, evoke for people. I have chosen the word "significance" rather than "values" to emphasize the inclusion of knowledge and meaning as well as feelings and values—an important point related to the conclusion of this chapter. It's also important to note that the expression "spiritual and cultural significance of nature" refers to nature in its broadest sense, not just sacred natural sites, although it includes the latter.

In 2001, the National Parks Conservation Association convened focus groups "to identify the single most compelling idea that people find to be a motivating message regarding the plight of the parks" (Wirthlin Worldwide 2001). Respondents considered the following seven concepts:

"Our National Parks provide us with some of the most beautiful, majestic
and awe-inspiring places on Earth."
"Our National Parks are the legacy we leave our children."
"Our National Parks provide a home for America's diverse populations of
plants and animals."
"Our National Parks provide an educational experience unlike any other."
"Our National Parks are the pride of America's history and heritage."
"Our National Parks provide us with opportunities for personal challenge,
adventure, fun and pleasure."
"Our National Parks are a symbol of democracy and the American way."

The concept with the greatest potential for developing a compelling
message for support of national parks was the first one—"our National
Parks provide us with some of the most beautiful, majestic and awe-
inspiring places on Earth." This concept succeeded on many levels. In
particular, it indicates people have primarily emotional and spiritual connections to parks, rather than rational or intellectual ones (Wirthlin Worldwide 2001). I would add that it also shows an *experiential* connection between people and parks.

The report included quotes from respondents that point to the importance of the spiritual and cultural significance of nature in inspiring connections between people and parks:

"I can envision my last trip. It makes me want to experience it again."
"It's visual, poetic, it places you there."
"It's the feeling I have when I'm sitting in the park. I want to preserve it. It's
more personal."
"It hits you on an emotional level, pulls at your heart strings."
"It gave me a vision of peace and tranquility. . . . How could you possibly
lose that?"
"It's not someplace far away. . . . It's someplace within your spirit."
"It paints the picture of what the parks should be about, not adventure, but
peace and tranquility . . . a spiritual thing."
"I think of the scenery, it's so overwhelming, so majestic, it's hard to
describe."

Nature has deep spiritual and cultural significance in cultures around
the world that can help address the second challenge of diversifying the
limited visitor base of parks. People throughout Latin America look to

mountains as sacred sources of water and healing (Bernbaum 1997). The graceful cone of Mount Fuji has come to symbolize the country of Japan and the quest for beauty and perfection that lies at the heart of Japanese culture (Bernbaum 1997). The sacredness of trees in cultures as diverse as those of India and Ghana has inspired people to maintain the biodiversity of sacred groves around the world (Barrow 2010). *Shanshui*, the term for landscapes and landscape painting in China, means "mountains and rivers," pointing to the importance of these two basic components of nature in Chinese art as well as life (Bernbaum 1997).

By highlighting the spiritual and cultural significance of nature in cultures around the world, programs of outreach and interpretation can establish links with the cultural backgrounds of diverse ethnic groups, such as African Americans, Hispanic Americans, and Asian Americans, and interest them in visiting parks. As the case studies below clearly indicate, the cultural and spiritual significance of nature also provides an important way of addressing the third challenge of engaging and involving Native Americans, Native Hawaiians, and Native Alaskans in interpretation and management of parks that include places and natural features of sacred, cultural, and historical significance for them.

Four Case Studies of Building Spiritual and Cultural Connections to National Parks

A program that I directed at The Mountain Institute (www.mountain.org) from 1998 to 2008 provides case studies of various ways of using the spiritual and cultural significance of nature to engage people with parks. The Mountain Institute works in the Andes, Himalayas, and mountain ranges in the United States to conserve mountain environments and benefit mountain peoples. Since mountains viewed from base to summit include a great diversity of environments and ecosystems, ranging from jungles and deserts to tundra and glaciers, lessons learned from mountains may have wide applicability to other, nonmountainous areas and regions. The program we initiated in 1998 worked with a number of US national parks, including Yosemite, Mount Rainier, Great Smoky Mountains, and Hawai'i Volcanoes. We developed interpretive and educational materials and activities based on the evocative spiritual, cultural, and aesthetic meanings and associations of natural features in mainstream American, Native American, Native Hawaiian, and other cultures around the world.

Yosemite National Park Exhibit on Major National Parks

At Yosemite National Park, we collaborated with interpretive staff on an exhibit on the 58 major national parks (at the time there were 58 parks with "National Park" at the end of their names) organized around the theme of the inspirational value of nature and wilderness (fig. 14.1). Each park had a panoramic picture by Stan Jorstad, and a plaque with a brief paragraph describing the park and the dates it was first established as a protected area and then designated as a National Park, if there was a difference. The Mountain Institute provided an inspirational quote appropriate to the park, ranging from the voices of conservationists, such as John Muir, to Cherokee storytellers and Native Hawaiian elders. The quotes were highlighted just below the descriptions of each of the parks.

To set the tone for the exhibit, I wrote an introductory panel with the following text:

> The unspoiled sanctuaries of wilderness and nature preserved in our national parks have an extraordinary power to awaken a profound sense of wonder and awe. The ethereal rise of a peak in mist, the smooth glide of an eagle in flight, the bright slant of sunbeams piercing the depths of a primeval

14.1. Exhibit on 58 major national parks at Yosemite National Park. Photo by Chris Stein.

forest—such glimpses of natural beauty can move us in inexplicable ways that open us to a reality far greater than ourselves. There, outside the artificial routines of routine existence, lies an awe-inspiring realm of wild mystery, governed by forces beyond our control. In coming to national parks, many seek to transcend the superficial distractions that clutter their lives and experience something of deeper, more enduring value. Indeed, these sanctuaries of wilderness and nature represent places of spiritual renewal where we can return to the source of our being and recover the freshness of a new beginning. (Bernbaum 2006)

As this introduction to the exhibit demonstrates, the spiritual experience of nature does not need to imply a belief in a deity or divine creator. It is open to everyone, be they religious, agnostic, or atheist. All that is necessary is a sense of wonder and awe, of being in the presence of something greater than oneself, such as the vastness of the star-filled sky or the beauty of a flower.

As an example of the brief descriptions of each park with dates of establishment and designation, the plaque for Yosemite National Park reads at the top:

Yosemite Grant, California   1864

Glacier-carved granite peaks and domes rise high above broad meadows, while groves of giant sequoias dwarf other trees and wildflowers in the heart of the Sierra Nevada Mountain Range. Lofty mountains, alpine wilderness, lakes, and waterfalls, including the nation's highest, are found here in this vast tract of scenic wildland. 761,266 acres

Later Designations

  Yosemite National Park—1890
  World Heritage Site—1984
  Wilderness (93%)—1984

To bring out the inspirational nature of Yosemite National Park, The Mountain Institute provided the following evocative quote from John Muir, the conservationist and naturalist most closely associated with its creation:

I invite you to join me in a month's workshop with Nature in the high temples of the great Sierra Crown beyond our holy Yosemite. It will cost you nothing save the time and very little of that, for you will be mostly in Eternity.

John Muir to Ralph Waldo Emerson, 1871

The quote from a letter Muir wrote to Emerson evokes the sense of time-lessness experienced by many in the quiet contemplation of nature that can fill one with spiritual feelings of wonder and awe.

Many of the quotes chosen for other parks are from native peoples reflecting their deep feelings for and connections to nature and the place of each park. For Great Smoky Mountains National Park, we chose a quote by the Cherokee elder and storyteller Jerry Wolfe, who lives right next to the park in the ancestral lands of the Cherokee people:

> The Great Smoky Mountains are a sanctuary for the Cherokee people. We have always believed the mountains and streams provide all that we need for survival. We hold these mountains sacred, believing that the Cherokees were chosen to take care of the mountains as the mountains take care of us.
>
> Jerry Wolfe, Cherokee Elder, 2000

In addition to expressing beautifully the Cherokee relationship to the mountains of the park, the quote carries an inspiring message of environmental stewardship that viewers in general can easily understand and appreciate.

Hawai'i Volcanoes National Park Painting Competition and Radio Program

A park where our work with interpretive staff generated a great deal of excitement and community participation was Hawai'i Volcanoes National Park located on the Big Island of Hawai'i. As the seat of two of the world's most active volcanoes, Kilauea and Mauna Loa, the park is connected in Native Hawaiian tradition and in the public imagination with the fire goddess Pele. For Native Hawaiians, she is associated with many natural features, ranging from the fiery lava to various species of flora and fauna native to the area (Spoon 2005). When we came to Hawai'i Volcanoes, a group of Native Hawaiian elders known as the Kupuna Committee had been working with the superintendent for a number of years advising on cultural matters. They expressed concern that the painting of Pele in the main visitor center did not portray the fire goddess in a culturally appropriate manner (fig. 14.2). A *haole* (non–Native Hawaiian) had painted it in the 1920s and had depicted her without reference to Hawaiian culture. Pele had a Western-looking face and hair blazing yellow, so that she looked like a blonde surfer from California. The elders wanted to replace this painting of her with a painting of Pele more in accord with their traditions. The Mountain Institute had funds from a grant from the Ford Foundation to make it possible. We worked with the Kupuna Committee and interpretive staff to put out a call for people to submit paintings of Pele for the elders to judge and to choose a winning entry. Originally, organizers wanted to restrict the contest to Native Hawaiians, but since the park was a federal agency, the contest had to be open to everyone. We did, however, insert the wording "in consultation with your Kupuna [elder]" to make Native Hawaiian artists the primary target of the message.

The park sent the call out in a news release, and the two main newspapers in Hawai'i, headquartered in Honolulu on a different island, published front-page articles on the contest (Thompson 2003; Wilson 2003). Soon after, all the art stores on the Big Island of Hawai'i were sold out of supplies. The interpretive staff was expecting a few entries, perhaps 14 or so. Instead, they were inundated with what they called a "tsunami of art"—more than 140 paintings. Park staff had to work 12-hour days to process all the entries. Many of them said afterward that it was one of the most meaningful things they had done in their entire National Park Service careers. The Kupuna Committee chose the winning entry for its depiction of Pele with a serene, compassionate expression on her face and two objects in her hands representing important stories connected with her activities (fig. 14.3). For Native Hawaiians, rather than being a wrathful deity associated with volcanic eruptions, she is a benevolent, life-giving goddess who creates new land with her lava. The artist used as his model for Pele a Native Hawaiian neighbor, making the painting all the more authentic in appearance. Interestingly, it didn't matter to the Kupuna Committee that the artist himself was not Native Hawaiian—although, in his favor, he was born in Hawai'i, had lived all his life with Native Hawaiians, and had a deep knowledge of the culture. What mattered was that he had got the painting right.

The park had originally planned to display the remaining entries in the Volcano Art Center, but that venue had space for only about 14 paintings. The various partners in the project chose 67 paintings, from among the more than 140 submitted, and spread them throughout the park in the Jaggar (geology) Museum, Volcano House (hotel on the rim of Kilauea crater), and the Volcano Art Center in an exhibit titled *Visions of Pele*. The exhibit remained up for five weeks, and the artists had a chance to expose

14.2. Painting of Pele by D. Howard Hitchcock (1927) that had been on display
at the Kilauea Visitor Center, Hawai'i Volcanoes National Park, until 2003.
Courtesy of the National Park Service, Hawai'i Volcanoes National Park.

their work to the general public and sell their art. The call for the painting
of Pele, the ceremony around installing the winning entry, and the *Visions
of Pele* exhibit galvanized the communities surrounding Hawai'i Volcanoes
National Park and generated a great deal of enthusiasm for and interest in
the park (Spoon 2005, 2007).

14.3. Painting of Pele by Arthur Johnsen (2003) selected by Native Hawaiian elders to replace the painting by Hitchcock. Courtesy of Arthur Johnsen.

In another issue of concern, the Kupuna Committee wanted to let visitors know before they even entered the park that they were entering a special place sacred to Native Hawaiians, so they would not treat it disrespectfully as a mere recreation area or outdoor amusement park. I attended a meeting with Native Hawaiians and the park's interpretive staff in which they were talking about conveying this message by installing large signs and striking Polynesian sculptures outside the entrance to Hawai'i Volcanoes National Park. I had driven that morning past a sign well before the park that said something like "Tune into 640 AM on your radio for park information." I suggested they add an introduction about the special importance of Hawai'i Volcanoes National Park to the existing radio program that almost everyone entering the park listened to for information on where to see lava flowing and what else to do and see. Since the cars had to wait in line at the entry station and most people spent a lot of time driving around the park, the staff had a captive audience. They liked the idea. The interpreter in charge of the radio program was Native Hawaiian, and he composed the following introduction to the radio program—preceded by the music of a traditional Hawaiian nose flute—that blended together in a particularly sensitive way the spiritual and physical characteristics of the park and linked the concept of *wahi kapu*, or sacred area, to the more familiar idea of a World Heritage Site:

> Aloha and welcome to Hawai'i Volcanoes National Park. You may notice a change in the plant and animal life, climate, or maybe the way you feel as you enter the park. Don't be surprised; this is a common occurrence. For centuries people have felt the power and uniqueness of this place. Hawaiians call it a wahi kapu, or sacred area. You are in the domain of Pele, the volcano goddess. She is embodied in everything volcanic that you see here. This is also home to a forest full of species that are found nowhere else on earth and two of the world's most active volcanoes. Hawai'i Volcanoes National Park is now a World Heritage Site, a modern term for a wahi kapu, recognizing its importance to all of us.

The introduction to the radio program provides a useful model of an inexpensive way parks can use the spiritual and cultural significance of nature for native peoples to engage a large number of visitors and promote support for treating the environment with respect.

Mount Rainier National Park Traveling Exhibit

At Mount Rainier National Park, we used the inspirational value of mountains to reach out beyond the borders of the park to interest members of ethnic minority groups in coming to visit the park. The lofty, glacier-clad volcano of Mount Rainier occupies a special role as a prominently visible symbol of place and identity for millions of people living in the Pacific Northwest. Based on the iconic status of "The Mountain," as Mount Rainier is known to many, The Mountain Institute worked with interpretive staff at the park to develop a traveling exhibit titled "Mountain Views" to take to fairs, community centers, conventions, and other off-site venues in the Seattle–Tacoma area. The completed exhibit had three panels: "The Mountain," "Mount Rainier National Park," and "Mountains of the World" (fig. 14.4). Each panel had pictures of people and evocative quotes by them. The first panel, on the right, showed how Mount Rainier inspired park staff in different divisions of the park ranging from interpretation to maintenance. The second panel, in the middle, featured pictures and inspirational quotes about Mount Rainier by a conservationist, members of a Native American tribe, a mountain climber, a geologist, and a poet.

The third panel, on the left, used celebrated mountains around the world that function as icons, like Mount Rainier, to reach out to various ethnic groups in the Pacific Northwest who have not been visiting in large numbers. A picture of Mount Kailas in Tibet, the most sacred mountain in the world for more than one billion people in Asia and followers of at least four different religions, appeared at the top with the following quote from the Puranas, ancient works of Hindu tradition: "In the space of a hundred ages of the gods, I could not describe to you the glories of the

14.4. Off-site traveling exhibit on Mount Rainier National Park with interpretive ranger Ted Stout. Photo by E. Bernbaum.

Himalayas. . . . There are no other mountains like the Himalayas for there are found Mount Kailas and Lake Manasarovar. . . . As the dew is dried up by the morning sun, so are the sins of humankind by the sight of the Himalayas" (quoted in Bernbaum 1997).

To connect with Japanese Americans and others with Asian roots and interests, the panel juxtaposed a cloud-capped view of Mount Fuji, one of the most famous mountains in the world, with a haiku poem by Basho (1966), one of Japan's greatest poets:

> Delightful, in a way,
> to miss seeing Mount Fuji
> In the misty rain.

Reinforcing the connection with Mount Rainier National Park, the caption added: "Residents of the Pacific Northwest can relate to similar sentiments on 'not seeing' Mount Rainier in frequent mist and cloud."

For African Americans who have an interest in or feel a connection with their African heritage, the panel displayed a picture of Kilimanjaro, a mountain that in the minds of many stands out as an icon of the continent as a whole, with a quote by Julius Nyere, the founder and first president of Tanzania, written on the occasion of his country's independence: "We will light a candle on top of Mount Kilimanjaro which will shine beyond our borders, giving hope where there is despair, love where there is hate, and dignity where before there was only humiliation" (quoted in Hutchinson 1974). The quote appears on a metal plaque placed on the summit of Kilimanjaro and speaks to the concerns and aspirations of many African Americans as well as the general public. A picture of Martin Luther King Jr. and a refrain from his famous "I Have a Dream" speech used the symbolism of mountains in American culture to make a closer connection to African Americans and their history and dreams: "From every mountainside let freedom ring!"

The bottom of the exhibit posed in large white letters the question: "What does the mountain mean to you?" When park staff took the exhibit to various venues, they spread out a large sheet of paper and asked people to write down what the mountain meant to them. Since Mount Rainier occupies a prominent place in the region, the park got numerous, deeply felt responses, which they used in other interpretive exhibits and projects. At the same time, by encouraging this kind of participation, interpretive staff got people thinking and inspired additional connections with Mount Rainier National Park.

Santa Monica Mountains National Recreation Area Wayside Exhibit

A prominent feature of Santa Monica Mountains National Recreation Area near Los Angeles is Boney Mountain, which rises above a range that runs along the Pacific coast. The park's interpretive staff had plans to put up a wayside sign about the geology of the mountain. When we came to work with the park, they decided to change the theme to "Spirit of the Mountain" and place the wayside sign next to a cultural center for the Chumash Indians of the area. We suggested that they find a contemporary representative of the tribe and use a picture of that person and a quote from him or her on what Boney Mountain means to the Chumash people. National parks and other parks have a tendency to put up signage and exhibits talking about Native Americans and Native Hawaiians in the past tense: "The Indians used to . . ." At Hawai'i Volcanoes National Park, Native Hawaiians scrawled graffiti on these kinds of wayside signs saying, "We're still here!"

The interpretive staff asked Charlie Cook, the hereditary chief of the Chumash tribe, for a quote about Boney Mountain. They placed a picture of him on the wayside along with his words about the spiritual meaning of the mountain for his people: "Boney Mountain is a sacred spiritual area, a shaman's retreat, and a place for vision quests. It is a place for meditation. From up there, you can see everything." The purpose of the wayside was not to encourage park visitors to go on vision quests and imitate Native American traditional practices—something that many Native Americans object to. Rather, it was to acquaint members of the general public with the Chumash view of Boney Mountain and to encourage them to experience the mountain in their own particular ways, authentic to each of them. In this spirit, the interpretive staff developed the following sensitively written text at the bottom of the wayside sign:

> Boney Mountain stands as a majestic beacon filling the day and night sky. The mountain's spirit pervades the plants, animals and sense of place around you. It is in the cycle of the seasons, and the past and present generations of people. Whether alone or with others, this place anchored by the mountain invites you to pause, reflect, and look inward. Taste the salt rolling in on the morning sea breeze. Smell the pungent sage warmed by the afternoon sun. Witness the magical interplay of dark and light shadows. What insights, ideas and feelings does the spirit of the mountain evoke for you?

The wayside is unusual for signs and exhibits in parks in that it seeks to evoke experiences rather than convey information.

Principles for Spiritual and Cultural Interpretation

Out of the lessons learned from these examples of diverse projects at various US national parks, nine principles or guidelines emerge for developing interpretation based on the spiritual and cultural significance of nature that has the power to inspire connections between people and parks.

1. Focus on inspiring and enriching experience rather than simply conveying information. The text at the bottom of the wayside sign on Boney Mountain illustrates this principle. It asks visitors, in light of the quote by the Chumash chief Charlie Cooke above it, to experience with their senses various features of the natural environment, and to see what insights, ideas, and feelings these features and the mountain evoke for them. Interpretation that evokes the experience of a primal, old-growth forest—or better yet, having people walk through it to experience its cathedral-like nature—can be a much more powerful way of engaging park visitors and motivating conservation than simply talking about the need to protect the forest as a habitat for an endangered species. Scientific facts need to be conveyed, but there should also be an emphasis on evoking strong experiences and involving people at a more than intellectual level so that they come away enriched and inspired, as well as better informed.

2. Engage people as active participants rather than passive recipients. The traveling exhibit at Mount Rainier National Park involved people by asking them to write what "The Mountain" means to them, thereby engaging and holding their interest. The Pele contest and exhibit owed much of their success to the active participation of artists and the excitement that generated in the local community. Active participation reinforces the kinds of experience inspired by interpretation in the first principle above and leads to involvement in projects that support parks and conserve the environment. By its nature, active participation is also more fun and engaging than simply receiving information in a passive way. People come away enriched and remembering more of what they learned.

3. Make interpretation as personal as possible. People respond to people. Adding a personal element, such as what a feature or place means to someone that conveys the person's passion for that feature or place, can vastly increase the effectiveness of interpretation. The traveling exhibit at Mount Rainer was built around this principle, using pictures of people and their quotes and stories about what Mount Rainier and other mountains meant to them. This applies as well to interpretive materials conveying scientific information: their impact can be strengthened by quoting the scientists doing the research discussing how their subject matter inspired them. Stories about how they made their discoveries are particularly efficacious in sparking interest and engagement. Effective interpretation using the spiritual and cultural significance of nature should also engage visitors as much as possible on a personal level, as the Boney Mountain wayside did, addressing the viewer directly.

4. *Promote mutual respect and appreciation for different points of view.* The Mount Rainier traveling exhibit, the Yosemite exhibit, the Hawai'i Volcanoes radio program, and the Boney Mountain wayside all presented different points of view in a respectful manner, not privileging any particular point of view over the others. This makes interpretation as inclusive as possible, so that everyone feels welcome at the park and more open to exploring new ways of viewing nature. Presenting different points of view also promotes mutual understanding and cooperation, especially in the area of working together on supporting parks and conservation. People feel more inclined to appreciate and support others' views and concerns if they feel their own are being acknowledged and respected.

5. *Leave the final interpretation up to the visitor: "What meaning does it have for you?"* At the parks with which The Mountain Institute worked, most of the examples of interpretation ended with or included a question asking visitors to consider what meaning the natural feature under consideration had for them—or what feelings, insights, or ideas it evoked for them personally. It's important that interpretation does not impose a particular view on visitors, but rather leaves it up to each person to examine his or her own experience and decide what's meaningful or significant. This principle, along with the previous one, also avoids problems that the US National Park Service as a government agency might have over issues of separation of church and state. When interpretive materials present views that are derived from religious traditions found in cultures around the world, it ensures that the beliefs of no particular religion are privileged or advocated.

6. *Work closely with representatives of indigenous peoples and traditions.* In addition to examples from The Mountain Institute's work with US national parks, this principle draws on guidelines for sacred natural sites that the Sacred Natural Sites Initiative developed and guidelines on cultural landscapes developed by the UNESCO World Heritage Centre (Wild and McLeod 2008; Mitchell, Rössler, and Tricaud 2009). Native Americans and Native Hawaiians have lived in and used for centuries many of the lands and natural features within US national parks. They have close ties to them and need to be included from the beginning in interpretation and management of these lands and sites. In particular, interpreters need to respect traditions and sacred places that many Native American tribes want

to keep secret and to present what they want known. At the same time, Native Americans and Native Hawaiians have a great deal to contribute to interpretation and ensuring that what is said about their beliefs and practices is accurate and reflects their views. We can see this principle illustrated in the Pele painting project at Hawai'i Volcanoes National Park where the Kupuna Committee was involved from the beginning, and at the Boney Mountain wayside at Santa Monica Mountains National Recreation Area where interpretive staff consulted with Charlie Cooke, the hereditary chief of the Chumash.

7. *Make interpretation of indigenous views and traditions contemporary by using the voices of living traditional elders and storytellers when possible.* The plaque with a quote by the Cherokee elder Jerry Wolfe on Great Smoky Mountains National Park in the exhibit at Yosemite and the Boney Mountain wayside with a quote by the hereditary Chumash chief Charlie Cooke illustrate both aspects of this principle. They highlight the fact that the Cherokee and the Chumash are contemporary tribes, and that their traditions are alive and important to people today. Putting interpretation in the voices of traditional elders and storytellers ensures accuracy and makes their words speak with greater authority and resonance for park visitors. Too many signs and exhibits at parks focus on archaeology and talk about Native Americans, Native Hawaiians, and Native Alaskans in the past tense. Interpretation needs to acknowledge and highlight the fact that many of these groups are not only present but have living traditions with deep connections to lands and sites within parks.

8. *Generate multiple messages for different audiences rather than a single message.* US national parks draw visitors from around the world. In order to speak to this diversity and to attract members of different domestic ethnic groups who have not been visiting parks in large numbers, interpretation needs to employ multiple messages that welcome everyone and speak to diverse audiences. This also holds true for members of the general public in the United States who have varied interests and backgrounds. To create a single message for everyone is like creating a news release in English for readers of other languages. By incorporating views of natural features and what they mean to people in other parts of the world, as the Mount Rainier traveling exhibit did, interpretation can help visitors relate more easily to what they are seeing and experiencing in parks. This is especially important for inspiring feelings and connections that will motivate support for parks and conservation among a broad audience of people.

9. *Appeal to the cultural and historical backgrounds of diverse ethnic groups as a means of interesting them in coming to parks and feeling comfortable there.* The

Mount Rainier traveling exhibit used a picture of Mount Fuji and a haiku by Basho to appeal to Japanese Americans through the relationship many of them experience between the iconic status of Japan's national mountain and that of Mount Rainier, which they sometimes refer to as "Tacoma Fuji," using the Native American name of this icon of the Pacific Northwest. To reach out and connect with the concerns and aspirations of African Americans, the exhibit used a picture of Kilimanjaro, Africa's highest mountain, and a stirring quote by Julius Nyere found on its summit.Interpretive materials and activities, such as traveling exhibits, visits to schools, and ranger walks, can appeal to the interests of Hispanic Americans by relating natural features in parks to the importance those features have in the traditional cultures from which they have come in Latin America. For example, interpretation can highlight the fact that indigenous peoples from Mexico in the north to Argentina and Chile in the south have traditionally viewed mountains as sacred sources of water and healing, functions that mountains can also serve in parks in the United States. Strategically chosen sister park relations, such as the one between Yosemite National Park and Torres del Paine National Park in Chile, can be used to implement this principle as well as the previous one.

Using the Cultural and Spiritual Significance of Nature in Protected Area Management and Governance

In 2014, the International Union for Conservation of Nature (IUCN) group on Cultural and Spiritual Values of Protected Areas (CSVPA) initiated a project that extended the spiritual and cultural significance of nature to the conservation, management, and governance of protected areas in the United States and abroad.[1] Under the title "Recognition and Promotion of the Cultural and Spiritual Significance of Nature in Protected Area Management and Governance," workshops were held at the IUCN World Parks Congress in Sydney, Australia, in November 2014 (1) to bring protected area managers together with representatives of indigenous traditions and local communities, mainstream religions, and organizations representing the general public; (2) to gather ideas and start to develop a training module to promote the role of the cultural and spiritual significance of nature in the conservation, management, and governance of protected areas; and (3) to establish a network of people interested in lending support and sharing experiences and ideas for working together on projects and

1. See http://csvpa.org/cultural-spiritual-nature-programme/ (accessed 23 March 2016).

activities that integrate the cultural and spiritual significance of nature in protected area management and governance.

This project builds on the work the CSVPA and its affiliates, the Sacred Natural Sites Initiative (www.sacrednaturalsites.org) and the Delos Initiative (www.med-ina.org/delos/), have done with sacred natural sites, but broadens the scope to include the spiritual and cultural significance that nature has for people in both traditional and modern societies. The experience of The Mountain Institute developing interpretive materials with US national parks will be crucial in this regard. By being as inclusive as possible, including the general public and mainstream religions as well as indigenous traditions and local communities, the project has the potential for reaching a wide audience and could have great impact on a large number and variety of protected areas, including national parks. Mainstream religions (e.g., Christianity, Islam, and Buddhism) have millions of followers who can be and have been inspired by their religious leaders to support measures that protect the environment. As the history of environmental organizations like the Sierra Club demonstrates, the general public can be galvanized by inspirational messages to influence government policies and private companies affecting parks and other protected areas (Cohen 1988). The project plans to develop training modules, e-modules, workshops, best practice guidelines, and other products and activities designed to help protected area managers and key stakeholders use the cultural and spiritual significance of nature to work with diverse groups to make the management and governance of their protected areas more effective, equitable, and sustainable.

The main workshop at the World Parks Congress was attended by 80 to 100 people. The participants broke out into small groups to work on developing ideas for a training module. They saw a need for a training module that (1) made use of experiential learning and emphasized the role of experience as a means of connecting people to nature, and (2) imparted knowledge through stories, art, feelings, and values rather than just facts. These points highlight the importance of immediate, direct experience not only in interpretation but also in the management and governance of protected areas. They also point to another kind of knowledge that protected area managers need to take into consideration in addition to the factual knowledge provided by science—the subject of the concluding remarks of this chapter.

Workshop participants also suggested ways of developing a network that would enable people to keep in touch and share ideas and experiences. In addition to sustaining interest, such a network will play an important role in supporting the training modules that the CSVPA has begun to develop and test as the first step in the long-term project. People come away from workshops and training programs excited about what they plan to do, but they often encounter disinterest and apathy and gradually lose enthusiasm when they go back to their places of work. A network can provide crucial support and sharing of experiences that helps them maintain their excitement and implement new ways of doing things in their protected areas.

Concluding Remarks

The examples from The Mountain Institute's program with the US national parks and the points made in the workshop initiating the CSVPA project indicate that we have two kinds of complementary knowledge that need to be incorporated in the interpretation, management, and governance of protected areas. They are suggested by two forms of the verb "to know" in Spanish—*saber* and *conocer* (*savoir* and *connaître* in French). We can see these two kinds of knowledge reflected in the sentence "Yo se que el es un biólogo, pero no lo conozco," which means "I know that he is a biologist, but I don't know him personally." The first kind of knowledge, which corresponds to *saber* in Spanish, is to know about someone or something—in the case of our sentence, what I know about him is that he is a biologist. This kind of knowledge, which we might call objective knowledge, corresponds with scientific knowledge. It is descriptive and explanatory, and it tends toward generalization and theoretical abstraction. It focuses on the object of knowledge and seeks to remove the observer or subject, so that feelings, values, beliefs, and other subjective factors won't interfere with the accuracy of recording data and theorizing. It strives to be as value-free and objective as possible.

The second kind of knowledge, which corresponds to *conocer* in Spanish, is to know someone or something directly or intimately—I know him personally. This kind of knowledge, which we might call subjective knowledge, is knowledge that a person gets through direct experience or through deeply felt experiences evoked by stories, poetry, art, music, or traditional ways of knowledge. When written down or otherwise expressed, it is evocative rather than descriptive. Instead of tending toward abstraction and theory, it emphasizes the concrete uniqueness and immediacy of what we see and experience. A powerful poem, story, or work of art can heighten our perceptions and make us acutely aware of features of nature and our relationship to them that we have overlooked or taken for granted. The metaphors and symbols that many of these works employ may have no direct descriptive correspondence to what they reveal in nature; that lies in the experience they evoke, not in the story, poem, or work of art itself. This is important because without taking into account what they are revealing, it's all too easy to dismiss such works as fanciful creations of the imagination that have no bearing on the real world. Although we are calling it subjective, this kind of knowledge is not a matter of being merely subjective, but rather one of evoking subjective experiences of an objective reality. It reveals aspects of what is actually there that are not accessible to a purely objective approach to knowledge.

Subjective knowledge is important for conservation because it establishes an intimate connection with nature that motivates people to care for and protect the environment. Stories, poems, works of art, and traditional views of natural features help overcome the subject-object dichotomy that separates us from nature and rationalizes environmental destruction and desecration in today's predominantly economic world. Objective scientific knowledge, for all its great uses and benefits, tends by its very nature to separate the observer from the observed, placing a distance between people and nature. Subjective knowledge compensates for this tendency and complements scientific knowledge, so that we get a fuller and richer understanding of the natural world and our relationship to it. Park and protected area managers can improve their key mission of conserving the environment by including the expertise and skills of poets, writers, artists, traditional knowledge holders, and scholars in the humanities, as well as natural and social scientists, in their programs of interpretation, management, and governance. Both kinds of knowledge, subjective and objective, are needed to know nature in its fullest sense and to establish connections that ensure sustainable, long-term support of parks and protected areas.

Literature Cited

Armendariz, A. 2011. California minorities now the majority. Huffington Post, 1 September. Accessed June 2015. http://www.huffingtonpost.com/2011/09/01/california-minority_n_945181.html.

Barrow, E. C. G. 2010. Falling between the 'cracks' of conservation and religion: the role of stewardship for sacred trees and groves. Pages 42–52 in B. Verschuuren, R. Wild, A. McNeely, and G. Oviedo, eds. Sacred natural sites: conserving nature and culture. Earthscan, New York, New York.

Basho 1966. The narrow road to the north and other travel sketches. Translated by N. Yuasa. Penguin Books, New York, New York.

Bernbaum, E. 1997. Sacred mountains of the world. University of California Press, Berkeley, California.

———. 2006. The spiritual and cultural significance of national parks. Pages 10–11 in

M. J. Saferstein and C. E. Stein, eds. America's best idea: a photographic journey through our national parks. APN Media, New York, New York.

Cohen, M. P. 1988. The history of the Sierra Club, 1892–1970. Sierra Club Books, San Francisco, California.

Delos Initiative. Accessed June 2015. www.med-ina.org/delos/

Dolesh, R. J. 2012. Public support may save some California State Parks. Parks and Recreation, 1 June. Accessed June 2015. http://www.parksandrecreation.org/2012/June/Public-Support-May-Save-Some-California-State-Parks/.

Floyd, M. F. 2001. Managing national parks in a multicultural society: searching for common ground. The George Wright Forum 18:41–51.

Hutchinson, J. A., ed. 1974. Tanganyika notes and records. Rev. ed. Tanzania Society, Dar Es Salaam, Tanzania.

Leaman, G. 2013. Co-managing parks with Aboriginal communities: improving outcomes for conservation and cultural heritage. The George Wright Forum 30:287–294.

Mitchell, N., M. Rössler, and P.-M. Tricaud. 2009. World heritage cultural landscapes: a handbook for conservation and management. World Heritage Papers, No. 26. UNESCO World Heritage Centre, Paris, France.

Sacred Natural Sites Initiative. Accessed June 2015. www.sacrednaturalsites.org.

Spoon, J. 2005. Gazing at Pele. Unpublished MA paper, University of Hawai'i at Manoa, Honolulu, Hawai'i.

———. 2007. The "Visions of Pele" competition and exhibit at Hawai'i Volcanoes National Park. CRM: The Journal of Heritage Stewardship 4:72–74.

Thompson, R. 2003. Rendering Pele. Honolulu Star-Bulletin, 13 July. Accessed June 2015. http://starbulletin.com/2003/07/13/news/index2.html.

Wild, R., and C. McLeod, eds. 2008. Sacred natural sites: guidelines for protected area managers. IUCN, Gland, Switzerland.

Wilson, C. 2003. Volcanoes Park seeks new portrait of Pele. Honolulu Advertiser, 27 May. Accessed June 2015. http://the.honoluluadvertiser.com/article/2003/May/27/ln/ln15a.html.

Wirthlin Worldwide. 2001. National Parks Conservation Association qualitative research report. Unpublished report to the National Parks Conservation Association.

FIFTEEN

Strategic Conversation: Engaging and Disengaging People in Parks

EDITED BY EMILY E. KEARNY, AUDREY F. HAYNES, AND CARRIE R. LEVINE

The mission of the US national parks includes the preservation of natural and historical beauty and the enjoyment of that beauty by visitors. The recognition of this relationship between people and parks has been a guiding force in the management and development of national parks for the past century. Over that century in the United States, however, urbanization has increased and the demography and racial balance of its citizenry has changed dramatically, casting the future of this relationship into uncertainty. Most citizens now live in urban areas and their children grow up apart from nature. Visitation of national parks has been declining and use by ethnic minorities, which will soon outnumber Caucasians, lags behind other user groups. Nevertheless, some parks are heavily visited during summer months and impacts on sensitive park resources can require disengaging visitors from some areas.

This strategic discussion, which transpired at the Berkeley summit "Science for Parks, Parks for Science" on 27 March 2015, focuses on how the relationship between people and parks will change in the coming century and what the National Park Service can do to evolve with it. The discussion panel consists of four members. Justin Brashares is professor of wildlife ecology and conservation in the Department of Environmental Science, Policy, and Management at the University of California, Berkeley. Cyril Kormos is vice president for policy at The WILD Foundation, as well as the IUCN World Commission on Protected Areas (WCPA) regional vice-chair for world heritage. Christine Lehnertz is the Pacific West regional director of the National Park Service. Nina Roberts is a professor in the Department of Recreation, Parks, and Tourism at San Francisco State University and director of the Pacific Leadership Institute. This conversation was moderated by Jennifer Wolch, professor of city and regional planning and dean of the College of Environmental Design at the University of California, Berkeley.

JENNIFER WOLCH: *I wanted to ask how you thought the uses of national parks, both in this country and elsewhere, will be different in the coming century compared to the last century? Let's take a prospective view looking forward 100 years. Will use of the parks be different?*

CHRISTINE LEHNERTZ: The uses of parks in the future, if the National Park Service is successful, are going to be different because the users are going to be different demographically and experientially. First, over the past couple of days we've heard people talk about their childhood experiences. When I grew up, I went out with my brothers after school and we played with frogs, we looked at tadpoles, we played in the mud, and we floated things down the creek behind our yard on leaves. We had those kinds of experiences with nature. Many kids today don't go outside their backyard because of the dangers of being kidnapped, at least that's what their parents think. They don't have those kinds of experiences with nature at a young age. Second, they have the digital experience. So users, as they come to parks, are going to define their experiences in different ways than they did in the past, and the National Park Service is going to have to change and adapt from within and with its partners to those new kinds of uses.

JUSTIN BRASHARES: I would just add, and this is well known to this audience, but a great wildcard in all of this is the political climate. Are we going to continue to see, as we've seen here in California, an unwillingness to continue to fund our public places? What is that going to mean for the future of parks?

CYRIL KORMOS: One use that is already important, but needs to be scaled up substantially, is education. I'm surprised that even in a very progressive place like Berkeley, where the environment is a very important issue, many people are not well informed about parks and their many functions. That's something that needs to change very quickly over the next decade.

WOLCH: I think that's something that we've heard about quite a bit at this meeting. Professor Roberts, I wanted to ask you a question about the commitment the National Park Service has to engage an ever-broadening range of visitors and diversify the National Park Service staff as well. Under Director Jarvis, this effort has been redoubled. He has formed special committees looking at how to engage urban visitors. Most of us think about diversity in terms of race and ethnicity and perhaps gender, but it strikes me that identities are increasingly multifaceted. These facets intersect and depend on contexts and circumstances to shape our overall everyday experience. *How should we be thinking about diversity in parks and in the Park Service?*

NINA ROBERTS: That's a question we could spend the next two hours on! First, I want to give kudos to Director Javis for the work he's done, and acknowledge and recognize the progress that's been made. Lots of amazing work has also been done by former directors as well. But as far as how people define what they consider diversity to be or what it to means to them, we also need to recognize that this occurs park by park. That's where the magic happens, in the parks. There's been decades of focus on race and ethnicity, which is important, no question. We also need to talk about age, because we face an exodus of baby boomers, of which there are 78 million people. So we have an intergenerational workforce right now and that's pretty challenging. Even from a gender standpoint, you still have few women that are superintendents or managers of parks. And the park system is finally moving into a new level of embracing the gay, lesbian, bisexual, transgender community. Stonewall, where a series of spontaneous, violent demonstrations by members of the gay community took place in 1969 in New York City, is a great example. In 2002, it became a National Historic Site. But why did it take so long? Thinking about the intersection of all these different facets is complex, but we also have to look at the geography in terms of diversity. Geography matters. There are compelling reasons to believe that the geography of one's residence and geography of one's workplace have impacts on park uses, and that these impacts differ among racial and gender groups. Perspectives are important, but so are policies because many people are resistant to change. Educating the workforce around these factors is really valuable. Looking at the intersection of different identities, what they mean, and how they interact, is a huge and very complex question, and that's a direction that we need to start to explore.

LEHNERTZ: I think this is a great question, and it may be the question for the National Park Service that is going to be the difference in its future. Over the last two days, we heard that connecting with communities is critical to the future of the National Park Service and the national parks, and it may be the most difficult thing and the most challenging thing to crack. If we talk about how we identify, I identify as a woman of privilege, as a lesbian who is a white woman surrounded by white people in the National Park Service. I love my colleagues. The last report I saw is that about 90% of the National Park Service employees are white. How many people in the audience identify as white or Caucasian? Look right and left, perhaps 70% are men in this audience, which is about what the National Park Service looks like. We saw great pictures here of youth who were looking at park rangers, and a lot of the time we saw white park rangers looking at kids of color who got to go to parks and it's like "Ain't it great? Don't you want to be like me?" That's awesome because we want kids to be park rangers. But wouldn't it be fun if it's "Ain't it great? I look like you." And that is a part of what the National Park Service must be challenged to correct. It's going to take social science to help us get there. Dr. Roberts said it's going to be hard for us to change it. It's going to be the hardest for the National Park Service to change from the inside out.

WOLCH: As you say, it's an extremely persistent aspect of most park systems, but not all, and it is difficult to change from the inside. I want to pick up on the question of values, which is related to some extent. We've heard quite a bit at this meeting about the kinds of values that are promoted by parks and especially the National Park Service. *Is it the job of the National Park Service to promote a specific set of values to visitors, scientific, spiritual, or otherwise, or values around landscape, conservation, history, and culture? Or should the Park Service also be thinking about broadening its values to encompass those of the communities that it seeks to engage?* We just heard Edwin Bernbaum's presentation about listening to different histories. *Cyril, what do you think the balance should be and do we have it right?*

KORMOS: I think it has to be balanced. It's got to be a two-way street. From purely a policy standpoint, we have the Organic Act and a set of values that have to be transmitted through the National Park System. This has to be nationally relevant and significant. On the other hand, we just had a presentation by Edwin Bernbaum, which was really excellent, and explained the need to integrate a broad range of values, both from the community immediately around the park and more broadly. I think one way you could reframe that question is to ask, How do you take the sets of values that are discussed in the Organic Act and make them relevant to a lot more people and transmit them more effectively? It can't be that difficult. For example, if you live in San Francisco, you rely on the water that comes from Yosemite National Park.

We're living in an era of global change. Parks are important for a broad range of health, spiritual, and cultural values that we are beginning to understand better. It shouldn't be that difficult to make these messages relevant. But it's really disconcerting to me, as a conservationist, when I see so many people who still don't understand the value of parks. We've gotten our message out, so we need to try harder and we need to try differently. But it's got to be a dialogue, it's got to be a two-way street.

ROBERTS: I'm going to jump into this dialogue. Those are typical Eurocentric wilderness values. Those are what I often hear, and what I've studied and explored. Those are white values. Those are values that we, as professionals, try to impose on the general public. That's what has got to stop, right? We keep hearing about this two-way street. This reciprocity is not really happening because we forget about how we have come to protect these values by eradicating the public who has valued outdoor activities very differently. If people want to go fishing, sometimes it's not for sport—it's for dinner. People want to go out and do certain things on park lands; it may be to maintain their livelihood. But we tell them "no," because "that's not how you should value these lands." I have to ask, according to whom and what beliefs? We need to challenge how to accept and embrace the history and traditions, but not forget to continue to ask the questions stated in Edwin Bernbaum's presentation—"What does it mean to you?" and "How can we meet halfway?"—versus conveying those messages of "Here's the way you should value this wilderness or these national parks or historic sites." That's something that is very important to convey.

WOLCH: I think that's a good counterpoint, Nina. One thing that is so striking about the national parks is that we think about the iconic parks, but in fact there are nearly 500 units, and they have cultural missions, historical missions, and nature conservation as their foci. Even within one national park like Golden Gate, there's a huge array. *Do you think such extreme heterogeneity can actually play a role in engaging different kinds of people and attracting different kinds of workers?*

LEHNERTZ: It's a great question and it's one of the strengths of the National Park Service. I have a disorder that I'm going to share with you, and it's not one of those disorders that there are embarrassing ads about during nighttime television! I have a "Wayfinding Deficit Disorder," so I have a hard time finding my way. I've been with the Park Service just eight years, and even before I was with the Park Service, when I went to a park it was often difficult for me to find my way. I would get lost, and the Park Service's strength is not always "Here's how you find your way through a park." So, when you go to a park that has a lot of different features or a lot of different elements, it is not easy to find your way, particularly if you're in the wilderness. A park like Golden Gate National Recreation Area, which is in an urban setting, can be intimidating. The National Park Service needs to start thinking about starter parks, as 80.7% of Americans live in urban areas. Urban parks are going to be the Park Service's entrée to people now and into the future. That often starts with a cultural park rather than a large natural park, like Yellowstone or Yosemite or Rainier or Olympic. I think the Park Service has an opportunity to reach people through cultural parks in urban areas. If the Park Service can tell its stories better, if it can reach people with some confidence, if it can do it digitally before they get to the park, it's going to be able to use that heterogeneity in a way that may be simpler. I think it's a strength rather than a weakness.

WOLCH: One of the things Christine just said, which is really important, is that

80% of the US population lives in cities. We know that visitation rates to parks in general, but certainly also to the US national parks, are often very low in poor urban communities and communities of color. My research team conducted a focus group with immigrant Latinos in Los Angeles who did not go to the Santa Monica Mountains National Recreation Area, even though they lived fairly close by. Participants were asked why they didn't go. Some thought that maybe the park was only for white people, or that maybe you would be asked for papers to get in. This is a pretty profound perceptual barrier. *What do you think the National Park Service should do in urban areas to increase knowledge about and access to parks? Do we need a different model for outreach and especially local collaboration?*

ROBERTS: It's key for people to realize that this topic has been explored since the early 1960s, so this is not a new conversation. But are there new models? Should we be doing something differently? Yes. As we think about partnerships, collaborations, and our current knowledge base, we need to consider how we impart knowledge. And we also need to seek knowledge from the communities and the cultural icons whose voices are heard less often. A lot of us have PhDs but we don't have PhDs in the streets, like some of the communities we're trying to reach. How do we tap into those communities? For example, the health professionals are coming to the table more than before and it's about time. We've known about the health benefits of parks, but health providers, park professionals, and scientists have not been at the same table for many years. It's about looking at who our partners are. How many people are working with gang intervention programs? How many people are working with those and other nontraditional communities? We have to think about tapping into a variety of other communities that could really benefit from the resources that we are trying so hard to protect, to preserve—resources that these communities have an equal right to experience and explore. Often, it is just showing up, taking the risk. There's strategy after strategy, plan after plan. The standing joke in the National Park Service is that you have to have a plan to implement the plan, but then when you implement the plan, what's the plan to manage that plan? And so on.

LEHNERTZ: We'll plan about that later!

ROBERTS: We have heard a bunch of strategies, but we need to ask, Who are the people at the table? Who are the voices that are not being heard? That's where the real change will occur.

KORMOS: One thing that I feel could help would be more urban visitor centers. We have visitor centers in parks, but that presupposes you're already at the park. In the Bay Area, for example, there isn't one place where you can go and get a sense or a vision of the greenbelt that's being assembled here. The greenbelt is absolutely amazing and pulls together the regional parks, the state parks, the federal parks, and all the different land management agencies in the area. Maybe the Oakland Museum could be a location for this type of urban visitor center. Downtown San Francisco would be better, if there were the funding. But there's got to be places where we could present this in a dynamic way to a lot of people to show the relevance and demonstrate the options for access. That's a local example. There are probably others.

WOLCH: I think the idea is really interesting—stepping-stones or starter parks that give people a much broader outlook and future places to visit. We heard E. O. Wilson and others talk yesterday about the extreme threats to biodiversity, and we also heard how many millions of people come to the US national parks every year. *On the one hand, that's great and it's certainly crucial for political support, but does this also mean that some parks get used too heavily and that visitation can actually threaten biodiversity? Do we have to think about how to sensitively limit, as well as promote, access?*

BRASHARES: I think the answer is absolutely. We need to be thinking about balancing access and impacts. There's a pattern that we have observed across the planet with regard to protected areas: where protected areas are well managed, they start to look less and less like the areas around them, which become increasingly developed. So what does that mean over time? It means that's a wonderful thing and that our protected areas are serving a purpose. Whether it's biodiversity banks or spiritual places, they are serving that purpose and they are holding on to those resources. But it also means that more people are turning to those protected areas for resources, whether it's because land values are higher around our national parks and state parks since those areas are sought after, or whether it's because people are increasingly looking to our protected areas for ecosystem services and natural resources. So we need to continue thinking hard about balancing use and sustainability. I think that the National Park Service has done an excellent job. More and more we see places like Zion National Park where the traffic flow is limited and the human impact is being concentrated in certain areas. We need to recognize that these are places that are meant to preserve biodiversity and nature in all forms. Maybe it isn't as simple as "Hey, we're getting visited more than Disney World and the NFL and all the rest," but instead a matter of "Well, that's one measure of success. The other measure of success is to do the monitoring that's necessary to ask if we are meeting our goals. Are we doing our job in maintaining biodiversity in these areas?"

WOLCH: I want to turn to the question raised by E. O. Wilson about nature needing half. I know, Cyril, this is something that your organization is very much behind. The WILD Foundation is pursuing this goal at a global scale.

Does this initiative actually conflict with the goal of engaging more and more diverse people with national parks? How do you reconcile conservation biology and our need to increase access, diversity, and visitorship?

KORMOS: It's a big topic. It's an aspirational goal. The thinking behind it was that there are a number of studies indicating ecosystems need 30% to 70% of their land area to be protected in order to persist and to maintain their functions. If you take an average, that's roughly half. We're talking about designating half in protected areas as defined by IUCN's protected area classification system, of which there are seven categories. Only one of those categories actually excludes people; those are strict nature reserves or biological control areas. All the other categories, from national parks all the way to sustainable development areas, involve human use. The issue is not excluding people. The issue is what kind of uses are compatible with what category of protection. That's really important and it applies to wilderness as much as the other categories. At The WILD Foundation, we've taken the position that wilderness is not a place that excludes people, though that's the way it is sometimes perceived here in the United States as a result of the Wilderness Act. Wilderness is a place that has wild natural values with which people interact. It is actually about a human relationship. That relationship can be recreational. It can also be about indigenous cultures that make no separation between themselves and the land. It's a spectrum of relationships. The "nature needs half" vision is aspirational and supposes we need to protect a whole lot more of the planet because we're in really, really bad shape. The way it's done is through a range of protected area classifications that will be culturally and biologically meaningful at local scales as well as regionally and globally.

WOLCH: It sounds to me like you're really talking about a coexistence strategy so that humans can successfully coexist with other kinds of sentient life on the planet.

KORMOS: Yes, that's what it comes down to. Either you believe that the biosphere is in very bad shape, which has a serious impact on people all over the planet, or you don't. We feel that it is in bad shape and that's a serious problem. We are confronting climate change, and it is going to be a humanitarian crisis for millions of people around the world that needs to be addressed. Our premise is that we're going to need to do a lot more for people and for nature. How you get there is the key, obviously. You have to do it in ways that will benefit people and that are culturally sensitive, but we have to, in our view, be thinking about protecting a lot more areas than we have historically. It's difficult for the conservation movement to be claiming success at this point. We've been talking about what we need to do for three decades. Biodiversity is declining and climate change accelerating, so we're not winning. We need to be thinking clearly about what it's going to take to win because we are not there yet.

WOLCH: We are out of time. But one thing that strikes me about this conversation, particularly given what Cyril just said, is that it's more important than ever to have people understand this issue, to be in touch with diverse types of parks at different spatial scales, and to feel like they have a stake in what happens in these incredibly important spaces.

Future of Science, Conservation, and Parks

Study the past, if you would divine the future.

—Confucius (551–479 BCE), *Chinese teacher, writer, and philosopher*

The past, the present and the future are really one: they are today

—Harriet Beecher Stowe (1811–1896), *American author and political activist*

Well, I woke up this morning, and I got myself a beer.
The future's uncertain, and the end is always near.

—From "Roadhouse Blues" (1970), *by The Doors, American rock-and-roll band*

Prediction is very difficult, especially about the future.

—Niels Bohr (1885–1962), *Danish physicist and philosopher*

The future belongs to those who prepare for it today.

—Malcolm X (1925–1965), *African American minister and human rights activist*

When the second centennial of the US National Park Service (NPS) is celebrated in 2116, what will parks, science, and conservation look like? Should we have confidence that the park concept and national parks themselves will persist for another century, even as their features and constituencies are changing? What are the megachanges already underway—technological, social, and environmental—and how will they affect parks, the ways we do science, and our approaches to conservation?

This book closes with chapters that provide glimpses of the future, musings about science for parks, and contemplation about what conservation for parks may be like in the future. These and other related issues are considered from three different perspectives in the final section of this book. The authors, like many since Confucius who have tried to "divine the future," employ past patterns and recent trends as a basis, recognizing our limited ability to dissect past, present, and future, à la Stowe. Certainly, the "future is uncertain," which causes the authors to shy away from predictions, but in a tone that is different from many environmental forecasts, none of these contributors suggest "the end is always near." Instead, the chapters are offered, paraphrasing Malcolm X, for the purpose of preparing today for the futures of tomorrows.

This section begins by examining the kinds of futures that national parks may encounter over the next 50 to 100 years in a thought-provoking contribution by Jamais A. Cascio. He chooses to eschew the doomsday attitude that characterizes many environmental forecasts. Instead, taking the view that humanity and parks will persist through the next century of changes as sustainable societies eventually emerge, Cascio examines three scenarios for the coevolution of technology, culture, and nature. Each scenario is built on perspectives about the way that a global sustainability crisis is overcome—through increased *control* of the economy by global institutions, through *collaboration* among new institutions and open-source tools, and through *creation* of radical technological developments that result in social, economic, and environmental transformations. The three scenarios provide context for envisioning the role, composition, and context of national parks. A surprising fate may be in store for Earth's long-term future if we celebrate the quincentennial anniversary of the NPS in 2416.

To understand the future of science for parks and parks for science, we need to consider the future of science. A perspective on how science will be done for national parks in the near horizon—the next 40 years—is provided by Gary E. Machlis. While important science can be accomplished with no more than a pencil and a piece paper, science often progresses hand-in-hand with technological advances. In this chapter, Machlis envisions how recent trends in technology and science will combine to drive knowledge about natural and cultural resources in parks. He describes how environmental DNA, biocuration, artificial intelligence, and big data analytics may change the way we do science and gain inference for parks. These advances are likely to generate controversies, such as de-extinction, human-assisted evolution, and tension between data collection and surveillance. Machlis concludes by discussing the roles of science in reducing uncertainty and informing decisions related to stewarding parks through the next century of change, as envisioned by the *Revisiting Leopold* report of 2012.

In the final chapter, Steven Beissinger and David Ackerly provide a broad view of how science, conservation, and management of park resources have changed since the birth of the NPS, and how climate change may shift not only species and ecosystems but our conservation and management paradigms by the sesquicentennial of the NPS. Although early conservation controversies had arisen even before the 1916 Organic Act was signed into law, science was not incorporated into the NPS until more than decade later. Beissinger and Ackerly trace the repeated periods of growth and decline of science in US national parks over the past century, arriving at recent expansion of capabilities over the past decade. Perspectives on conservation in national parks have changed over that time and may be shifting again, given that climate change affects how the concept of "location" as well as "what to conserve" is viewed. Beissinger and Ackerly then review the three dominant paradigms that shaped the past century and will shape the future century of thought and actions by park managers and conservation biologists. They argue that, if parks are to be stewarded through the rapid environmental and cultural changes occurring, successful conservation may require a paradigm shift. The dominant conservation paradigm today advocates management to maintain current and historic baseline conditions; other emerging paradigms view conservation goals as managing for natural processes or managing to projected, future conditions. Determining which paradigm to embrace, or the best combination of paradigms needed to support valued natural and cultural resources, will be a major challenge facing scientists and managers stewarding parks worldwide.

We stand at a temporal crossroads, where the past and present may be of little help in understanding the conservation challenges that lay ahead of us in a rapidly changing world. Friedrich Nietzsche, German-Swiss author and philosopher of the 20th century, stated, "The future influences the present just as much as the past." We would be wise to keep the wisdom of the three chapters of this section in mind as we work toward a sustainable future for parks, people, and biodiversity.

SIXTEEN

A New Kind of Eden

JAMAIS A. CASCIO

Introduction: The Work of a Futurist

Ask someone to describe a "futurist," and you're likely to get references to crystal balls or jet packs, or maybe something about the latest technologies. One word you're almost certain to hear is "predict." But foresight professionals today have little interest in trying to predict what the coming years will hold. Forecasts are created not to offer a pinpoint-accurate description of tomorrow, but to provide some insight into the complexity of today by illustrating possible consequences. The goal is to focus on uncovering the unexpected but plausible implications of today's choices, especially when seemingly unconnected systems intersect.

Over the last generation, futurism as a discipline has become something strongly resembling the social humanities of history, political science, and especially anthropology. Foresight looks at the combination of economic, political, social, technological, and environmental dynamics as drivers of change, and does so with a sharp focus on how these dynamics affect the lives of everyday people. Futurists studying a particular industry or region pay close attention to changes in cultural norms, values, and traditions. Futurism is an attempt to understand how human systems connect and clash. But instead of doing so retrospectively, futurism tries to anticipate the connections and clashes yet to come while informed by the best available information about the ideas and innovations on the near horizon.

Futurists are certainly not alone in this endeavor. Economists, demographers, energy specialists, and more are all quite comfortable with making multidecade projections based on their best current understanding of the dynamics of their respective fields. Where futurists differ is that they don't just focus on a single category of system—they're more interested in what happens at the littoral zone where systems overlap. They focus on the interdependence and conflicts between arenas like information technology, demography, ecology, bioscience, political movements, and much more. One of my recent projects illustrates this: my set of scenarios on the possibility of global nuclear disarmament for a security nongovernmental organization wove in climate projections, the political impact of social media, the race between state-funded and privately funded space travel, institutional responses to global pandemics, and the evolution of international rivalries (Cascio 2015).

It should come as no surprise, then, that foresight professionals have long paid close attention to the work of environmental and climate scientists. In some ways, the environmental sciences offer another kind of parallel to what foresight specialists do. Much of the work done in both of these disciplines is anticipatory, attempting to describe the environment to come, not just the conditions of the past. Moreover, a useful understanding of the dynamics of ecosystems or ocean-atmosphere systems depends on our comprehension of how different domains connect. The converse is also true: our understanding of the changes underway in nearly every complex human system requires us to examine how these systems interact with environmental systems.

One of the more difficult aspects of foresight work is an unhappy tendency that many people have—not just futurists, but thinking people in general—to zero in on the myriad ways in which human civilization is trying to doom itself. I sometimes refer to this as "apocaphilia," an attraction to the concept of the end of the world. Catastrophic stories are easy to tell: there are just so many ways in which we seem to be digging our own grave, from collapsing fisheries (Gaines and Costello 2013) and agriculture (Lal 2013), to accelerated evolution of antibiotic resistance (Carlet et al. 2012), to unstable regimes with nuclear weapons (Wilson 2013), and much, much more. The sheer diversity of apocalyptic futures is mesmerizing.

All of this makes finding positive scenarios of tomorrow difficult, which is especially true of climate and environmental futures. This doesn't just arise from simple apocaphilia. For many people, whether foresight specialists, climate scientists, or the general public, focusing on the possibility of failure rather than on the potential for success can feel wiser and more responsible. In my experience, forecasts of the world to come are often used as dire warnings, or even anticipatory scoldings.

It's important, then, to face this problem directly, and to look intentionally at what kinds of "positive" futures might emerge. These need not be utopian—even our wisest decisions can have undesired side-effects—but they should describe futures in which our choices are no less constrained than they are today, and in which the overall quality of life for humanity and Earth's environment has arguably improved over the present. This isn't meant to say that everything is okay, but (arguably, more importantly) that everything *can be* okay, if we work to get it right.

This is doubly true when thinking specifically about the future of the US National Park Service (NPS). In most crisis-driven, disaster-laden scenarios, organizations like the NPS are quite vulnerable. As California is currently witnessing with its massive drought, when resources that are required for both the ongoing protection of the environment and society's food needs and economic stability become scarce, the environment inevitably suffers. Scenarios that explore the future of the NPS amid various dire possible futures facing us would be dismal and likely painfully repetitive.

Positive scenarios, ones with broad, sustainable improvement over time, are actually more plausible than most people expect. Much of my work as a futurist involves collaborating with organizations seeking to make a significant difference in the world, and as a result, I frequently run into groups and individuals working on projects that have a very real potential to make our lives immeasurably better. Many of these projects are technological in nature (e.g., see Cascio 2010; RepRap 2015), but not all of them. Some of the most important developments now underway focus on creating new social and economic models, using cutting-edge science, big data, and real-world experiments to create new systems of sustainable human interaction (McCoy 2014). If even a small fraction of these projects are successful, we will have within our grasp an array of tools and concepts that will allow us to create a future that's not just worth living in; it will be worth celebrating.

This doesn't mean that success is guaranteed, or that it will be easy. The dramatic warnings that turn into catastrophes in the more dystopian scenarios remain threatening. What differs in the more successful scenarios is how we decide to work against them. The dystopias expect us to be overwhelmed and to fail. The comparatively positive scenarios that will follow here start from a simple set of questions: What would the world look like if we succeeded? What would it take to build a livable, plausible tomorrow?

The following are 50-year scenarios. I chose this timescale for several closely related reasons. The first is that, in human terms, 50 years is distant but fits within the lifetime of a single individual; most of us can either remember 50 years past or can expect to be around for 50 years hence (and increasingly, both are possible). Second, it's near enough in time that many of the core aspects of lived experience will still be recognizable; much of how we live today has a distinct lineage stretching back beyond 50 years, and it's reasonable to expect that the same would be true for the people living in the mid-2060s. Third (and to counterbalance the second), it's far enough ahead of us that there's sufficient time for a number of important developments in technology and culture to emerge.

Such developments could in many cases be described as transformative. Each of the three scenarios takes a different path of transformation, and the nature of these paths may be surprising. We should take to heart the insights of a couple of people long practiced at thinking about tomorrow. Roy Amara, past president of the Institute for the Future, observed that we constantly overestimate short-term changes and underestimate the long term (Morrison 2001). And Jim Dator, founder of the Futures Studies Department at the University of Hawai'i, notoriously asserted that "any useful idea about the future will at first seem ridiculous" (Dator 1993).

Scenario 1: Walking the Tightrope

The first scenario is one of *control*, in which we manage to build a relatively sustainable world through top-down measures, allowing us to maintain some of our core institutions and behaviors, and changing just what is necessary in order to achieve environmental stability. This world could easily fall back into unsustainability—or worse—so ongoing success demands strict oversight. Control results in the kind of world that many people who care about the environment would love to see: cleaner, healthier, but still eminently familiar. A person from 2015 dropped into this version of 2065 would likely be as confused as someone from 1965 visiting 2015—that is, definitely perplexed by what he or she sees and experiences, but not catastrophically so. There are big technological differences, but most day-to-day social and cultural behavior remains quite recognizable.

This scenario comes about from asking some basic questions: What would a successfully sustainable future look like if our key economic and political institutions truly got serious about making changes? If our governments considered climate change an existential risk? If, in particular, the tools of ubiquitous observation we've deployed in the name of national security were instead used for environmental security? How might we build a sustainable future if the tools of state and corporate power were turned to deal directly with the challenge?

The resulting future doesn't demand fundamental changes to our way of life, only to how we make that way of life happen. At its best, it's a world that's focused on replacing the dirty, inefficient elements of our lives with cleaner, greener alternatives. It wouldn't be easy; there's plenty of potential here to fall off the tightrope. As a result, it's also a world where any actions taken that threaten the well-being of the planet can be punished swiftly and severely.

Among the present-day enablers of this scenario (what futurists often refer to as "signals"), the accelerating push toward greening industry and infrastructure stands out. We can see signs that many large corporations and powerful nations have begun to recognize the need for faster change (e.g., see Bloomberg Business 2015; General Motors 2015; Apple 2015; Walmart 2015). Environmental economists have long argued that a more sustainable world is ultimately more profitable (Hawken, Lovins, and Lovins 1999) and that an increasingly unsustainable world will cut profits dramatically over time. In this scenario, that concept has taken root. As the costs of renewable energy and high-efficiency materials and products drop, we will be able to make large-scale changes to how we organize our society and economy, changes that would have seemed unthinkable just a few years earlier. The breakthroughs and experiments that we see today in building and product design, transportation, and efficiency snowball in the Walking the Tightrope world, becoming an engine of radical—but ultimately beneficial—transformation.

Another critical driver of this world is the power of the financial services industries, particularly the reinsurance market, to push changes on reluctant government and corporate actors. Reinsurance is an industry that isn't often discussed but holds enormous power: the companies offering this service essentially provide insurance for the insurance companies. When a massive disaster proves too much for conventional insurers, the reinsurance agencies are there to back them up. Two of the largest global reinsurance companies, SwissRe and MunichRe, have been at the forefront of demanding that their clients (and, by extension, the governments in the countries in which the clients operate) include anthropogenic global warming in their models of future risk. In the Walking the Tightrope scenario, such demands have borne fruit.

The core logic of the Walking the Tightrope scenario is this: *Building a sustainable 21st century that is a clear descendent of our present-day world and one that emerges swiftly enough to escape the more catastrophic environmental scenarios requires significant top-down coordination within and across governments and corporations.* Such coordination, in turn, requires abundant, detailed information. There would be sensors and cameras everywhere, all making sure that nothing is being done that could endanger the world's health. One might think of this scenario as "Sustainable Disneyland"—it looks beautiful, perhaps even perfect, but behind the scenes it's highly controlled and closely watched.

On the surface, our day-to-day lives in this future would effectively be zero-carbon parallels of how we presently live. Most of the big changes would be structural, but not systemic. It's a world of (1) better-than-LEED buildings and all-electric cars, but not completely redesigned urban and transportation models that eliminate suburbs and personal vehicles; (2) big solar and wind farms (or even cheaper, more resilient forms of nuclear power), but not radically distributed energy production; and (3) robot agriculture and carefully bioengineered foods, but little use of permaculture, aquaponics, or similarly disruptive models of producing food. Everything is recyclable, everything is high efficiency, and everything is tagged and traceable for maximum transparency. Depending on how societies evolve in this scenario, such transparency may or may not be available to citizens as well as administrators.

These kinds of changes would not happen overnight. Even at the accelerated rate posited in the scenario, the necessary evolution of energy and industrial systems will happen too slowly to avoid significant climate warming. So to head off a global catastrophe, this future relies on solar radiation management geoengineering as a stop-gap measure. Solar radiation management geoengineering uses radical, large-scale techniques to alter climate systems (principally insolation and cloud patterns) in order to hold down temperatures (Govindasamy and Caldeira 2000). The goal of solar radiation management geoengineering isn't to "fix" global warming, but to slow its harmful consequences in order to give us the necessary time to shift to a zero-carbon economy. Because of its global impacts, geoengineering requires international cooperation, and the policies introduced to forestall both private geoengineering efforts and popular backlash against these projects reinforce this future's top-down structure.

Economics are as closely watched and controlled as politics and the environment in the Walking the Tightrope scenario. Although this is a strongly market-driven scenario, it's a heavily regulated market. Government policies seek to reduce the impact of externalities, especially those arising from potentially disruptive technologies. Robots are common, for example, but haven't completely replaced human labor; 3D printers lower the cost of manufacturing, but don't lead to desktop factories. Tools improve, but their use is often strictly controlled. Research and development remains prolific, but it's largely directed toward getting to zero carbon as quickly and as efficiently as possible. It's not so much that technological advances themselves are restricted or limited, but that they are focused.

The result is that, over time, life becomes better for a significant portion of the world. As the global economy reinvents itself, changes become dramatic, especially in developing nations. Up-and-coming countries and regions *leapfrog*, bypassing legacy industries and infrastructure in order to take advantage of the better technologies of the 21st century. Rapidly developing societies are eager to adopt a better-than-Western lifestyle without Western levels of environmental harm. Social and political tensions sporadically emerge, but the rise of the "Green South" becomes an engine for faster economic and technological evolution, as the previous recipients of eco-development assistance become exporters of eco-development innovation.

Ongoing support for this scenario depends on a delicate balance of systems working together smoothly, and a capacity to avoid any big surprises. The primary challenge is to keep that balance right: (1) to maintain a growing world economy without overheating (figuratively and literally); (2) to encourage the speedy development of the Green South without undermining the postindustrial world; and (3) to use geoengineering to constrain temperatures without triggering conflicts over who controls the global thermostat. Simply put, it's a very top-down scenario because *success requires deliberate action by forward-looking leadership as well as an international community willing to cooperate.*

Getting to this kind of future would need a level of foresight and collaboration around the world that is not often in evidence. And as comfortably familiar as the resulting future might be, taking this path would necessitate significant trade-offs in national sovereignty, in global power, and especially in privacy.

Environmentally, this is a future of managed recovery. Ecosystem-critical species are protected, and restrictions on ocean and land use are enforced aggressively. It's a triage scenario, but one that ends up working better than feared; within two decades, the health metaphor most often used is no longer "triage" but "intensive care." By the 2060s, global biodiversity has diminished compared to the beginning of the century, but the rate of extinctions has also dropped considerably. Ecosystem management and operation is a popular field of study, with graduates overseeing former wildlands, now officially called "Administered Bioregions."

National parks in this scenario emphasize their preservation and protection role, serving as bioregion role models. Initially, some parks will be effectively closed, and will serve more as ecosystem service providers than as entertainment; others will be opened to limited visits or virtual tours, as a reminder of what we're trying to recover. These parks are not so much museums as archives, to be tentatively reopened once we restore the environment. Assuming no unpleasant surprises, by the NPS bicentennial most if not all the parks should be reopened to visitors. Swarms of nanosensors will keep a close watch on the ongoing evolution of the ecosystems, as well as on our behavior.

Scenario 2: Recovery as Reinvention

The second scenario is one of *collaboration*. In this world, change is driven from the bottom up, and sustainability comes not from ponderous national policies and global policing, but through the rapid spread of ideas and solutions across social networks and around the world. While in the first scenario the world just manages to avoid disaster, in this scenario things *do* fall apart and need to be fixed.

At first glance, this future may seem like a hippie nirvana of bicycles and vegetarianism, and while both of those features can be found in abundance, they don't tell the whole story. Recovery as Reinvention is actually an environment of intense competition between regions and among individuals as we work to rebuild the world. If, superficially, hippie behaviors are common, it's not because of changing ideologies, but because—in this kind of world—they *work*.

Recovery as Reinvention is very much a crisis-driven scenario. Historically, confronting and recovering from large-scale crises nearly always results in some fundamental changes to system dynamics. Power long accumulated may have been spent; institutions charged with our protection may have failed; citizens' expectations of their roles and the duties of the state may shift dramatically. In the Recovery as Reinvention future, we get hit, hard, by near-term crises in our environment and economy. As a result, we fundamentally rethink how we live. A populace demanding to recover not just quickly, but wisely, questions and very often discards many present-day institutions. The fundamental driver here is *adaptation*, with different parts of the world adapting at very different speeds and in very different ways.

The present-day signals of this scenario are fundamentally social, even if enabled by information and communication technologies. We've seen the potential of tech-powered social movements for years, from open-source software to citizen science to Facebook- or Twitter-driven revolutions. Such movements aren't always successful, but even in failure they can alter the scope and narrative of the larger socioeconomic structure in which they operate. These methods and tools of social disruption gather strength just as the combination of economic, environmental, and societal pressures start to undermine the already shaky legitimacy of traditional systems of governance.

In a future reshaped by crisis, big changes will often focus on the practical. Recovery as Reinvention is a world of abandoned suburbs and omnipresent mass transit, distributed energy and water recycling, permaculture farming and vegetarianism by necessity (if not always desire). Bicycles fill the streets and dryer lines abound in backyards. Yet it's not a "return to simplicity" or a similar neo-Luddite fantasy. It's still high-tech, but widely used technologies are social and resilient. In this future, we give a lot of thought to the impacts of our tools; you might even call it "cyber-Amish." Our 2015 person dropped into the future of Recovery as Reinvention would find a mix of the familiar and the baffling that roughly parallels what we saw in Walking the Tightrope, but here it's the technology that remains surprisingly familiar; the social, economic, and political structures would be much more foreign.

The environmental and economic disasters leading to this scenario are global in scope; no part of the world is left untouched by the "Long Crisis." The European Union splits; China is overwhelmed by environmental disasters and demographic instability, driving large-scale unrest; the United States sees its global dominance evaporate, and may well see its own separatist movements. Well before 2065, there's no longer any such thing as a leading world superpower. Nations dependent on either international trade or international aid face tremendous difficulties. If you were to tell citizens of Recovery as Reinvention in 2040—that is, midway through this scenario—that they were living in a "successful" future, they would have little visible cause to believe you.

It's not just the citizens who face unexpected difficulties. Many of the institutions we presently take for granted become unrecognizable as they evolve to cope with the new demands and expectations of this future. Economic, environmental, and technological catalysts replace globalized trade with strong regionalism. With apologies to Tom Friedman, the world isn't *flat*—it's *bumpy* (Friedman 2005). Many of the 20th-century nation-states become little more than administrative bodies, with citizenship often more closely tied to city-states and megapolitan areas. Money is either local cash or a global digital currency.

In 2065, manufacturing uses both traditional methods and ultra-high-tech tools, the distant descendants of the 3D printers of today. These fabrication devices are increasingly *general purpose* technologies, easily and swiftly reconfigured to print anything from electronics to clothing to (in some cases) organic products such as food and biomedical implants. There are present-day 3D printers able to do each of these tasks; what changes here is the breadth of function. As a result of all of this, the largest remaining global markets are in digital blueprints for physical products and food design.

The most commonplace technologies are immersive and social. These are the distant great-grandchildren of Facebook and Skype, providing instant and ongoing connections in a manner that's often indistinguishable from an in-person conversation. As is the case with fabrication devices, these kinds of technology are in development today and would be cheap and commonplace by the 2060s.

Biotechnology and related fields go through a fairly significant crash and recovery cycle. The loss of trust in most traditional institutions means that organizations already facing heightened public skepticism (like many biotech firms) suffer a nearly total collapse of political and economic support. At the same time, the weakening of governments' ability to enforce regulations opens up greater opportunities for the careless use of potentially dangerous technologies. After a small number of accidents (and one intentional near disaster), prohibitions on research into biotechnologies, nanotechnologies, and artificial intelligence technologies pop up everywhere. It isn't until the late 2040s that these restrictions begin to be loosened.

All of this results in a rather tepid advance in technology over 50 years. Where there are significant advances, they are either in tools for human interaction (social and economic) or in the elimination of carbon emissions and other environmental hazards. Imagine "hybrid" cargo ships that mix high-efficiency electric motors and massive sails; permaculture farms replacing suburban megastores; and rooftops covered in gardens, offering food, cooling, and a small bit of carbon sequestration. For most people, the most disruptive everyday "decline" compared to the present would be reduced mobility. Fewer people travel long distances, and far fewer travel overseas.

But deep connections between regions and communities remain present, even if the links are largely virtual. We maintain the social ties between friends and family thousands of miles away. Citizens may even feel more solidarity with people of a geographically distant but culturally familiar region than they do with citizens closer in space but not in belief. This has important political implications; geographically large states become increasingly difficult to govern, even as new regional ecology-based configurations emerge. Young people in 2065 are much more likely to refer to themselves as citizens of London, California, or Amazonia than England, the United States, or Brazil.

These persistent and widespread interconnections become fundamental to the adaptation strategies of Recovery as Reinvention. We readily copy and improve on the most successful approaches, sometimes through trade, sometimes through the commons, and occasionally through intellectual property "piracy," learning from each other's mistakes. While there can be an intense competition in this scenario, much of the work to adapt better, to be more resilient, and to create a more livable environment ends up readily shared. Where there's a need for oversight, it's more often by neighbors and fellow citizens than by governments.

The environment in this scenario faces dramatic challenges, initially serving as the nightmare version of the "hands-off" or "rewilding" ecosystem management model. The rate of extinction worldwide actually *increases* in the first decades of the Long Crisis. Where species do grow, they're typically the opportunistic species best able to adapt quickly; there's a reason that some observers refer to this as a "rats and kudzu future." Over time, however, this serves to strengthen ecosystem health.

This "hands-off" model isn't intentional, at least initially. Vast stretches of land across the United States are abandoned by people. Internal displacement due to economic or environmental factors reshuffles the population dynamics of mid-21st-century America. This will lead to "de facto rewilding," in which nature retakes land no longer managed by people. Over time, the NPS or its successors (whether local authorities or a revitalized national group) eventually enclose and watch over these regions.

As for the US national parks themselves, the results are mixed. During the Long Crisis, superficially noncritical government programs like national parks take a significant hit. Funding declines or even disappears, boundaries are no longer respected, and a need for survival can overwhelm a desire to protect our planet. Rising temperatures, alongside fewer resources for firefighting, will make wildland fires widespread and devastating. Fortunately, restoring the parks is very much part of the postcrisis agenda, in the recognition that parks are both symbols of and manifestations of the world we're trying to build. The abundant wildland fires of the 2030s and 2040s end up as ultimately beneficial, clearing out the accumulated underbrush and dead trees that resulted from aggressive firefighting, and that made the fires that did happen far more severe. By 100 years out, we come to think of parks as our human-friendly gateways to our increasingly extensive wildlands. There is even discussion of a "wild corridor" linking the Pacific to the Atlantic, a continent-wide stretch of parks and protected lands.

Scenario 3: New Mythologies

The last scenario is about *creation*, a concept and term loaded with cultural implications and meaning. This isn't "Creation" in the Biblical sense, but it's certainly richer than the strict "act of making" definition. Creation, in this scenario, is the use of tools and knowledge to shape the continued evolution of ourselves and the planet, along with the wisdom to do so mindful of the possibility of unintended consequences. It's a scenario in which Stewart Brand's famous line—"we are as gods, and we might as well get good at it"—hits unsettlingly close to home (Brand 1968).

Brand's observation (itself paraphrasing an observation by British anthropologist Edmund Leach) is too often thought of as being a celebration of hubris, an invocation of our growing ability to do anything we'd want. In this interpretation, "get good at it" means "learn how to do more and more amazing things, consequences be damned." That's a misreading. What Brand meant by "get good at it" was "figure out how not to mess up when we do more and more amazing things, because the consequences would damn us." It's this latter version of Brand's aphorism that sits at the heart of this scenario.

New Mythologies explores what happens to sustainability when we feel the full impact of a set of radical—but actually surprisingly plausible—technological and social developments. Some of these are now in active development, while others are waiting for the right combination of economic, scientific, and social dynamics to emerge. In many cases, the innovative tools or philosophies aren't individually transformative, but become so when combined with other catalytic changes.

As a result, New Mythologies describes a world where some of the most complex and wicked economic and environmental issues of the present day have been successfully resolved, and by 2065 active steps are being taken to restore the most damaged regions. Ecosystem science is in its heyday, and our understanding of how complex systems interact allows us to balance long-term changes and unintended consequences. We make few decisions of significance without deep consideration of downstream implications. This isn't a utopia by any means, however, and new challenges will arise.

This future undergoes a degree of social, economic, and technological upheaval akin to the Industrial Revolution, but at a franticly accelerated pace. A key element of this scenario is that the transformation is still ongoing—in many ways, the big changes of the first 50 years are just preparation for truly staggering changes over the next. Recall how, in the Walking the Tightrope scenario, the reaction of a 2015 person dropped into 2065 would be akin to a 1965 person dropped into the present. In New Mythologies, the parallel would be more like a citizen of 1865 being pulled into the present day.

On the surface, New Mythologies may seem like a primarily technology-driven scenario, but that's not quite right. This future is a noisy mix of radical industrial and technological developments and major social and political earthquakes, with the dials all turned up to 11. These cycles of upheaval feed into each other: economic and environmental drivers lead to cultural disruptions, which inspire the development of new technologies that in turn serve as triggers for new waves of environmental, economic, and ultimately social change.

The key driver of this scenario is *experimentation*. In this future, human civilization confronts challenges not by trying to hold on to the status quo, or by focusing solely on immediately practical responses, but by exploring new systems of governance, work, and even play. Many of these new systems do make use of cutting-edge technologies, but the mere presence of new tools isn't itself enough: people—communities, organizations, and individuals alike—need to be willing to investigate what the technologies offer. The challenge of experimentation is that the more significant the leap forward, the more likely (and potentially the more dangerous) the failure. In the New Mythologies scenario, the benefits of this wave of experimentation outweigh the risks. All of this results in dramatic shifts in power; notions of citizenship, commerce, even human identity are over time all called into question.

Technologically, the most significant development is the proliferation of molecular-scale fabrication systems. Superficially, these systems would work like extremely precise 3D printers, able to build physical objects of nearly any kind, using basic elements like carbon as inputs. At present, these exist largely as concept, but in this scenario the key technologies work. This development redefines many of our economic and environmental challenges.

For someone in our present, these kinds of technologies can seem almost magical. Waste products we now call trash and pollution—including carbon—are just another resource in a world of molecular nanotechnology. Materials become "smart," able to sense and respond to their surroundings, as well as to convert incidental sunlight into a flow of energy. It's a world where every surface is active, almost alive.

The ability of nanofabricators to cleanly and cheaply construct most physical objects overturns conventional market forces. The cost of production is close to nil when you can take yesterday's garbage and turn it into today's gadget at the push of a button. Most importantly, the ability to operate at a molecular scale enables us to reshape the physical environment, and even allows us to stop, and then reverse, global climate disruption. By 2065, some scientists are even warning of the potential to draw too much carbon from the atmosphere.

The other key technological advance is the continued evolution of autonomous machine intelligence, usually called artificial intelligence (AI). This realm of technology doesn't require a Rapture-like "Singularity" to be highly disruptive. By the 2030s, advanced AI systems have moved well beyond repetitive tasks, and by 2065, they perform much of the "knowledge work" now performed by highly trained people. These systems are still considered tools, but the assistance they provide is orders of magnitude more sophisticated than today's computers. Intelligent machines are everywhere, but they're usually working invisibly, becoming part of the infrastructure, or even the environment.

This scenario holds big advances in medical treatment, energy production, urban design, and more. It's a world of biomimetic buildings and robot transportation networks, food fabricators and ultra-clean water, engineered microbiomes and vertical farms. Environmental, social, and economic benefits and harm can be readily visualized and modeled, and the use of advanced simulations allows for extraordinarily detailed analysis.

Economics and politics see their own massive disruptions. With the increasing capabilities of autonomous machines, the traditional notion of "work" is becoming largely extinct, outside of the highly creative arts and jobs demanding deep empathy for others, such as schoolteacher or nurse. In a growing proportion of the world, people work not because their survival depends on it, but because they enjoy what they do. As long as the world remains primarily market driven, a guaranteed basic income model ensures continued economic growth and allows for even more entrepreneurial experimentation.

With AI-enabled planning and declining scarcity, politics increasingly focuses on conflict resolution. Most societies remain functionally democratic, although a growing number have adopted a "snap election" model allowing for immediate public response to proposals. This doesn't happen solely because of the assistance of intelligent machines. Advanced neurotechnologies allow us to understand, even shape, how the brain processes ideas and forms opinions.

And as all of this suggests, while the benefits are great, so too are the risks. New Mythologies is easily the most dangerous of the three scenarios—there's quite a bit of new power to play with, and not everyone plays nice. Moreover, the speed at which these changes have come has left a large number of people—especially but not exclusively older generations—suffering from what amounts to PTSD (posttraumatic stress disorder). One big role for governments is to provide cognitive and emotional protection; it's a world where mental health is civil defense.

But it's also the only scenario of the three that gets us out of the sustainability trap, in which efforts to keep things "sustainable" can mean living on a constant precipice of becoming unsustainable. This is the inherent dilemma of a sustainability focus: sustainability is treated as an end goal, rather than as a baseline. In short, sustainable is not enough.

The first scenario, Walking the Tightrope, remains (as its name suggests) vulnerable to a sudden loss of sustainability. An unfortunate combination of setbacks could fatally harm the entire system. As an example, solar radiation management geoengineering without the rapid elimination of carbon emissions leads to a need for increasing amounts of particulate material to be put into the stratosphere, leading to greater disruption of rainfall patterns and a growing vulnerability to "rebound shock," in which the elimination of aerosol particles used for solar radiation management—whether intentionally or by accident—results in a dangerously rapid jump in temperatures.

The second scenario, Recovery as Reinvention, is primarily focused on keeping society's figurative head above water. The 2065 of a postdisaster scenario remains brittle, and further changes to economic, social, and environmental behavior would be scrutinized for potential threats to the ongoing recovery. Such trepidation isn't a permanent feature, but the need to "sustain" would override a desire to "thrive" for much of the century.

In contrast, in this last world, New Mythologies, sustainability is only the beginning—the goal isn't simply to stop making things worse, but to actively make things better.

In New Mythologies, the overall rate of extinction has dropped dramatically. According to multiple studies, biodiversity (both within the United States and globally) is actually starting to increase, as the careful reintroduction of recently extinct species begins to reverse some of the worst ecological sins of the past two centuries. Proposals to "re-Pleistocene" North America by restoring mammoths and giant sloths (along with appropriate predators) go nowhere—this isn't restoration for entertainment or novelty, but a sober attempt to fix what we have broken.

In this future, parks are everywhere, almost literally. In this world, because of our greater understanding of complex systems and more nuanced view of our long-term impacts, we treat all of our physical environment as our heritage to be protected. Some parts of the nation and the world may have less overt human activity than others, but everything is curated, watched over, and gently steered, not just with the goal of transient stability, but mindful of the need to pass it along to our descendants. After another 50 years pass, the only difference will be the nature of the caretakers—instead of being monitored by easily distracted people, these parks will be, as in Richard Brautigan's 1967 poem, "all watched over by machines of loving grace."

Gazing at the Long-Term Future: The Fate of Earth

But 50 years, even 100 years, isn't actually that far into the future. As a final step, let's take a look at what 500 years might hold.

We're unlikely to be celebrating the 600-year anniversary of the founding of the NPS, because the governmental entity of the NPS (and very likely the United States itself) will likely no longer exist at that point. Political configurations tend not to have that kind of longevity, and a society empowered by the technological and social developments witnessed since the onset of the 21st century is unlikely to be satisfied with political institutions developed in the 18th, 19th, and 20th centuries. But the foundational ideals of the NPS will continue to shape our culture: the desire for knowledge, the need for preservation, and the interweaving of memory and legacy.

Our planet will be far healthier than it is today. Increasingly, the careful monitoring and interventions that had characterized the latter half of the 21st century and beyond will be reduced or eliminated, no longer being needed. In fact, I believe that Earth itself, the entirety of it, will be on a path to becoming the future society's equivalent of a national park.

What better way to pay our respect to the planet on which we were born than to let it return to its natural course of evolution? To carefully remove our footprint from the oceans, the sky, and the land? Perhaps we look again at the possible return of species that we had driven to extinction in the preindustrial or even prehistoric past. Or perhaps we just let the world rewild in a way that treats our historical impact as just another passing natural phenomenon. In whatever way we transform Earth into a park, it is with a light touch, constant observation and study, and most of all with reverence.

Where are the humans? We haven't killed ourselves off, not in this future, but we have begun to leave the nest. Some of us may be exploring the galaxy on starships accelerating to close to the speed of light. Others may be on vast arks, taking the seeds of Terra to new worlds welcoming to—but as yet untouched by—biology. Or maybe we stay near home, turning Mars, Venus, Ganymede, and Europa into new realms of earthly life. But in every case, we don't leave Earth behind; we expand it: "Earth" becomes wherever we live, a growing sphere of DNA and mind spreading through the cosmos, and at its center, our first home.

Not forgotten. Not ignored. Celebrated.

Literature Cited

Apple. 2015. Environmental responsibility. Apple.com. Accessed 2 July 2015. https://www.apple.com/environment/.

Bloomberg Business. 2015. China adds solar power the size of France in first quarter. Bloomberg.com. Bloomberg Business, 19 April. Accessed 2 July 2015. http://www.bloomberg.com/news/articles/2015-04-20/china-adds-solar-the-size-of-france-s-capacity-in-first-quarter.

Brand, S. 1968. Whole Earth Catalog 1968 Purpose. Whole Earth Catalog 1.2.

Brautigan, R. 1967. All watched over by machines of loving grace. The Communication Company, San Francisco, California.

Carlet, J., V. Jarlier, S. Harbarth, A. Voss, H. Goossens, and D. Pittet. 2012. Ready for a world without antibiotics? The Pensières Antibiotic Resistance Call to Action. Antimicrobial Resistance and Infection Control 1:11.

Cascio, J. 2010. World Water Day 2010: Three projects that are changing the future. Fastcompany.com. Fast Company, 22 March. Accessed 2 July 2015. http://www.fastcompany.com/1593806/world-water-day-2010-three-projects-are-changing-future.

———. 2015. Five scenarios of giving up on nuclear weapons. Reinventors.net. Reinventors Network, 24 March. Accessed 2 July 2015. http://reinventors.net/five-scenarios-of-giving-up-on-nuclear-weapons/.

Dator, J. 1993. From future workshops to envisioning alternative futures. Futures Research Quarterly. 9:108–112.

Friedman, T. 2005. The world is flat. Farrar, Straus and Giroux, New York, New York.

Gaines, S., and C. Costello. 2013. Forecasting fisheries collapse. Proceedings of the National Academy of Sciences USA 110:15859–15860.

General Motors. 2015. Innovation: environment. GM.com. Accessed 2 July 2015. http://www.gm.com/vision/environment1/general_motors_isfueledbythesun.html.

Govindasamy, B., and K. Caldeira. 2000. Geoengineering Earth's radiation balance to mitigate CO_2-induced climate change. Geophysical Research Letters 27:2141–2144.

Hawkins, P., A. Lovins, and H. Lovins. 1999. Natural capitalism. Little, Brown, and Company, Boston, Massachusetts.

Lal, R. 2013. Food security in a changing climate. Ecology & Hydrobiology 13:8–21.

McCoy, S. 2014. Citizen science: fire recovery monitored by crowdsourcing photos. Gearjunkie.com. Gear Junkie, 23 May. Accessed 2 July 2015. http://gearjunkie.com/fire-recovery-monitored-crowdsourcing-photos.

Morrison, I. 2001. The rise and fall and rise of e-health. Ianmorrison.com. Reprinted essay from Health Forum Journal January/February 2001. Accessed 2 July 2015. http://ianmorrison.com/the-rise-and-fall-and-rise-of-e-health/.

RepRap. 2015. Welcome to RepRap.org. Reprap.org. Accessed 2 July 2015. http://reprap.org/wiki/Main_Page.

Walmart. 2015. Renewable energy. Walmart.com. Accessed 2 July 2015. http://corporate.walmart.com/global-responsibility/environment-sustainability/energy.

Wilson, W. 2013. Five myths about nuclear weapons. Houghton Mifflin Harcourt, Boston, Massachusetts.

SEVENTEEN

The Near-Horizon Future of Science and the National Parks

GARY E. MACHLIS

Introduction

In 1942, C. C. Furnas published a book entitled *The Next Hundred Years: The Unfinished Business of Science*. A professor of chemical engineering at Yale University, he cheerfully predicted that after World War II there would be food pellets for everyone and that the technological limit of wireless receivers for personal use would not shrink beyond the two-pound backpack. But he also lamented the "shortcomings of science and society" that allowed poverty, war, lack of education, and environmental destruction to maintain their hold on the nation and the world. He saw science as "unfinished business" and strove to identify the near-horizon future of science. He noted: "We cannot see the goal, but we can see the nearer sections of the road leading to it. . . . If we look along the roadways of scientific thought and accomplishment that we have already passed, we should be able to prognosticate a bit and tell something about the road of the future" (Furnas 1942).

There is something both cheerfully off base about Furnas's book (food pellets did not soar in popularity after World War II) and on target: there is potential to constructively consider the near-horizon future of science. Near-horizon science can be defined as science that is just now emerging, its applications beginning to be tried, tested, and outlined, and its future potential far exceeding its past scientific or applied contributions.

Given that the US National Park System is at a historic benchmark of centennial progress in 2016, and that science has played a critically important role in the preservation, conservation, and management of the system, an assessment of near-horizon science and the national parks may be useful. The purpose of this chapter is to do so for science and the US national parks.

The chapter is organized as follows. First, several emergent disciplines and fields of science are briefly presented, with examples of how these new approaches to science may be applied to national park conservation and management. Second, select emerging methods, tools, and data are presented—again this represents only a sample of near-horizon science, but enough to demonstrate that it is a common occurrence for new methods or tools to drive new scientific ideas or applications. Third, a series of knowledge frontiers and scientific challenges are discussed—not only for the issues they raise within the sciences, but for the controversies they create for society.

Emerging Disciplines, Fields, and Subfields

In the early decades of the 21st century, science has exponentially increased in its complexity, primarily in two directions—toward narrower specialties and subspecialties, and toward interdisciplinary syntheses of selected disciplines, fields, and subfields. These are simultaneous and not inconsistent trends. Science is becoming both increasingly reductionist and pandisciplinary; specialization and consilience both characterize contemporary science. The following are examples of near-horizon disciplines and fields relevant to science and national parks.

Quantum Biology

Quantum biology applies quantum mechanics, particularly wave particle transitions, to biological phenomena (Ball 2011). An example is the ability of birds to navigate using magneto-reception linked to Earth's magnetic field (Al-Khalili and McFadden 2014). Quantum biology could provide deeper understanding of these mechanisms that in turn could help inform park scientists and managers in their life-cycle stewardship for migratory species. For instance, the four North American migration flyways (Atlantic, Mississippi, Central, and Pacific) cross hundreds of parks and are followed by hundreds of species of birds that use these parks as stopover sites. Protected areas harbor large acreage in which flora and fauna exist, making parks (and networks of parks) strategic places for researchers to study migratory mechanisms over a great variety of species and large-scale movements.

Conservation Paleobiology

Conservation paleobiology is the application of paleological, ecological, and geochemical techniques to the analysis of biotic remains of species threatened with extinction (Flessa 2002). Using the near-time and deep-time fossil record, the ecological and evolutionary responses of species to changes in their environment can be better understood (Deitl and Flessa 2011). This emerging field has broad opportunity to help scientists in parks redefine species extinctions over longer timescales and document wide ranges of environmental variability. Conservation paleobiology can inform species reintroduction efforts by expanding options beyond those supported by contemporary or historical records. An example is the effort to reintroduce the California condor (*Gymnogyps californianus*) in the Vermillion Cliffs area of northern Arizona, guided by knowledge garnered from late Pleistocene cave deposits that showed the species once occupied the area (Flessa 2002).

Conservation paleobiology can also be used to more accurately estimate the natural range of environmental variability of ecosystems. An example is work on the Colorado River Delta that examined the effects of freshwater diversion and provides target parameters for its restoration based on the fossil record (Flessa et al. 2001). Understanding the past through conservation paleobiology can help parks plan and manage for the future.

Reconciliation Ecology

The field of reconciliation ecology studies and promotes strategies to encourage biodiversity in human-dominated landscapes (Lundholm and Richardson 2010). Rosenzweig articulated the concept in his 2003 book *Win-Win Ecology* based on the assessment that there is not enough area within designated nature reserves for Earth's biodiversity to be effectively protected. Reconciliation ecology's focus on heavily human-dominated and small-scale landscapes distinguishes it (somewhat) from the 1990s concept of ecosystem management. Proponents often call reconciliation ecology "win-win ecology" because it aims to increase biodiversity in anthropogenic ecosystems while not decreasing their human utility (Rosenzweig 2003). Critics have called it "rose-tinted ecology," arguing that reconciliation ecology may be naively optimistic and that biodiversity conservation may be low on the list of priorities in regions that are dealing with pressing social issues such as poverty and civil conflict (Brooks 2003).

Parks could serve as control sites for reconciliation ecology research. Additionally, park scientists can utilize reconciliation ecology to invent, establish, and maintain new habitats to conserve species diversity in areas of parks that have been historically human-dominated, such as lodges, visitor centers, and surrounding areas.

Reconciliation ecology can be particularly useful in US National Park Service (NPS) units that do not have the explicit goal or theme of conserving biodiversity, such as historical and cultural sites. An example of successful reconciliation ecology shows that shrikes (Laniidae) thrive where wooden fence post perches facilitate easy pouncing on prey, and decline where steel fence posts do not provide perch advantage (Francis and Lorimer 2011). Another example is that longleaf pine (*Pinus palustris*) in the southeastern United States thrives when wildfires are allowed after timbering and declines when fires are prevented (Francis and Lorimer 2011).

Parks can also serve to disseminate knowledge and lessons from reconciliation ecology to park visitors in order to influence implementation of these strategies beyond park boundaries. Reconciliation ecology stands to become more effective in the near term, given its small spatial scale and low bar for resilient biodiversity. In some situations, it also presents a significant challenge to current strategic approaches for biodiversity conservation.

Cliodynamics

It is not only within the biophysical sciences that new disciplines and fields are emerging. Cliodynamics combines historical macrosociology, economic history, and modeling to understand long-term social processes (Turchin 2010). For example, recent comparative research demonstrates that agrarian societies experience long periods of instability—oscillations termed "secular cycles" (Turchin and Nefodov 2008). This new area of work could be used to more accurately predict park visitation over longer time periods and to identify key drivers of long-term visitor trends. Since many parks have extensive historical records, these parks could provide cliodynamic researchers with case examples to aid their advance in theory and cliodynamic methods.

Methods, Tools, and Data

It is not only emerging disciplines and fields that are near-horizon trends for park science. Science often advances when new methods, technologies (or tools), and data sets become practical, accessible, and applied. Such advances may not always come directly from basic scientific research; an example is how 1960s Cold War intelligence-gathering using photogrammetry led the way to remote sensing tools for natural resource applications from forestry to fisheries management. The following are examples of near-horizon methods, tools, and data relevant to science and national parks.

Environmental DNA (eDNA)

Environmental DNA (eDNA) is genetic material from whole microbial cells or shed from organisms via metabolic waste, damaged tissues, or sloughed stem cells (Kelly et al. 2014). Environmental DNA collects where organisms have passed through or spent time. Recently developed and rapidly expanding technology utilizing eDNA allows species detection within meters to kilometers of a monitoring site (Kelly et al. 2014). As the cost of gene sequencing declines, the use of eDNA becomes both more efficient and potentially a core of noninvasive inventory and monitoring. Environmental DNA samples may complement and someday replace traditional and current park inventory and monitoring methods, such as direct observation and diurnal area searches, which can be more invasive and require relatively more personnel, time, energy, and/or money.

In addition, eDNA collected in parks can be used to address larger applied research questions by park scientists and other researchers. For example, eDNA from entire communities across taxonomic groups within a park could potentially be analyzed simultaneously. Ancient eDNA could be comparatively studied to learn about a park's history and inform its future. Historical, current, and comparative data could be incorporated into park interpretation programs and used in resource decision making.

Biocuration

Biocuration is the activity of organizing, representing, and making biological information accessible to both humans and computers (Howe et al. 2008). The methodological area of study is relatively new: the Fourth International Biocuration Conference took place in 2010, with an international participation encompassing 30–50 scientists.[1] Data access enables scientific advance: new questions can be posed and new hypotheses can be

1. See http://hinv.jp/biocuration2010/ (accessed 21 March 2016).

tested. Uploading park-related data and tagging it with controlled, universally agreed-on variables (sometimes called "advanced tagging") is not yet widely available, but biocuration tools will make this advantageous and practical in the near future. Expanding access (including data, metadata, and analytic tools) can help create a new generation of citizen scientists doing science in parks, and can contribute to interpretive programs that expose visitors (both real and virtual) to park-related data and trends. It can (particularly when paired with advanced analytics and decision support tools) empower park managers via evidence-based decision making. Because second-generation biocuration tools are largely in development, parks can serve as proving grounds, support the visibility of biocuration, and promote it as a professional career.

Artificial Intelligence (AI)

The term "artificial intelligence" (AI) was coined in 1956 by computer scientist John McCarthy. Initially, it was used to describe the intelligence exhibited by a computer programmed with essential features in order to study human intelligence. AI now refers to intelligence shown by all technology and the research to create such technology (Gil et al. 2014). AI methods that extend beyond research and statistical analysis to advanced semantic analysis will create the opportunity for AI research assistants. Such AI assistants could scout the literature, identify metadata patterns, construct hypotheses, and eventually test and report results. With this advancement comes the inevitable probability of a completely AI-generated, hoax scientific paper that could buffoon the scientific community.

AI may allow park scientists to more robustly and productively find hypotheses worthy of study and to use the steps of strong inference (focused on null hypotheses testing) to advance theory and create usable knowledge. AI methods could also be used to free park professionals to attend to other work. If used at park entrances, employees previously tasked with visitor intake could focus on other park operations. Intelligent machines could answer park visitors' questions and offer a list of detailed suggestions tailored to their interests, time constraints, and abilities. Such machines could provide visitors with the collective experience of all park employees (and eventually all past and present employees), not just a "greeting ranger." The implications of AI for natural and cultural history, inventory and monitoring, long-term field studies, and visitor education are significant.

Big Data Analytics

Big data is a broad term used to describe data that is either too large or too complex to be processed by traditional applications (Li and Chen 2014). Big data analytics is the process of uncovering hidden patterns, unknown correlations, and other useful information by examining structured, semi-structured, and/or unstructured big data sets (Howe et al. 2008). Big data analytics merge large data sets, biocuration, and AI methods, and have significant value for near-horizon science in parks.

If used correctly, big data analytics can help park scientists examine scientific hypotheses and advance theory—particularly in the fields of landscape ecology, coupled human-natural systems, and climate science. Where traditional data analytics are reactive, big data analytics can enable park managers to make proactive decisions by utilizing in new ways established methods such as optimization, predictive modeling, text mining, and forecasting. Park data scaled up from individual parks to regions or biomes can be made more accessible through biocuration for analysis (including analysis by AI methods), using the steps of strong inference to identify patterns and correlations worthy of study. These processes can be used to make better decisions based on data unavailable (or inaccessible) to conventional analytics and solutions. If misapplied or abused, as these methods can be, they can create spurious correlations. An inventory of poor hypotheses, data, and noise must be carefully evaluated, rejected, and ultimately pushed aside for science to advance (Fan, Han, and Liu 2014). Hence, big data analytics is both an opportunity and a challenge for science in national parks.

Sensor Technologies

Sensor technology machines detect some characteristic(s), event(s) or change(s) in quantities or qualities in their environment. A new wave of public environmental monitoring is emerging with small-scale, affordable tools such as the crowd-funded Air Quality Egg and Smart Citizen Kit. The Air Quality Egg collects high-resolution readings of nitrogen dioxide (NO_2) and carbon monoxide (CO) concentrations, two of the gases that are the most related to urban air pollution (Austen 2015). The Smart Citizen Kit measures everything the Air Quality Egg does, as well as light intensity and noise (Austen 2015). These and other potential sensor technologies are low-cost, provide useful monitoring functions, and have data upload capabilities.

As these tools become more accurate, reliable, sophisticated, and widespread, they will be able to provide managers with high-value, low-cost, real-time data on park resources. For example, park rangers (particularly those in the backcountry who follow similar routes repeatedly) can become environmental monitors that efficiently collect essential monitoring data without in situ equipment, particularly appropriate for wilderness areas. Air and water quality, light intensity, and noise (and in the future other variables) can be measured by individual sensors given to visitors or park rangers and uploaded to managers and park scientists who can use the data even before the actual sensor technology machine is returned to their hands.

Another useful emerging sensor technology is CubeSats. CubeSats are 10-centimeter miniature satellite boxes that have been released from the International Space Station to collect data for education (Hand 2014). Their relatively cheap price is altering the way remotely sensed environmental data is collected because it alters the risk calculus of using space-based sensor equipment. In the near-horizon future, government agencies such as the NPS and organizations such as The Nature Conservancy could commission their own CubeSats. Data from NPS CubeSats could be used to inform resource management decisions, and environmental organizations with CubeSat capability could be armed with proprietary and distinctive data sets at global scales. The convergence of such near-horizon sensor technology, big data sets, biocuration, and AI support tools suggests that conservation science is poised for an explosive increase in available and accessible data on park resources, conditions, and trends.

Near-Horizon Frontiers and Challenges

In addition to advances in scientific disciplines and fields, as well as methods, tools and data, the near-horizon future includes a wide range of opportunities (frontiers) and controversies (challenges). In many cases, the advances (such as in big data analytics) directly or indirectly lead to the concerns (such as overreaching surveillance). Several of these near-horizon frontiers and challenges are described below.

The Next Generation of Citizen Science

Citizen science refers to the involvement and engagement of the nonscientist public in scientific activities (Miller-Rushing, Primack, and Bonney

2012). In 2009, Bonney et al. described public participation in scientific research as being in three categories: contributory, collaborative, and cocreated; each involves a rising level of citizen engagement and integration into the research enterprise. Haklay (2013) separated citizen science into four categories: crowd-sourcing (citizens as sensors), distributed intelligence (citizens as interpreters of science), participatory science (citizens involved in problem definition and data collection), and collaborative science (citizens help plan the research and conduct analyses of results).

For park science, citizens have largely been engaged in crowd-sourcing, collecting data, and serving as "sensors." Examples are the hugely popular Audubon winter bird counts and NPS BioBlitzes (park-level species counts) in many units of the National Park System.

In the near future, opportunities to engage citizen scientists in participatory and even collaborative science will become available. Distributed technology and broad data access will help create conditions in which citizens could participate in collaborative science and directly contribute to problem definition, data collection, and analysis. Parks can encourage visitors, virtual visitors, and individuals not visiting parks but interested in science to utilize data (potentially made more readily available by biocuration) and conduct their own science for parks in parks or at home.

De-extinction

De-extinction is the use of methods such as cloning and genetic engineering to re-create extinct species (Ogden 2014). For example, scientists in Spain are close to cloning the Pyrenean ibex (*Capra pyrenaica pyrenaica*), which went extinct in 2000 (Minteer 2014). In April 2015, an international team of scientists completed sequencing the entire genome of the wooly mammoth (*Mammuthus subplanifrons*), and is attempting to study characteristics of the long-extinct animal by inserting its genes into Asian elephant (*Elephas maximus*) stem cells (Ghosh 2015). For de-extinction to be successful, suitable habitats and large landscapes (depending on the resurrected species) will need to be stocked and protected—and national parks represent obvious candidates.

While there are scientific and ethical arguments in favor of de-extinction, such as that of "righting the past wrongs" of human-caused extinction (Sherkow and Greely 2013; Brand 2014), there are also challenges related to concerns about limited genetic diversity and the high probability of unintended consequences, as well as critiques of its feasibility (Ehrlich 2014;

Minteer 2014; Ogden 2014). Aldo Leopold provides a cautious reminder: "Our tools are better than we are and they grow better and faster than we do" (Leopold 1991).

Human-Assisted Evolution

Another emerging frontier with intertwined scientific, public policy, and ethical issues is "human-assisted evolution" (van Oppen et al. 2014). Scientists are currently experimenting with heat-hardy coral, like that found in American Samoa, in order to combat the threat climate change poses to coral reefs elsewhere (Mascarelli 2014). In controlled nurseries, plant husbandry specialists are choosing selected kinds of variants and hybrids in order to create plants with desirable traits for Earth's changing environment.

Parks can serve as laboratories and in situ testing grounds for human-assisted evolution and controlled plant husbandry, with the objective of creating organisms adapted to current and future environments and resilient to natural or human-induced environmental changes. Genetically modified plants and animals can be grown and/or introduced into parks in the future. This is controversial; there is an intense and important debate about the help or harm associated with genetically modified organisms in general (e.g., Hails 2000) and their introduction into natural ecosystems (like selected parks) in particular (Beringer 2001).

The Triple Helix

The convergence of interest and organizational structures that link universities, government, and industry (Ranga and Etzkowitz 2013) has sometimes been called "the triple helix." An example of the triple helix includes the Massachusetts Institute of Technology, the federal government, and the high-tech companies that surround the university campus. The triple helix has become one of the characteristics of contemporary science. The convergence of economic relationships can result in both significant science applications and legitimate concerns about power, transparency, accountability, academic freedom, proprietary control, and bias-driven science. In addition, nongovernmental organizations (NGOs) with strong advocacy objectives can substitute for industry in similar conditions and with similar concerns.

To date, the triple helix of government, university, and industry/NGO has only just begun to exert influence on science in parks. But as partnering becomes increasingly the modus operandi of park science, and on-demand scientific expertise replaces institutional careerists, the potential for the triple helix bonds to create problems in the conduct and outcomes of park science (and in extreme cases, scientific misconduct) is likely to increase. Even more prevalent may be bogus claims of scientific misconduct aimed at impeding, halting, or reversing science-informed decisions regarding park resources. Protections include clear common missions; vastly increased training on science ethics; transparency of methods, data, analysis, and results; vigorous peer review; and open access to the results of partnered park science.

Data Collection versus Surveillance

The near-horizon future may see a heightened tension between useful data collection for science and aggressive surveillance. Oversurveillance represents an intrusion into personal privacy as well as the potential for misuse and abuse of information about citizens, from personal preferences to private property conditions. The line between acceptable and appropriate research data collection and surveillance is unclear, and the ability to create immense and intrusive data sets under the original use for science and transfer this information to industry, advocacy, or security uses is pervasive and expanding. Collecting personal data at unforeseen scale and detail and using the emerging tools of big data analytics, biocuration, and AI to transform the data into actionable information runs up against the dangers of creating a surveillance society (Lyon 2004). For example, the National Security Agency's program that collected the metadata on phone records of millions of Americans in bulk as a counterterrorism measure has been criticized as an invasion of privacy (Lyon 2014), and its legal provisions have been (as of summer 2015) restricted.

It has been and will remain useful for park managers, often working with elements of the triple helix described above, to collect information about visitors and the wider general public. Protections are in place (such as Office of Management and Budget approval and university institutional review boards) to reduce potential misuse; they are likely to be and need to be strengthened in the near future.

Decisions on such data collection do not hinge only on existing requirements, protections, and administrative processes. Public policy, informed by professional judgment, will need to respond to the tension of data collection *versus* surveillance. For example, parks (particularly wilderness parks) have the potential to serve as refuges from surveillance by the state, but also face the plausibility of intrusions, such as face recognition systems installed at trailheads under the moniker of safety purposes. Data collected about visitors in parks by cameras or other technology might help managers answer questions about visitors or promote visitor safety. However, the use of technology in parks may become invasive, and may degrade the ethos of parks and the values they represent. Park policies will need to balance the desire for and use of visitor information to support management with visitor rights and privacy concerns.

The Divide between the Scientific Community and the Public

The demonstrated and growing divide between the knowledge and beliefs of the scientific community and those of the general public suggests a near-horizon weakening of public support for science, and lack of public support poses a threat to the advancement of science in the United States. The Pew Foundation in cooperation with the American Association for the Advancement of Science recently completed a major study of public and scientific community attitudes about science (Pew Research Center 2015). The results are revealing. For example, the public does not share the scientific consensus about evolution. While 87% of scientists believe life has evolved over time owing to natural processes, only 32% of the general public shares this belief. These and other results highlight public indecision and confusion about evolution and other scientific understandings (such as climate change), and demonstrate the increasing disconnect between the scientific community and the public.

For park science—including research on evolution, ecosystems, climate change, wildlife behavior, geology, and more—such disconnection has important implications. If scientists in general and park scientists in particular are dismissive of public attitudes, or even more inappropriately treat citizens with disguised or open contempt over their views (on everything from genetically modified organisms to climate change to evolution), it is at the scientific community's and the parks' peril. Instead, parks should serve as platforms for knowledge dispersion and public forums for controversial topics like de-extinction and human-assisted evolution. Parks can expose the public to scientific tools and processes by engaging them in all four levels of citizen science, and can employ their interpretive programs to help visitors experience and understand science and scientific concepts. In doing so, parks can reduce the belief gap and increase the likelihood of citizen support for science generally and in parks.

The Horizon Past and the Near-Horizon Future: *Revisiting Leopold*

In 1963, A. Starker Leopold (son of Aldo Leopold and brother of Estella Leopold) chaired a committee of scientists charged with examining wildlife policy in the US national parks. Their report, *Wildlife Management in the National Parks* (Leopold et al. 1963), was soon known as the Leopold Report. The report, first resisted by the NPS, became foundational to NPS natural resource policy, and its description of mission objective became (and is currently) a core part of NPS management philosophy: "A national park should present a vignette of primitive America."

Yet much has changed since 1963: (1) the National Park System has grown considerably in acreage, number of units, and kinds of parks; (2) there has been a fourfold increase in visitation, and the demographic mix of Americans (and park visitors) has diversified; (3) climate change, drought, sea-level rise, and biodiversity loss (all possibly related) have created major shifts in ecosystems and biological communities; (4) exotic species and human development in the form of cities, suburbs, and gateway communities have led to biodiversity loss; and (5) new scientific fields and techniques have emerged directly relevant to park science.

Hence, in 2011, nearly 50 years after the Leopold Report was released, NPS Director Jarvis commissioned a prestigious group of scientists (including Nobel Prize winners, National Academy of Science members, and others) to reexamine the Leopold Report, consider the near-horizon future, and provide recommendations on the future of park stewardship and science for the national parks in their second century. The report, *Revisiting Leopold: Resource Stewardship in the National Parks* (Colwell et al. 2012), provides a very different foundational paradigm for park resources: "The overarching goal of NPS resource management should be to steward NPS resources for continuous change that is not yet fully understood."

The proposal that the near-horizon future of the national parks is one of "continuous change that is not yet fully understood" is both challenge and opportunity for park science. Conducting science in parks undergoing dynamic environmental change requires new disciplines and fields of study that can generate and test new hypotheses. New methods, tools, and data are necessary, along with new ways of organizing and delivering science for parks. It requires (as the report notes) "broad disciplinary and interdisciplinary scientific knowledge and scholarship . . . necessary to manage change while confronting uncertainty."

"Confronting uncertainty" is the very essence of science, and a necessary strategy for advancing our scientific understanding of parks, how they function, and how best to protect them for future generations. From quantum biology to cliodynamics and from biocuration to big data analytics, the emerging new areas of science will likely have significant impacts on science in the parks, often with unintended or unforeseen consequences. C. C Furnas's depiction of science as "unfinished business" still holds true and always will. The near-horizon future of park science will most likely be extraordinary and, the modest predictions in this chapter aside, surprise us all.

Acknowledgments

The author would like to acknowledge and thank the National Park Service, the University of California, Berkeley, and Madeline Duda, graduate student at Clemson University, who assisted in the preparation of this chapter. The views and opinions of the author do not state or reflect those of the US government. Any reference within the author's work to specific commercial products, processes, or services by trade name, trademark, manufacturer, or otherwise, does not constitute or imply its endorsement, recommendation, or favoring by the US government.

Literature Cited

Al-Khalili, J. S., and J. J. McFadden. 2014. Life on the edge: the coming of age of quantum biology. Bantam Press, London, United Kingdom.

Austen, K. F. 2015. Environmental science: pollution patrol. Nature 517:136–138.

Ball, P. 2011. Physics of life: the dawn of quantum biology. Nature 474:272–274.

Beringer, J. E. 2001. Releasing genetically modified organisms: will any harm outweigh any advantage. Journal of Applied Ecology 37:207–214.

Bonney, R. E., C. B. Cooper, J. L. Dickinson, S. Kelling, T. B. Phillips, K. V. Rosenberg, and J. L. Shirk. 2009. Citizen science: a developing tool for expanding science knowledge and scientific literacy. BioScience 59:977–984.

Brand, S. 2014. The case for de-extinction: why we should bring back the woolly mammoth. Yale Environment 360, 13 January. Accessed 3 June 2015. http://e360.yale .edu/feature/the_case_for_de-extinction_why_we_should_bring_back_the_wooly _mammoth/2721/.

Brooks, D. B. 2003. Rose-tinted ecology. Review of Win-win ecology: how the Earth's species can survive in the midst of human enterprise. PLoS Biology 1:e73.

Colwell, R., S. Avery, J. Berger, G. E. Davis, H. Hamilton, T. Lovejoy, S. Malcolm, et al. 2012. Revisiting Leopold: resource stewardship in the national parks. A report of the National Park System Advisory Board Science Committee. http://www.nps.gov/ calltoaction/PDF/LeopoldReport_2012.pdf.

Dietl, G. P., and K. W. Flessa. 2011. Conservation paleobiology: putting the dead to work. Trends in Ecology and Evolution 26:30–37.

Ehrlich, P. R. 2014. The case against de-extinction: it's a fascinating but dumb idea. Yale

Environment 360, 13 January. Accessed 3 June 2015. http://e360.yale.edu/feature/the_case_against_de-extinction_its_a_fascinating_but_dumb_idea/2726/.

Fan, J., F. Han, and H. Liu. Challenges of big data analysis. National Science Review 1:293–314.

Flessa, K. W. 2002. Conservation paleobiology. American Paleontologist 10:2–5.

Flessa, K. W., D. L. Dettman, B. R. Schone, D. H. Goodwin, C. A. Rodriguez, and S. K. Noggle. 2001. Since the dams: historical ecology of the Colorado Delta. Poster presented at United States–Mexico Colorado River Delta Symposium, International Boundary and Water Commission, Department of Interior and the Mexican Secretariat of the Environment and Natural Resources, Mexicali, Baja California, Mexico, 11–12 September.

Francis, R. A., and J. P. Lorimer. 2011. Urban reconciliation ecology: the potential of living roofs and walls. Journal of Environmental Management 92:1429–1437.

Furnas, C. C. 1942. The next hundred years. World Publishing Company, Cleveland, Ohio.

Ghosh, P. K. 2015. Mammoth genome sequence completed. BBC News: Science & Environment, 23 April. Accessed 24 April 2015. http://www.bbc.com/news/science-environment-32432693.

Gil, Y. A., M. F. Greaves, J. A. Hendler, and H. Hirsh. 2014. Amplify scientific discovery with artificial intelligence: many human activities are a bottleneck in progress. Science 346:171–172.

Hails, R. S. 2000. Genetically modified plants: the debate continues. Trends in Ecology and Evolution 15:14–18.

Haklay, M. E. 2013. Citizen science and volunteered geographic information: overview and typology of participation. Pages 105–122 in D. Sui, S. Elwood, and M. Goodchild, eds. Crowdsourcing geographic knowledge. Springer Netherlands, Dordrecht, Netherlands.

Hand, E. 2014. The rise of the CubeSat. Science 346:1449.

Howe, D. E., M. C. Costanzo, P. Fey, T. Gojobori, L. I. Hannick, W. A. Hide, D. P. Hill, et al. 2008. Big data: the future of biocuration. Nature 455:47–50.

Kelly, R. P., J. A. Port, K. M. Yamahara, R. G. Martone, N. E. Lowell, P. F. Thomsen, M. E. Mach, et al. 2014. Harnessing DNA to improve environmental management. Science 244:1455–1456.

Leopold, A. S. 1991. Engineering and conservation. Pages 249–254 in S. L. Flader and J. B. Callicott, eds. The river of the mother of God and other essays by Aldo Leopold. University of Wisconsin Press, Madison, Wisconsin.

Leopold, A. S., S. A. Cain, C. M. Cottam, I. N. Gabrielson, and T. L. Kimball. 1963. Wildlife management in the national parks: the Leopold report. Unpublished report to the US Secretary of the Interior.

Li, Y. S., and L. Chen. 2014. Big biological data: challenges and opportunities. Genomics, Proteomics & Bioinformatics 12:187–189.

Lundholm, J. T., and P. J. Richardson. 2010. Habitat analogues for reconciliation ecology in urban and industrial environments. Journal of Applied Ecology 47:966–975.

Lyon, D. 2004. The electronic eye: the rise of surveillance society. Polity, Cambridge, United Kingdom.

———. 2014. Surveillance, Snowden, and big data: capacities, consequences, critique. Big Data & Society, July. doi:10.1177/2053951714541861.

Mascarelli, A. L. 2014. Designer reefs: biologist are directing the evolution of corals to prepare them to fight climate change. Nature 508:444–446.

Miller-Rushing, A. J., R. B. Primack, and R. E. Bonney. 2012. The history of public participation in ecological research. Frontiers in Ecology and the Environment 10:285–290.

Minteer, B. A. 2014. Is it right to reverse extinction? Nature 509:261.

Ogden, L. E. 2014. Extinction is forever . . . or is it? BioScience 64:469–475.

Pew Research Center. 2015. Public and scientists' views on science and society. Washington, DC.

Ranga, M., and H. Etzkowitz. 2013. Triple helix systems: an analytical framework for innovation policy and practice in the knowledge society. Industry and Higher Education 27:237–262.

Rosenzweig, M. L. 2003. Win-win ecology: how the Earth's species can survive in the midst of human enterprise. Oxford University Press. Oxford, United Kingdom.

Sherkow, J. S., and H. T. Greely. 2013. What if extinction is not forever? Science 340:32–33.

Turchin, P. V. 2010. Arise 'cliodynamics.' Nature 467:18–21.

Turchin, P. V., and S. A. Nefedov. 2008. Secular cycles. Princeton University Press. Princeton, New Jersey.

Van Oppen, M. J. H., J. K. Oliver, H. M. Putnam, and R. D. Gates. 2014. Building coral reef resilience through assisted evolution. Proceedings of the National Academy of Sciences USA 112:2307–2313.

EIGHTEEN

Science, Parks, and Conservation in a Rapidly Changing World

STEVEN R. BEISSINGER AND DAVID D. ACKERLY

Introduction

In March 1915, Stephen Mather and Horace Albright gathered 75 park administrators and rangers, businessmen, scientists, politicians, and conservationists for a three-day conference on US national parks in Berkeley, one of the first gatherings of its kind. Some arrived after a week of train travel, others after trips in motor cars or on horseback, and one Berkeley resident, Mrs. E. T. Parsons, walked in off the street after having read about the conference in the local newspaper (Albright and Albright Schenck 1999). They met on the University of California, Berkeley, campus for two days in California Hall, the building now occupied by the university's chancellor and administrative staff. Led by UC Berkeley graduates Mather, Albright, and Mark Daniels, the nation's first superintendent and landscape engineer for parks, they discussed the conditions and management problems facing the national parks, which at that time numbered about a dozen parks and several national monuments, each being managed independently. As described by Albright, the conversations centered on issues related to development of park facilities and problems facing the management of parks and their wildlife (Albright and Albright Schenck 1999). One and a half years later, the US National Park Service (NPS) would be enshrined in legislation (the Organic Act of 1916), a landmark event that spurred the growth of parks and protected lands in the United States and around the world over the next century.

In this chapter, we conclude the book by considering how science for parks and conservation has changed over the past century since the birth of the NPS, and how the present rapidly changing world may demand changes to conservation and management practices during the second century of the NPS. We begin by examining the state of science and conservation at the time the Organic Act was written and the NPS was launched. Then we consider how conservation science and park management have changed over the past century. We finish by discussing the key issue facing the future of national parks in the United States and parks throughout the world—how to steward them through the rapid environmental and cultural changes taking place in the world—and by considering three paradigms for park stewardship.

Science and Conservation at the Birth of the National Park Service

Attendees spent the third day of the Berkeley national parks meeting in March 1915 at the Panama-Pacific International Exposition in San Francisco (fig. 18.1), where they were greeted as dignitaries (Albright and Albright Schenck 1999). The Pan-Pacific Fair, as it was called, was a 10-month exposition of technology and culture that celebrated the recent completion of the Panama Canal and the rebirth of San Francisco following the 1906 earthquake (Ackley 2014). Called a "world university" by its director, James Barr (Nolte 2015), the fair's exhibits provide a view of the state of "progress" and show a world on the cusp of globalization and innovation. At the fair, the first transcontinental phone call was made from New York City to San Francisco by dignitaries including Alexander Graham Bell, inventor of the telephone and cofounder of AT&T. The airplane was a new thrill and crowds were treated to aerial displays until one crashed into San Francisco Bay. A 15-foot (4.6 m) tall, 28,000-pound (12,700 kg) Underwood typewriter was awarded the grand prize. Advertising described it as "the machine you will eventually buy," and it went on to revolutionize the workplace, not only with the QWERTY keyboard still in use today, but by increasing the number of jobs and demand for women in the workplace. The fair also launched a ukulele and Hawaiian music craze that swept across the United States (Ackley 2014).

Science at this time was primarily focused on advancing human health and welfare, producing efforts to eradicate such fatal diseases as leprosy, tuberculosis, typhoid, and hookworm. However, 1915 was also a year of important advances in basic science. The first genetic mutations were discovered by T. H. Morgan (Morgan et al. 1915). In 1915, Einstein's field equations of general relativity (Einstein 1915) and Wegener's theory of Pangaea (Wegener 1915) were first published. The former provided a framework for understanding time and space, while the latter eventually

18.1. On the third day of the conference at Berkeley on 19 March 1915 attendees took a ferry to San Francisco where they visited the Panama-Pacific International Exposition. Photo courtesy of the Library of Congress.

changed how we think about our world today and as it once was in paleo times. Also in that year, the Ecological Society of America was formed; forestry and entomology along with plant and animal ecology were among its subjects of interest. Henry Cowles, one of the society's founders, urged its members to embrace experimentation to advance the study of ecology beyond descriptions of species, their natural history, their geographic ranges, and their associations with other species (Kingsland 2015).

While conservation was still a relatively new concept in 1915 and had yet to develop into a science, important lessons had already been learned that would affect American perspectives on conservation throughout the next century. Experiences with the Passenger Pigeon (*Ectopistes migratorius*), American bison (*Bison bison*), and wading birds provided glimpses of the effects of unregulated exploitation, albeit with different outcomes. Martha, the last Passenger Pigeon, died in the Cincinnati zoo in 1915 (Blockstein and Tordoff 1985), about two decades after wild populations throughout the eastern United States, of what had been perhaps the most abundant land bird in North America, were wiped out to feed an urbanizing America. The American bison, which probably numbered 20–60 million individuals in the 19th century, would have suffered the same fate at the hands of market hunters and government agents if conservation and reintroduction efforts had not been started in 1902, when around 100 individuals remained (Hedrick 2009). Egrets and herons (Ardeidae) had been indiscriminately slaughtered to provide plumes for lady's fashionable hats in the late 19th century. But widespread efforts by concerned citizens who had formed some of the first conservation nongovernmental organizations (NGOs), including the National Audubon Society, resulted in legislation that outlawed the slaughter and employed wardens that patrolled the swamps to enforce it (Doughty 1975). Overexploitation of forests had also occurred from unregulated harvest, in part prompting the founding of the US Forest Service to regulate cutting on federal lands. For instance, deforestation of the northern woods of Wisconsin had been so severe that in 1896, with the decline of the logging industry, unemployed loggers invented the Hodag—a mythical forest creature—in an attempt to attract tourists to visit the region (Kearney 1928).

By 1915, however, the effects of introduced species or diseases on native flora and fauna were not yet fully understood. The Chestnut blight (*Cryphonectria parasitica*) had arrived in the eastern United States by 1915 on ornamental shrubs and was spreading throughout eastern hardwood forests at a rate of 24 miles (39 km) per year, but the effects of this fungus accidentally introduced from China were not yet well known (Freinkel 2007). Indiscriminant introductions of wildlife outside native ranges were already common. Game fish, such as rainbow trout (*Oncorhynchus mykiss*) and largemouth bass (*Micropterus salmoides*), were being moved around the United States (and soon the entire world), and the European starling (*Sturnus vulgaris*), a songbird, had been introduced from Europe to Central Park in the 1890s, beginning a westward range expansion that would eventually cover all of North America by the 1940s (Cabe 1993). Although Swedish scientist Svante Arrhenius suggested in 1896 that fossil fuel combustion would eventually cause global warming (Arrhenius 1896), global climate change had yet to be envisioned as a problem for park management or conservation.

Evolution of Science and Conservation in National Parks

Early Management Controversies Precipitated by the Organic Act

The Organic Act of 1916—the enabling legislation that created the NPS—consists of four short sections, and contains only brief statements to guide the growth and management of parks. The main dictate in the Organic Act—"to conserve the scenery and the natural and historic objects and the wild life therein and to provide for the enjoyment of the same in such manner and by such means as will leave them unimpaired for the enjoyment of future generations" (or the shorthand, "to conserve unimpaired")—has provided both a clear mission and a major challenge for those entrusted with managing US national parks. Moreover, conflicts sometimes arose between the two major goals articulated above in the Organic Act: "to conserve unimpaired" and "to provide for the enjoyment of" the natural and cultural resources.

There were already plenty of management controversies by 1916 facing the newly created NPS that needed science to be conducted in parks and a scientific approach to management to be enacted. Many of these issues were articulated by Joseph Grinnell in an essay he wrote with his former student Tracy Storer that was published in the journal *Science* in September 1916, within a month of the passage of the Organic Act. They wrote that "without a scientific investigation" of national park wildlife, "no thorough understanding of the conditions or of the practical problems they involve is possible." In other words, there was a strong need for "science for parks."

Grinnell and Storer advocated that the highest purpose of parks should be the preservation of their natural conditions as free as possible from all human interference for the purpose of "retaining the original balance in plant and animal life." Grinnell, having surveyed vertebrates throughout California and the western United States since the early 20th century to build the collections of UC Berkeley's Museum of Vertebrate Zoology, had seen the extent and impacts of land-use and environmental change in the region on biodiversity. He recognized that the national parks would "probably be the only areas remaining unspoiled for scientific study." In other words, an important goal of the US national parks should be "parks for science."

To Grinnell, preservation of parks' natural conditions meant that no trees (living or dead) should be cut, no understory vegetation should be removed to reduce fire hazards, no fires should be suppressed, no predators should be killed as part of control programs, no "pest" animals should be removed, and no nonnative species should be introduced (Grinnell and Storer 1916). Instead, national parks should be places where ecological processes are permitted to occur in the absence of human influence and where people can visit to recreate with nature. In contrast with this vision, all the above activities were taking place in and around US national parks in 1916 (Sumner 1983). Indeed, some national parks, including Yosemite, were displaying animals in cages or promoting human interactions by feeding wildlife. These activities expressed Mather's view of national parks as places of peace and beauty, free of fires and predators, and classrooms for the teaching of American values (Sellars 1997).

Wildlife management crises in US national parks had already arisen by 1916. Albright writes, "There was deep concern that some animals, the bison and the antelope in particular, might die out as a species. Years later this problem would be called one of endangered species" (Albright and Albright Schenck 1999). Large die-offs were happening in winter, especially in Yellowstone National Park, and some were advocating that park wildlife should be maintained with artificial feeding. The problem was exasperated by predator control in parks: "By 1916 mountain lions appeared to have been wiped out, and only a handful of wolves were left in Yellowstone. Coyotes roamed in abundance even though hundreds were shot or poisoned each year" (lbright and Albright Schenck 1999).

Thus, within a year or two of enacting the Organic Act, debates had already emerged about how parks and their wildlife should be managed, and about the impacts of "designed development" of infrastructure on wilderness.

The Fall and Rise of Science in Parks

Yet, as Mather began to build the NPS in 1916, he ignored the advice of Grinnell, Storer, and others, and de-emphasized science in favor of tourism and development (Sellars 1997). Mather chose to invest funding and resources in infrastructure, such as roads, and the development of facilities, such as grand hotels, to attract and serve tourists. A scientist was not hired until 1928, when in response to large forest fires in Glacier National Park, John Coffman moved from the US Forest Service to work under Ansel Hall, then the NPS chief naturalist, as part of the newly created NPS Division of Education and Forestry located at UC Berkeley (Sellars 1997). In 1929— two years after George Melendez Wright informally began work with his own funds—the Wildlife Division of the NPS was established with offices at UC Berkeley (Sumner 1983).

This marked the start of a century of cycles of the fall and rise of science in the US national parks (table 18.1). Sumner (1983) provides context for the first 50 years, beginning in the late 1920s with the rise of science. At the recommendation of an advisory committee, the NPS created the Branch of Research and Education in 1930, which was headed by Dr. Harold C. Bryant, who had been trained at UC Berkeley with Grinnell. At about the same time, the earliest scientific investigations were initiated in the form of inventories of fauna and flora in national parks. Wright wrote *Fauna of the National Parks of the United States* (hereafter *Fauna No. 1*) in 1933 with Joseph Dixon and Ben Thompson, which analyzed the ecological health of each park and its problems in the late 1920s and early 1930s. *Fauna No. 1* became the "bible" for park biologists and was in many ways the predecessor of the 1963 Leopold Report on wildlife management in national parks (Leopold et al. 1963). All three authors had been students in UC Berkeley's Museum of Vertebrate Zoology, where Grinnell taught.

Fauna No. 1 interpreted what "unimpaired" in the Organic Act meant in relation to protecting predators, artificial feeding of threatened ungulates, removal of exotic species, and restoration of extirpated native species, suggesting these were inappropriate activities in US national parks (Sellars 1997). Wright, Dixon, and Thompson (1933) recognized the inherent conflict between managing national parks to sustain natural conditions and the presence of large numbers of visitors in national parks. In Wright's view, the appropriate objective for management was identified as the ecological conditions that occurred between "the arrival of the first whites and the entrenchment of civilization," which included removal of exotic species. Furthermore, the report argued that NPS resource management poli-

Table 18.1 Summary of major actions resulting in the fall (↓) and rise (↑) of science in the US national parks over the past century

Direction	Year	Action
↓	1915	Mather invests mostly in infrastructure, little in science.
↑	1928	NPS Education and Forestry Division is established.
↑	1929	George Wright establishes the NPS Wildlife Division.
↑	1930	Harold Bryant heads new Branch of Research and Education.
↓	1939	NPS Wildlife Division is transferred to US Fish and Wildlife Service.
↓	1939	"Research" is dropped from the Branch of Research and Education.
↓	1941	World War II further decimates number of park scientists.
↑	1963	National Academy of Sciences report on research in parks stimulates expansion.
↑	1967	NPS Office of Natural Science Studies is created.
↑	1970	Cooperative Park Study Units (CPSUs) are initiated with universities.
↓	1993	Scientists are moved from the NPS to the new National Biological Survey.
↓	1993	CPSUs are phased out with the start of the National Biological Survey.
↑	1995	Position of associate director of NPS natural resource stewardship and science is created.
↑	1999	Cooperative Ecosystem Study Units (CESUs) open at universities.
↑	2000	NPS Inventory and Monitoring (I&M) Program is created.
↑	2007	NPS Climate Change Response Program is established.
↑	2009	Position of science advisor to the NPS director is established.
↑	2010	NPS Social Science Division is established.

Note: Based on descriptions in Sumner (1983), Sellars (1997), and the NPS archives (http://www.archives.gov/research/guide-fed-records/groups/079.html).

cies should be based on scientific research and that species should be not be actively managed unless threatened with extinction in a park.

Yet Wright, Dixon, and Thompson (1933) were already aware of the limitation of managing by noninterference: "Protection, far from being the magic touch which healed all wounds, was unconsciously just the first step on a long road . . . to restore and perpetuate the fauna in its pristine state by combating the harmful effects of human influence." Of particular significance was the realization that protection of many mammal species that migrated out of the high mountain parks required expanding park boundaries to protect wintering habitats. *Fauna No. 1* was followed by additional park inventories; by studies of wildlife management issues, such as Thompson's (1933) examination of conflicts between bird-watchers and fishermen over persecution of White Pelicans (*Pelecanus erythrorhynchos*) in Yellowstone National Park; and by early efforts to restore threatened wildlife, such as research to recover the highly endangered Trumpeter Swan (*Cygnus buccinator*) by Wright and associates in Yellowstone National Park and nearby Red Rock Lakes National Wildlife Refuge.

The rise of science for and in the US national parks reversed course and began an abrupt, steep descent in the late 1930s (see table 18.1). In 1939, the entire NPS Wildlife Division was transferred to the US Fish and Wildlife Service. Positions and duty stations mostly remained the same, but biologists now reported to a different agency with a different mission. Moreover, Sumner (1983) writes, "In 1939 the climate in Congress had grown so increasingly unfavorable to the concept of research that this word was dropped from the Branch of Research and Education." Biological research was further decimated by World War II, which reduced the number of wildlife biologists working in national parks "down to a vestige" (Sumner 1983). While park archaeology and history programs recovered rapidly after World War II ended, the number of scientists working in US national parks had not been restored by the time of the Leopold Report 20 years later (Sumner 1983). In the absence of the ecological knowledge necessary for good management, which was a result of the lack of support for science in parks from 1942 to 1963, the wildlife and ecosystems in many US national parks deteriorated (Sumner 1983; Sellars 1997). For example, saguaro cactus (*Carnegiea gigantea*) were disappearing from Saguaro National Monument, feral goats threatened the survival of native flora in the Hawai'ian national parks, diking and draining of upstream wetlands were drying Everglades National Park, and overpopulation of elk had occurred again in Yellowstone National Park.

The Leopold Report, *Wildlife Management in the National Parks*, in early 1963 was another turning point for science in parks (Leopold et al. 1963). Written by an advisory board appointed by Secretary of the Interior Stuart Udall, the report is a study of science and resource management in US national parks, especially as they relate to wildlife management. It took on the name of the committee's chair, A. Starker Leopold, who was a professor of wildlife ecology at UC Berkeley and the eldest son of famed conservationist Aldo Leopold. The Leopold Report independently reached many of the same conclusions about park science and management found in the long-forgotten *Fauna No. 1*, such as ending artificial feeding programs for ungulates and predator control, but it also emphasized the need for active management in parks (Sumner 1983; Sellars 1997). Similar to *Fauna No. 1*, it recommended that "the biotic associations within each park be maintained, or where necessary recreated, as nearly as possible in the condition that prevailed when the area was first visited by the white man." The Leopold Report recognized that this goal might not be fully achieved, given the extinction of species and the invasion of exotics, but it could be approached in many parks. The major management target identified in the

Leopold Report was to preserve or re-create a "reasonable illusion of primitive America" in US national parks for the "aesthetic, spiritual, scientific and educational values they offered to the public" (Leopold et al. 1963). The Leopold Report was followed soon after by a survey on research capacities of the NPS by a National Academy of Sciences Committee (Robbins et al. 1963).

As a result of these and other independent reviews resulting from recommendations of scientists not affiliated with the NPS, renewed interest grew in science-based management of national parks. This stimulated an expansion of funding for research in US national parks and hiring of scientists that began in the late 1960s (Sumner 1983). In 1967, the Office of Natural Science Studies was created. The number of scientists working for the NPS continued to increase with the establishment of the Air and Water Resources Division in 1978 and the creation of the Cooperative Park Study Units (CPSUs) located at a handful of universities in the 1970s (Agee, Field, and Starkey 1982). At its zenith in 1980, the CPSU network included 35 universities (Sellars 1997).

From 1993 to 2000, the cycle of fall and rise of science in the US national parks occurred yet again (see table 18.1). As in the 1930s, it began with a bureaucratic reorganization. In 1993, all NPS scientists were transferred, along with scientists and employees from six other agencies in the Department of the Interior, to staff a newly formed research agency, the National Biological Survey. It also absorbed all the scientists in the CPSUs and the units were closed. After congressional elections in 1994 resulted in a conservative Republican Congress that proposed to roll back environmental legislation, the National Biological Survey was transferred into the US Geological Survey's Biological Resources Division and its budget was cut by 15% (Wagner 1999). Once again, the science capacity in US national parks was reduced despite the best intentions of the US Geological Survey's Biological Resources Division to be the arm of science for many federal agencies.

Responses by the NPS at the turn of the century began the process of restoring its science capacity. First, the Cooperative Ecosystem Studies Units (CESUs) began forming at universities in 1999. Led initially by efforts of the NPS, the CESU network was a way to expand the science in the parks by replacing the CPSU model with a much larger network of collaborators. It quickly grew beyond park boundaries to become a national consortium of 371 partnering federal agencies, academic institutions, tribes, state and local governments, and conservation NGOs working together in all 50 states and the US territories to support research, technical assistance, education, and capacity building.[1] Second, based on a congressional mandate in 1998, the NPS established an Inventory and Monitoring (I&M) Program in 2000. Its goal is to track park conditions and resources in order to inform park managers about the status and trends of natural resources for making management decisions, working with other agencies, and communicating with the public.[2] The I&M Program is currently established in more than 270 parks, has more than 100 employees, and represents most of the science capacity that resides within the NPS. Science expansion over the past decade also included the creation of the Climate Change Response Program, the position of science advisor to the NPS director, and the establishment of the NPS Social Science Division (see table 18.1).

Thus, the NPS now has greater scientific capacity and greater access to scientists nationwide than at any time in its past. National parks will need all the science they can get given the challenges they now face, as discussed in the next section.

The Future of Science, Parks, and Conservation in a Rapidly Changing World

National Parks Today at a Pivotal Time in the Anthropocene

We live in a rapidly changing world. Over the past century that the NPS has been in existence, people have transformed land use at regional and life-zone scales through habitat loss, fragmentation, and defaunation (Vitousek et al. 1998; Ellis 2011; Dirzo et al. 2014), and overexploited fish stocks in the ocean, especially top predators, so that the average trophic level of the species now composing ocean communities has greatly declined (Pauly et al. 1998; Myers and Worm 2003). They have transported and introduced around the globe species and diseases that have invaded ecosystems and have reduced the ranges of native species, sometimes to the point of global extinction (Houlahan et al. 2000). Humanity has increased greenhouse gas concentrations through fossil fuel use and deforestation to cause climate change (Stocker et al. 2013) and ocean acidification (Tyrrell 2011). Some scientists have championed the label "Anthropocene" to describe this period in Earth's history in order to designate the rapid, large-scale changes to Earth and its ecosystems caused mainly by human actions (Crutzen 2002). Whether the Anthropocene began in 1610 when carbon

1. See http://www.cesu.psu.edu/default.htm (accessed 22 March 2016).
2. See http://science.nature.nps.gov/im/ (accessed 22 March 2016).

dioxide levels reached their lowest point in the Holocene owing to forest regrowth, in the late 18th century with the onset of the Industrial Revolution that would eventually shift the composition of atmospheric gasses causing climate warming, in 1950 when coal burning peaked and nuclear explosions left their stratigraphic mark, or at some other point in time continues to be debated (Steffen et al. 2011; Corlett 2015; Lewis and Maslin 2015).

The chapters in this book provide perspectives on the state of national parks and conservation at the start of the second century of the NPS. While the times have clearly changed since Mather and Albright, many of the problems from that era remain the same or have grown over the ensuing century; others are new. We depend on national, regional, and local parks to form the backbone of a biodiversity conservation network, which Edward O. Wilson (this volume, ch. 1) argues must be greatly expanded to be successful. There is an even larger need to create protected areas in the ocean where international collaboration is required (Grorud-Colvert, Lubchenco, and Barner, this volume, ch. 2). Yet how to design this system in a rational way to maximize returns from investments and engender public support is unclear, and could be a combination of designating new protected areas and restoring other lands to become parks (Possingham, Bode, and Klein 2015).

Change is a constant theme that resonates through most of the contributions in this book—biological, cultural, and technological change. Climate variation is natural, but climate change, caused by greenhouse gas emissions from human activities, involves increases in temperature and changes in precipitation that will rapidly exceed historical variation, with widespread consequences for natural and cultural resources in parks (Gonzalez, this volume, ch. 6). Novel climates and disturbance regimes are increasing, but whether they require novel management solutions or less intervention is debated (Turner et al., this volume, ch. 5). And it's not just climate that is changing. New species are arriving with greater frequency, successfully invading and transforming park ecosystems (Simberloff, this volume, ch. 8). Pollutants and wildlife of all kinds move across park boundaries (Baron et al., this volume, ch. 7; Berger, this volume, ch. 9). Boundary problems require thinking outside of the box and outside of the park, as the eminent ecologist Dan Janzen urged when he wrote about the "eternal, external threat" to parks (Janzen 1986). He advocated the need for park managers to work with landowners and governments outside park boundaries. It is a short step from this perspective to conceptualizing parks as coupled natural-human systems (DeFries, this volume, ch. 11) that engage a range of stakeholders as contributors to debates about park management (Dietz, this volume, ch. 12). Cultural change may be occurring even faster than ecological change. In the United States, Americans of African, Asian, or Latino heritage are increasing in proportion relative to the overall population and are rapidly gaining political and economic influence, but visit US national parks far less often than Americans of European heritage (Floyd 1999, 2001; Taylor, Grandjean, and Gramann 2011). Youth have also chosen to recreate in ways other than outdoor pursuits (Pergams and Zaradic 2008). Engaging youth and minorities as citizen scientists and inspiring them through cultural and spiritual connections to park features (Bernbaum, this volume, ch. 14) may be important ways to ensure a strong future for US national parks. For national parks throughout the developing world, conservation is most likely to be politically or culturally feasible and effective when the important roles those parks can provide for people's livelihoods are recognized and when alternatives to the "Yellowstone Model" of wilderness are considered (Enkerlin-Hoeflich and Beissinger, this volume, ch. 3). Finally, science itself is changing with rapid technological advances, and science in parks is likely to follow suit (Machlis, this volume, ch. 17).

Thus, the key issue facing the future of national parks in the United States and throughout the world is how to steward them through the environmental and cultural changes taking place in our world and societies. This was a theme in the most recent report on the state of stewardship in the US national parks by an external expert panel. The *Revisiting Leopold* report (Colwell et al. 2012) recognizes this new reality: "The overarching goal . . . should be to steward NPS resources for continuous change that is not yet fully understood, in order to preserve ecological integrity and cultural and historical authenticity, provide visitors with transformative experiences, and form the core of a national conservation land- and seascape." It also suggests the "precautionary principle" should be fully integrated into NPS decision making at all levels to avoid "actions and activities that may irreversibly impact park resources and systems" and enthusiastically requires that "stewardship decisions reflect science-informed prudence and restraint."

Stewarding parks for continuous change while at the same time embracing the precautionary principle to prevent impairment of historical authenticity may at times present a paradox. At the Berkeley summit that spawned this book, throughout its chapters, and in the scientific literature over the past decade, some voices have been urging that we need to move beyond the comfort zone of the precautionary principle by embracing proactive management to maintain ecological integrity, even using forward-looking management experiments, in preparation for the environmental changes that are coming, especially from climate change. Reconciling these views requires understanding how climate change becomes a game changer for management goals.

Management Paradigms and Goals for
National Parks in the Anthropocene

The 1963 Leopold Report that guided park management for the next half-century called for US national parks to capture "vignettes of primitive America" in its many forms. This vision aligns with the broader goal that has motivated much of 20th-century American conservation to preserve ecosystems in "pristine" conditions, as in wilderness areas, or to restore systems already affected by human activity to similar historical baselines. These are aspirational and difficult goals in the best of circumstances, but the continuing impact of human activity, especially the onset of rapid, global climate change, is forcing us to reevaluate the very goals themselves, as was implied in the *Revisiting Leopold* report (Colwell et al. 2012).

Two aspects of climate change, and its effects on biodiversity, set it apart from the threats that have been faced in the past, and require new approaches to conservation and resource management in the coming century. First, because greenhouse gases are thoroughly mixed in the atmosphere, the anthropogenic causes and impacts of climate change are shared globally, and the trajectory of change is "committed" for the next several decades. Unlike other environmental threats, resource managers and regional policymakers cannot effectively reduce the immediate exposure of local systems to changes in temperature, rainfall, and extreme events. Solutions that will significantly reduce the rate or eventual magnitude of change over the next several centuries depend on innovations in energy technology and policy far removed from the domain of local conservationists and park managers (although park agencies, along with individuals and organizations, all have a role to play reducing the carbon footprint of their own activities). Many of the management strategies developed over the years, from legal and regulatory restrictions to direct management interventions (e.g., invasive removal, toxin cleanups, etc.), have a very limited role in the response to climate change because of the global nature of the problem.

The second unusual feature is the nature of the "threat" itself. Climatic conditions define the fundamental physical context for natural processes influencing, and being influenced by, the composition and functioning of the biosphere. Over geological timescales, climate change is a ubiquitous aspect of Earth history, shaping the dynamics of biological evolution and the structure and function of ecological communities. What is exceptional in the current episode of anthropogenic climate change is the pace, global extent, and magnitude of change (Blois and Hadley 2009), setting it apart from all but the most unusual episodes in Earth history. The high rate of change will rapidly shift conditions beyond the range of historical variability observed over the last several thousand years (Parmesan 2006; Williams, Jackson, and Kutzbacht 2007). These novel conditions have already caused and will continue to precipitate biotic impacts, such as elevated tree mortality (Allen et al. 2010), as well as adaptive responses, including local adaptation and range shifts (Parmesan 2006). However, constraints on the rates of biological response (e.g., limited genetic variation, dispersal limitation) are likely to result in widespread ecological disequilibria and potentially irreversible tipping points for biodiversity, including species extinctions (Svenning and Sandel 2013).

From a human perspective, perhaps the most difficult challenge presented by climate change is the threat to our sense of place and the stability of nature. In a time of unsettling social and technological change, the natural world offers personal rejuvenation and reassurance that not everything is trampled and transformed by consumption and growth. Even scientists can feel a disconnect between the intellectual knowledge of the dramatic ecosystem changes that accompany episodes of climate change and the emotional sense of stability and timelessness provided by nature, especially in returning to places they have been in the past. *The conservation movement in many ways was built on the promise of sustaining that sense of stability; embracing and even facilitating change, especially in ecosystems that are not viewed as degraded, breaks that promise.* This emotional and psychological threat is further compounded by the sense of collective responsibility for the current episode of anthropogenic climate change and the resulting moral obligation to act. Even though ecosystems respond to climate change through a range of entirely natural processes, it is difficult to embrace the ecological changes underway as "natural" when the underlying changes in climate itself arise from profoundly "unnatural" processes of fossil fuel burning and land-use change.

The conservation literature is buzzing with ideas about new strategies and tactics to adapt to climate change, to enhance ecosystem resilience, and to conserve biodiversity while accepting the inevitability of changes that will result from life in the Anthropocene. As often noted, science is best equipped to answer the question of how to achieve particular goals, or to evaluate the consequences of alternative actions. *Perhaps the more difficult task is to reevaluate the goals themselves, and to articulate conservation objectives that are not tied to maintenance or restoration of particular communities or to a narrow focus on conservation of particular species (endangered or otherwise).* In recent years, a range of management objectives has been articulated, including naturalness, historical fidelity, ecological integrity, ecosystem resilience, and the maintenance of nature's autonomy (Hobbs et al. 2010). While each of these presents opportunities and challenges, we believe that a clear focus on the question of stability versus change and historic versus future conditions is an essential starting point in the discussion of goals.

In this context, we propose three paradigms (discussed below) that may be useful as a basis for continued discussion of management goals in national parks, and protected lands in general, in the 21st century. Figure 18.2 links these paradigms to a spectrum of management interventions commonly undertaken or proposed and the resulting targets that represent the desired outcome of intervention. Most interventions in the toolbox of resource management can serve a range of objectives, depending on the details of how, when, and where they are used, and how they are tailored to achieve specific objectives. We focus on terrestrial and freshwater ecosystems, and hope that this framework, while not comprehensive, will at a minimum be useful in framing ongoing discussions of these challenging problems.

1. Manage to preserve historical and current ecological communities. In a world of rapid change and degradation of places affected by human activity, parks have and will continue to provide a glimpse of the past, a place where people can see a bit of what the world used to be like. The ideal of preserving or restoring systems to pre-European conditions, as recommended by the 1963 Leopold Report, now appears to be a wistful goal to many observers. In the face of rapid environmental change, the baseline may even need to shift forward to the goal of preserving vignettes of the present, the world as it was at the dawn of the 21st century. Managing for persistence of contemporary biodiversity is at the core of the foundational commitment to prevent extinction, and focuses attention on conservation of climate refugia and resilient landscapes that provide suitable conditions for as many species and communities as possible. In the face of legal mandates to protect particular systems and species, a cautious approach that hews closely to historical practice may still present the fewest risks to managers and policymakers.

In this management paradigm, parks would serve the role of "ecosystem museums" (Tweed 2010), living repositories of ecosystems that perhaps are changing more quickly in places lacking intensive management. This

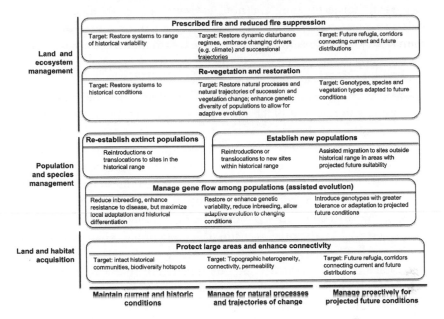

	Maintain current and historic conditions	Manage for natural processes and trajectories of change	Manage proactively for projected future conditions
Land and ecosystem management	**Prescribed fire and reduced fire suppression**		
	Target: Restore systems to range of historical variability	Target: Restore dynamic disturbance regimes, embrace changing drivers (e.g. climate) and successional trajectories	Target: Future refugia, corridors connecting current and future distributions
	Re-vegetation and restoration		
	Target: Restore systems to historical conditions	Target: Restore natural processes and natural trajectories of succession and vegetation change; enhance genetic diversity of populations to allow for adaptive evolution	Target: Genotypes, species and vegetation types adapted to future conditions
Population and species management	**Re-establish extinct populations**	**Establish new populations**	
	Reintroductions or translocations to sites in the historical range	Reintroductions or translocations to new sites within historical range	Assisted migration to sites outside historical range in areas with projected future suitability
	Manage gene flow among populations (assisted evolution)		
	Reduce inbreeding, enhance resistance to disease, but maximize local adaptation and historical differentiation	Restore or enhance genetic variability, reduce inbreeding, allow adaptive evolution to changing conditions	Introduce genotypes with greater tolerance or adaptation to projected future conditions
Land and habitat acquisition	**Protect large areas and enhance connectivity**		
	Target: intact historical communities, biodiversity hotspots	Target: Topographic heterogeneity, connectivity, permeability	Target: Future refugia, corridors connecting current and future distributions

18.2. Three paradigms for management of national parks (*x* axis) in relation to a spectrum of management interventions (*y* axis) commonly undertaken or proposed, and the targets that represent the desired outcome of intervention.

approach will appeal to those who think the potential impacts of climate change may be overstated, and maintains focus on the many other persistent threats to parks and biodiversity. For those who think that the biological impacts of climate change will be rapid and severe, it seems fundamentally unrealistic to even strive for this goal and impossible to achieve it, except on very limited spatial scales (e.g., the use of intensive irrigation to save individual groves of trees, or similar approaches). While a cautious approach that hews closely to historical practice may still present the fewest risks to managers and policymakers, inadvertent consequences can occur when attempting to restore systems to historical conditions, such as in the case of the emergence of giant fennel (*Foeniculum vulgare*) as a dominant nonnative invasive plant that displaced native vegetation after sheep (*Ovis aries*) had been removed from Santa Cruz Island (Dash and Gliessman 1994), or by taking no management actions.

Managing for the maintenance of current systems can require active intervention, and restoration of natural processes, such as fire regimes, can play a critical role in efforts to restore systems toward historical baselines. As Safford et al. (2012) have argued, historical conditions can serve as a "waypoint" for restoration, and not necessarily an "endpoint"; they provide a reference condition to guide restoration toward more resilient communities, although over time they may give way to more significant change, transitioning to our second paradigm.

2. Manage for natural processes and trajectories of change. A second management paradigm would be to manage ecosystems for "unimpaired" or restored functions and ecological processes, allowing the resulting trajectories of ecological change to unfold in response to changing conditions, especially changing climate. Allowing fires to burn and postfire succession to proceed without intervention is an important example, and paleoecological research can add to the collective understanding of the range of historical variability as context for current trajectories (Turner et al., this volume, ch. 5). In other cases, however, it may be more difficult to embrace the processes and trajectories as "natural"—for example, when witnessing a wildlife die-off from disease (Smith, Lips, and Chase 2009), or widespread tree mortality and potential vegetation conversion following drought or pest outbreaks (Allen et al. 2010). It is important to remember that species range shifts unfold as a series of local population extinctions at "trailing edges," which may be perceived as conservation failures, and/or establishment of new populations at "leading edges," where the arrival of novel species may raise concerns. Embracing changes of this sort will require significant shifts in conventional conservation thought.

The idea of managing ecosystems for enhanced resilience is broadly consistent with this paradigm, as it focuses on enhancing the capacity of systems to respond and sustain biodiversity and ecosystem function, while accepting and in some cases embracing the inevitability of change. Conservation of heterogeneous landscapes is a critical strategy in this regard, aiming to increase the probability that a species will encounter suitable conditions in the future in close proximity to current populations (Loarie et al. 2009; Lawler et al. 2015). Conservation or restoration of landscape connectivity, via corridors or enhanced permeability, is similarly important, increasing dispersal capacity for species to move across the landscape in response to climate change (Beier 2012). The conservation of large landscapes, in individual parks or integrated regional systems that are well connected, also greatly enhances the likelihood that components of the historical biodiversity occupying a region will persist somewhere within the park network, even if not in their original locations.

Managing for natural processes reflects the overarching goal for US national parks recommended in the *Revisiting Leopold* report (Colwell et al. 2012): "to steward NPS resources for continuous change that is not yet fully understood, in order to preserve ecological integrity and cultural and historical authenticity." The Canada National Parks Act, first passed in 1930, also established the goal of maintaining "ecological integrity" when reauthorized in 2000, which it defined as "a condition that is determined to be characteristic of its natural region and likely to persist, including abiotic components and the composition and abundance of native species and biological communities, rates of change, and supporting processes." The objectives of national parks in both Canada and the United States capture the tension inherent in embracing change and natural processes while seeking to maintain historical components deemed "characteristic" or "authentic" for a region or park. At times, this tension could lead to relaxing the standards set by historical baselines. While this tension may capture the ongoing reexamination of conservation goals in the face of environmental change, it may also provide the necessary ambiguity that allows managers on the ground to adapt general principles to specific situations and local values.

3. *Manage for change based on projected future conditions.* A third management paradigm would be to actively manage for change, facilitating changes that move systems toward projected future function or composition. The rationale for such an approach would result from confidence in scientific projections that the rates of climate change are going to be sufficiently high for a location that natural rates and mechanisms of response will be insufficient for communities to keep pace, resulting in widespread loss of biodiversity and reduced ecosystem services. These losses could be mitigated by active intervention, especially to remove barriers to dispersal limitation and enhance genetic diversity. Management actions could include restoration with novel genotypes or species (i.e., new "planting palettes"), managed relocation of threatened species, assisted evolution by introducing potentially adaptive genotypes, and other proactive interventions (Hobbs et al. 2011; Koralewski et al. 2015, van Oppen et al. 2015). Experiments in these types of interventions are already underway (van Oppen et al. 2014; Castellanos-Acuna, Lindig-Cisneros, and Saenz-Romero 2015).

Much has been written about the ethics, efficacy, and feasibility of interventions designed for future conditions, especially managed relocation (Richardson et al. 2009; Schwartz et al. 2012; Klenk 2015), and whether the projected benefits of intervention outweigh the costs and the burden of responsibility that accompanies active intervention. It is important that these debates be grounded in a clear discussion of conservation goals and the question of whether natural responses to climate change will be insufficient to mitigate undesirable effects, such that the severity of the consequences that will result from inaction is deemed unacceptable. It may also be increasingly important to evaluate the acceptability of the interventions themselves, in addition to the goals. The end will not always justify the means if the means themselves require unaffordable investments or actions that violate the integrity of systems that are objects of preservation themselves (Heller and Hobbs 2014).

Managing for the future also requires increased attention by park managers to the context of regional landscapes, shifting cultural and social values, and the multiple layers of local, regional, and national governance (Hobbs et al. 2011). In other words, it requires thinking strategically outside the park. Range shifts will rearrange species distributions across mosaics of private and public lands (and different public agencies). As a result, assessment of conservation outcomes may need to be evaluated at larger scales than in the past. Changing management regimes, especially in the case of fire, must be coordinated at larger scales, as ecosystem processes operate across park boundaries with contrasting impacts on different stakeholders. National parks have a key role to play as anchors in regional conservation plans, exemplified in the vision of the "Y2Y" (Yellowstone to Yukon) project and the linked corridor planning along the spine of the continent (Locke 1993; Chester 2015). This potential role is captured in the goal of "Scaling Up" outlined in the NPS "Call to Action" (National Park Service 2014), sowing the seed for an expansive view of the new directions that will be required to steward America's greatest places through the next century of great change.

Conclusions

Over the past century since the founding of the NPS, there has been dramatic growth in the number and size of protected areas in the United States and around the world—both on land (Enkerlin-Hoeflich and Beissinger, this volume, ch. 3) and more recently in the ocean near the shorelines of participating countries (Grorud-Colvert, Lubchenco, and Barner, this volume, ch. 2). These conservation gains, while probably not enough to secure the future for biodiversity conservation (Wilson, this volume, ch. 1), have been accompanied by relatively modest changes in the philosophy and strategies of park and protected area management. As science has fallen and risen in prominence within the NPS over the past century (see table 18.1), it has been accompanied by the fall and rise of science-based management of park resources (Sellars 1997). With the emergence of climate change as a threat to natural and cultural resources in parks and to biodiversity in general (Gonzalez, this volume, ch. 6), and with the continued expansion of human enterprise and population growth, it seems likely that the near-term future will be a very difficult time for parks and for conservation before a new approach to sustainability eventually emerges (Cascio, this volume, ch. 16).

Serious thought needs to be given to the adequacy of the dominant paradigm embraced by 20th-century conservation—manage to maintain current and historic baseline conditions—to consider other paradigms, including managing for natural processes and trajectories of change and managing proactively for projected future conditions (see fig. 18.2). In the absence of being able to effectively reduce the immediate exposure of parks and protected areas to changes in climate and other future environmental threats, embracing and sometimes facilitating change in ecosystems that are not degraded may be the best hope to maintain these systems somewhere on the landscape of the future. There is, however, no simple solution or set of management actions to accomplish this goal. We have presented the three management paradigms as core alternatives, not as prioritized recommendations. Stewards of protected areas will likely need to embrace all three paradigms, determining the best combination of approaches to support valued natural and cultural resources in their parks. For instance, the goal of maintaining elements of historic ecosystems may be combined with restoring natural processes and preparing for change, while at the same time undertaking some active interventions to seed resilience for desired future conditions.

Determining which paradigm and associated course of action is most appropriate for a given situation in a park won't be a simple task. "Science for parks," when done well, can illuminate the benefits and costs associated with various management actions, and can be used, in combination with societal desires, to make wise decisions.

Acknowledgments

Reviews by Patrick Gonzalez, Nicole Heller, and Gary Machlis greatly improved this chapter. Gary Davis and John Dennis provided help with table 18.1.

Literature Cited

Ackley, L. A. 2014. San Francisco's jewel city: the Panama-Pacific International Exposition of 1915. Heyday Books, Berkeley, California.

Agee, J. K., D. R. Field, and E. F. Starkey. 1982. Cooperative Park Study Units: university-based science programs in the National Park Service. Journal of Environmental Education 14:24–28.

Albright, H. M., and M. Albright Schenck. 1999. Creating the National Park Service: the missing years. University of Oklahoma Press, Norman, Oklahoma.

Allen, C. D., A. K. Macalady, H. Chenchouni, D. Bachelet, N. McDowell, M. Vennetier, T. Kitzberger, et al. 2010. A global overview of drought and heat-induced tree mortality reveals emerging climate change risks for forests. Forest Ecology and Management 259:660–684.

Arrhenius, S. 1896. On the influence of carbonic acid in the air upon the temperature of the ground. Dublin Philosophical Magazine and Journal of Science (5th series) 41:237–275.

Beier, P. 2012. Conceptualizing and designing corridors for climate change. Ecological Restoration 30:312–319.

Blockstein, D. E., and H. B. Tordoff. 1985. Gone forever: a contemporary look at the extinction of the passenger pigeon. American Birds 39:845–851.

Blois, J., and E. A. Hadley. 2009. Mammalian response to Cenozoic climatic change. Annual Review of Earth and Planetary Sciences 37:181–208.

Cabe, P. R. 1993. European starling (*Sturnus vulgaris*). *In* A. Poole, ed. The Birds of North America Online. Cornell Lab of Ornithology, Ithaca, New York. Retrieved from the Birds of North America Online, http://bna.birds.cornell.edu/bna/species/048.

Castellanos-Acuna, D., R. Lindig-Cisneros, and C. Saenz-Romero. 2015. Altitudinal assisted migration of Mexican pines as an adaptation to climate change. Ecosphere 6:art2.

Chester, C. C. 2015. Yellowstone to Yukon: transborder conservation across a vast international landscape. Environmental Science & Policy 49:75–84.

Colwell, R., S. Avery, J. Berger, G. E. Davis, H. Hamilton, T. Lovejoy, S. Malcolm, et al. 2012. Revisiting Leopold: resource stewardship in the national parks. A report of the National Park System Advisory Board Science Committee. http://www.nps.gov/calltoaction/PDF/LeopoldReport_2012.pdf.

Corlett, R. T. 2015. The Anthropocene concept in ecology and conservation. Trends in Ecology & Evolution 30:36–41.

Crutzen, P. J. 2002. Geology of mankind. Nature 415:23–23.

Dash, B. A., and S. R. Gliessman. 1994. Nonative species eradication and native species enhancement: fennel on Santa Cruz Island. Pages 505–512 *in* W. L. Halvorson and G. J. Maender, eds. The Fourth California Islands Symposium: update on the status of resources. Santa Barbara Museum of Natural History, Santa Barbara, California.

Dirzo, R., H. S. Young, M. Galetti, G. Ceballos, N. J. B. Isaac, and B. Collen. 2014. Defaunation in the Anthropocene. Science 345:401–406.

Doughty, R. W. 1975. Feather fashions and bird preservation: a study in nature protection. University of California Press, Berkeley, California.

Einstein, A. 1915. Die Feldgleichungen der Gravitation. Sitzungsberichte der Preussischen Akademie der Wissenschaften zu Berlin 1915:844–847.

Ellis, E. C. 2011. Anthropogenic transformation of the terrestrial biosphere. Philosophical Transactions of the Royal Society A 369:1010–1035.

Floyd, M. F. 1999. Race, ethnicity and use of the national park system. Social Science Research Review 1:1–24.

———. 2001. Managing parks in a multicultural society: searching for common ground. Managing Reacreation Use 18:41–51.

Freinkel, S. 2007. American chestnut: the life, death and rebirth of a perfect tree. University of California Press, Berkeley, California.

Grinnell, J., and T. I. Storer. 1916. Animal life as an asset of national parks. Science 44: 375–380.

Hedrick, P. W. 2009. Conservation genetics and North American bison (*Bison bison*). Journal of Heredity 100:411–420.

Heller, N. E., and R. J. Hobbs. 2014. Development of a natural apactice to adapt conservation goals to global change. Conservation Biology 28:696–704.

Hobbs, R. J., D. N. Cole, L. Yung, E. S. Zavaleta, G. H. Aplet, F. S. Chapin, P. B. Landres, et al. 2010. Guiding concepts for park and wilderness stewardship in an era of global environmental change. Frontiers in Ecology and the Environment 8:483–490.

Hobbs, R. J., L. M. Hallett, P. R. Ehrlich, and H. A. Mooney. 2011. Intervention ecology: applying ecological science in the twenty-first century. BioScience 61:442–450.

Houlahan, J. E., C. S. Findlay, B. R. Schmidt, A. H. Meyer, and S. L. Kuzmin. 2000. Quantitative evidence for global amphibian population declines. Nature 404:752–755.

Janzen, D. H. 1986. The eternal external threat. Pages 286–303 *in* M. E. Soulé, ed. Conservation biology: the science of scarcity and diversity. Sinauer, Sunderland, Massachusetts.

Kearney, L. S. 1928. The Hodag and other tales of the logging camps. Democrat Printing Company, Madison, Wisconsin.

Kingsland, S. 2015. The Ecological Society of America: founders, founding stories, foundations. Bulletin of the Ecological Society of America 96:5–11.

Klenk, N. L. 2015. The development of assisted migration policy in Canada: an analysis of the politics of composing future forests. Land Use Policy 44:101–109.

Koralewski, T. E., H.-H. Wang, W. E. Grant, and T. D. Byram. 2015. Plants on the move: assisted migration of forest trees in the face of climate change. Forest Ecology and Management 344:30–37.

Lawler, J. J., D. D. Ackerly, C. M. Albano, M. G. Anderson, S. Z. Dobrowski, J. L. Gill, N. E. Heller, et al. 2015. The theory behind, and the challenges of, conserving nature's stage in a time of rapid change. Conservation Biology 29:618–629.

Leopold, A. S., S. A. Cain, C. M. Cottam, I. N. Gabrielson, and T. L. Kimball. 1963. Wildlife management in the national parks: the Leopold Report. Report delivered to the US Secretary of the Interior.

Lewis, S. L., and M. A. Maslin. 2015. Defining the Anthropocene. Nature 519:171–180.

Loarie, S. R., P. B. Duffy, H. Hamilton, G. P. Asner, C. B. Field, and D. D. Ackerly. 2009. The velocity of climate change. Nature 462:1052–1055.

Locke, H. 1993. Yellowstone to Yukon. A strategy for preserving the wild heart of North America. Wild Earth 3:68–72.

Morgan, T. H., A. H. Sturtevant, H. J. Muller, and C. B. Bridges. 1915. The mechanism of Mendelian heredity. Henry Holt, New York, New York.

Myers, R. A., and B. Worm. 2003. Rapid worldwide depletion of predatory fish communities. Nature 423:280–283.

Nolte, C. 2015. S.F. took global bow with 1915 fair. San Francisco Chronicle, San Francisco, California.

National Park Service. 2014. A call to action: preparing for a second century of stewardship and engagement. National Park Service, Washington, DC.

Parmesan, C. 2006. Ecological and evolutionary responses to recent climate change. Annual Review of Ecology, Evolution, and Systematics 37:637–669.

Pauly, D., V. Christensen, J. Dalsgaard, R. Froese, and J. Torres. 1998. Fishing down marine food webs. Science 279:860–863.

Pergams, O. R. W., and P. A. Zaradic. 2008. Evidence for a fundamental and pervasive shift away from nature-based recreation. Proceedings of the National Academy of Sciences USA **105**:2295–2300.

Possingham, H. P., M. Bode, and C. J. Klein. 2015. Optimal conservation outcomes require both restoration and protection. PLoS Biology **13**:e1002052.

Richardson, D. M., J. J. Hellmann, J. S. McLachlan, D. F. Sax, M. W. Schwartz, P. Gonzalez, E. J. Brennan, et al.2009. Multidimensional evaluation of managed relocation. Proceedings of the National Academy of Sciences USA **106**:9721–9724.

Robbins, W. J., E. A. Ackerman, M. Bates, S. A. Cain, F. D. Darling, J. M. Fogg Jr., T. Gill, et al. 1963. National Academy of Sciences Advisory Committee on research in the national parks: the Robbins Report. A report by the Advisory Committee to the National Park Service on research. National Academy of Sciences, National Research Council, Washington, DC.

Safford, H. D., G. D. Hayward, N. E. Heller, and J. A. Wiens. 2012. Historical ecology, climate change, and resource management: can the past still inform the future? Pages 46–62 *in* J. A. Wiens, G. D. Hayward, H. D. Safford, and C. M. Giffen, eds. Historical environmental variation in conservation and natural resource management. John Wiley & Sons, West Sussex, United Kingdom.

Schwartz, M. W., J. J. Hellmann, J. M. McLachlan, D. F. Sax, J. O. Borevitz, J. Brennan, A. E. Camacho, et al. 2012. Managed relocation: integrating the scientific, regulatory, and ethical challenges. BioScience **62**:732–743.

Sellars, R. W. 1997. Preserving nature in national parks, a history. Yale University Press, New Haven, Connecticut.

Smith, K. G., K. R. Lips, and J. M. Chase. 2009. Selecting for extinction: nonrandom disease-associated extinction homogenizes amphibian biotas. Ecology Letters **12**:1069–1078.

Steffen, W., J. Grinevald, P. Crutzen, and J. McNeill. 2011. The Anthropocene: conceptual and historical perspectives. Philosophical Transactions of the Royal Society A **369**: 842–867.

Stocker, T. F., D. Qin, G.-K. Plattner, M. Tignor, S. K. Allen, J. Boschung, A. Nauels, et al., eds. 2013. Climate change 2013: the physical science basis. Contribution of Working Group I to the Fifth Assessment Report of the Intergovernmental Panel on Climate Change. Cambridge University Press, Cambridge, United Kingdom.

Sumner, L. 1983. Biological research and management in the National Park Service: a history. The George Wright Forum **3**(4):3–27.

Svenning, J. C., and B. Sandel. 2013. Disequilibrium vegetation dynamics under future climate change. American Journal of Botany **100**:1266–1286.

Taylor, P. A., B. D. Grandjean, and J. H. Gramann. 2011. National Park Service comprehensive survey of the American public, 2008–2009: racial and ethnic diversity of National Park System visitors and non-visitors. Natural Resource Report NPS/NRSS/SSD/NRR—2011432. National Park Service, Fort Collins, Colorado.

Thompson, B. 1933. History and present status of the breeding colonies of the white pelican in the United States. Contribution of Wild Life Division, Occasional Paper No. 1. National Park Service, Washington, DC.

Tweed, W. C. 2010. Uncertain path: a search for the future of national parks. University of California Press, Berkeley, California.

Tyrrell, T. 2011. Anthropogenic modification of the oceans. Philosophical Transactions of the Royal Society A **369**:887–908.

van Oppen, M. J. H., J. K. Oliver, H. M. Putnam, and R. D. Gates. 2015. Building coral reef resilience through assisted evolution. Proceedings of the National Academy of Sciences USA 112:2307–2313.

van Oppen, M. J. H., E. Puill-Stephan, P. Lundgren, G. De'ath, and L. K. Bay. 2014. First-generation fitness consequences of interpopulational hybridisation in a Great Barrier Reef coral and its implications for assisted migration management. Coral Reefs 33:607–611.

Vitousek, P. M., H. A. Mooney, J. Lubchenco, and J. M. Melillo. 1998. Human domination of Earth's ecosystems. Science 277:494–499.

Wagner, F. H. 1999. Whatever happened to the National Biological Survey? BioScience 49:219–222.

Wegener, A. 1915. Die Entstehung der Kontinente und Ozeane. Friedr. Vieweg & Sohn, Braunschweig, Germany.

Williams, J. W., S. T. Jackson, and J. E. Kutzbacht. 2007. Projected distributions of novel and disappearing climates by 2100 AD. Proceedings of the National Academy of Sciences of USA 104:5738–5742.

Wright, G. M., J. S. Dixon, and B. H. Thompson. 1933. Fauna of the national parks of the United States: a preliminary survey of faunal relations in national parks. Contribution of Wild Life Survey. Fauna Series No. 1—May 1932. US Government Printing Office, Washington, DC.

Historical Connections between UC Berkeley, the Birth of the US National Park Service, and the Growth of Science in Parks

STEVEN R. BEISSINGER AND TIERNE M. NICKEL

This book, and the summit at the University of California, Berkeley, from 25 to 27 March 2015 that spawned it, builds on the historic linkage between UC Berkeley and the US National Park Service (NPS). National parks and public education—America's "two best ideas"—grew up together at UC Berkeley, which was established in 1868. Much of the inspiration for and the effort that produced the NPS came from UC Berkeley and its graduates more than a century ago. This remarkable history is revisited here, as it set the stage for the summit and this book.

UC Berkeley and the Birth of the NPS

Even before the NPS was born, Berkeley graduates and faculty were leaders in conservation science. *Joseph LeConte* (fig. A.1) began his 30+ year career as a professor of geology, natural history, and botany at UC Berkeley in 1869. He was also a cofounder of the Sierra Club and a pioneering researcher and educator in Yosemite before it became a national park. *Teddy Roosevelt* (fig. A.2), who played a significant role in the creation of 5 national parks, 18 national monuments, and 140 national forests, among other works for conservation, delivered two speeches on the Berkeley campus. The first was the commencement speech in 1903, after which he went on a camping trip with naturalist *John Muir* in Yosemite that led to the transfer of Yosemite from the state of California to federal protection. The second was the Charter Day speech in 1911, which inspired a young *Horace Albright* (alum of 1912; fig. A.3). In 1914, Albright was hired as an administrative assistant by then Assistant Secretary of the Interior *Stephen Mather* (alum of 1887; fig. A.4), another conservation-minded Berkeley graduate. Together, Mather and Albright are widely crediting with helping launch the NPS.

If Mather was going to build a park service, he needed the support of his superior, *Franklin Lane* (fig. A.5), the Secretary of the Interior during the early years of the national parks. Lane had a strong philosophy about nature: "A wilderness, no matter how impressive and beautiful, does not satisfy this soul of mine, (if I have that kind of thing). It is a challenge to man. It says, 'Master me! Put me to use! Make me something more than I am.'" Lane had studied at UC Berkeley for two years, but dropped out to become a reporter, then a lawyer, and eventually a politician. Had he continued at UC Berkeley until graduation, perhaps he would not have championed a dam in Yosemite National Park in 1913 that flooded the Hetch Hetchy Valley to create a water supply for the city of San Francisco.

In March 1915, Mather and Albright gathered 75 park administrators and rangers, businessmen, scientists, politicians, and conservationists at the Berkeley campus for a three-day conference on national parks, one of the first gatherings of its kind (see fig. 18.1). They met for two days in California Hall, the present-day location of the chancellor's office, to envision and build a future for the country's existing and desired national parks. Mather arranged it all, housing them in his old fraternity house, Sigma Chi, on College Avenue.

Mark Daniels (alum of 1908; fig. A.6), the first general superintendent and landscape engineer for the US national parks, was the first speaker. He believed that "economics and esthetics go hand in hand," and his philosophy was to plan parks for the maintenance of "the inexhaustible commercial resources of scenery." Led by the three Berkeley graduates—Mather, Albright, and Daniels—the conversations centered on issues related to development of park facilities and problems facing the management of parks, which at the time included about a dozen national parks and several national monuments, each being managed independently. Appalled by what they learned about the condition of parks, Mather and Albright had a goal to develop and enact legislation that would enable the handful of individual parks and monuments in existence to be brought together under a single national agency that would be responsible for their management, infrastructure, promotion, and protection.

In June 1915, Mather invited a group of 15 influential Americans to join him for a two-week trip through the Sierra Nevada Mountains of California. Dubbed the "Mather Mountain Party," the group comprised promi-nent publishers, including *Gilbert Grosvenor* of the National Geographic So-ciety, politicians, businessmen, and railroad builders. The challenging trip through the high Sierra backcountry of Sequoia National Park and climb of 14,495-foot Mount Whitney bonded the group who then embarked on a campaign to create an agency to oversee the parks.

The result was legislation (the National Park Service Organic Act) es-tablishing the NPS in 1916, a landmark event that over the next century inspired the growth of parks and protected lands in the United States and around the world. Mather became the first NPS director (1917–1929) and presided over the professionalization and expansion of the NPS. Albright worked closely with Mather on the launch, and then served as superinten-dent of Yellowstone and Yosemite before becoming the second NPS direc-tor (1929–1933). UC Berkeley and other universities have been intimately involved with conservation, parks, and protected areas ever since.

UC Berkeley's Contribution to Science in Parks

But UC Berkeley's legacy was not just launching the NPS. Much of the early and influential research in national parks was done by Berkeley faculty and graduates. None of this would have been possible without the influence of a Hawaiian sugar heiress, *Annie Alexander* (fig. A.7). A 33-year-old Alexan-der attended lectures on the Berkeley campus in 1900 and was inspired to become a vertebrate zoologist and paleontologist. She founded UC Berke-ley's Museum of Vertebrate Zoology in 1908, chose its first director, *Joseph Grinnell* (fig. A.8), and financially supported work by its faculty as well as faculty in the Museum of Paleontology. With her companion, *Louise Kel-logg* (alum of 1901; fig. A.7), Alexander collected over 22,000 vertebrate, fossil, and plant specimens throughout the western United States, many in locations that would become national parks, to catalogue the disappearing indigenous flora and fauna for posterity.

Grinnell also realized that California was changing rapidly. He led early inventories of birds and mammals in Yosemite in 1915, eventually surveying all the large national parks in California while developing his influential ideas about how climate shapes the ecological niche of plants and animals. He was also an early proponent of the role of science in park management, as well as the value of parks for science.

Grinnell trained many Berkeley graduates who would go on to become prominent leaders and biologists in the NPS. *Tracy Storer* (BS, 1912; PhD,

1921) worked with Grinnell on surveys in Yosemite and elsewhere in the Sierras, arguing strongly against the common practice of the time of shooting predators in parks. He had a distinguished career as a faculty member at UC Davis. *George Melendez Wright* (BS, 1925; fig. A.9) started the NPS Wildlife Division with his own funds in 1927, and for many years operated it on the Berkeley campus out of the top floor of Hilgard Hall in what is now the College of Natural Resources. With *Joseph Dixon* (PhD, 1915; fig. A.10) and *Ben Thompson* (BS, 1928; fig. A.11), Wright conducted the first formal survey of wildlife in national parks, *Fauna of the National Parks of the United States* (*Fauna No. 1*), which became the bible for NPS biologists. Wright died early in his career in a tragic car crash, but Dixon and Thompson produced many subsequent surveys in their long and distinguished Park Service careers.

Harold Bryant (PhD, 1914; fig. A.12) was an outstanding researcher who held a number of posts in the NPS over his eminent Park Service career spanning three decades, but he was most proud of establishing its interpretive program. *Ansel Hall* (BS, 1917; fig. A.13) was trained in UC Berkeley's forestry school. He became the first park ranger in Sequoia National Park and the first chief naturalist and first chief forester of the NPS, and established the Yosemite Museum.

Finally, three other NPS giants had strong Berkeley connections. *Newton Drury* (alum of 1912; fig. A.14) was the executive director of the Save the Redwoods League before he became the fourth NPS director (1940–1951), where he was known for resisting demands for consumptive uses of park resources. He later went on to oversee the California Division of Beaches and Parks. After *Lowell Sumner* received a master's degree under Grinnell's tutelage in 1931, he spent nearly his entire career as a wildlife researcher with the NPS, conducting seminal wildlife studies, pioneering aerial surveys of wildlife, and rising to the position of chief research biologist for the entire system before retiring in 1967. He was known as an outspoken champion of keeping parks in their natural condition and was instrumental in establishing wilderness parks in Alaska. *A. Starker Leopold* (PhD, 1944; fig. A.15) was Aldo Leopold's eldest child and a giant in his own right in wildlife conservation. Starker obtained his PhD at UC Berkeley, and returned to become a professor of wildlife ecology for 30 years. His 1963 report guided the management of the US national parks for 50 years.

Today at UC Berkeley, we carry on this tradition, standing on the shoulders and sometimes even literally walking in the footsteps of our predecessors, as we have done in the Grinnell Resurvey Project, which has resurveyed birds and mammals throughout national parks in California to understand the effects of climate and land-use change on distributions of species. Berkeley faculty, students, and alumni continue to conduct research for parks that produce key data and insights related to conservation and management with local and global impacts. Research on the social, cultural, and health benefits of parks contributes to understanding the barriers to and benefits of park use.

A.1. Joseph LeConte lecturing on his 76th birthday at UC Berkeley, 1899. Photo courtesy of the Bancroft Library, University of California, Berkeley (BANC PIC 1960.010ser.1).

A.2. President Theodore Roosevelt (*left*) at the 1903 commencement address with UC Berkeley president Benjamin Ide Wheeler (*right*), for whom the lecture hall was named where the 2015 "Science for Parks, Parks for Science" summit was located. Photo courtesy of Joseph R. Knowland Collection, Oakland History Room, Oakland Public Library (Knowland Neg. 88).

A.3. Horace Albright on his first day as superintendent of Yellowstone National Park, 1919. Photo courtesy of the Library of Congress.

A.4. Stephen Mather at Glacier Point, Yosemite National Park, 1926.
Photo by George Stone, courtesy of the National Park Service Historic
Lantern Slide Collection (Image Number C567.7-2).

A.5. Portrait of Franklin Lane. Photo courtesy of the Library of Congress Harris & Ewing Collection.

PHOTOGRAPH BY BOYÉ

Mark Daniels, of California, General Superintendent and Landscape Engineer of National Parks, an office recently created for him by the Secretary of the Interior. Mr. Daniels will put into practice some new and definite ideas upon what he terms "the inexhaustible commercial resources of scenery"

A.6. Portrait of Mark Daniels. Photo courtesy of Troy Ylitalo, Period Paper.

A.7. Annie Alexander (*left*) and Louise Kellogg (*right*) at a farm exhibit with their dairy shorthorn cattle. Photo courtesy of the Historic Photo Collection, Museum of Vertebrate Zoology Archives, University of California, Berkeley (MVZ IMG 6022).

A.8. Joseph Grinnell with a skull at UC Berkeley's Museum of Vertebrate Zoology, 1930. Photo by G. Elwood Hoover, courtesy of the Historic Photo Collection, Museum of Vertebrate Zoology Archives, University of California, Berkeley (MVZ IMG 8421).

A.9. George Melendez Wright, assistant naturalist at Yosemite National Park, talking with Maria Lebrada (Totuya), one of the last surviving members of the Yosemite Indian nation, 1929. Photo by Joseph S. Dixon, courtesy of the National Park Service.

A.10. Joseph Dixon measuring a day-old fawn in Yosemite National Park, 1928. Photo by J. Dixon, courtesy of the Historic Photo Collection, Museum of Vertebrate Zoology Archives, University of California, Berkeley (MVZ IMG 5770).

A.11. Portrait of Ben Thompson. Photo courtesy of the National Park Service.

A.12. Harold Bryant at Yosemite National Park. Photo courtesy of the National Park Service.

A.13. Ansel Hall at the laying of the museum cornerstone at Yosemite National Park, 1924. Photo courtesy of the National Park Service.

A.14. Portrait of Newton Drury. Photo courtesy of the National Park Service.

A.15. Portrait of A. Starker Leopold. Photo by Norden H. (Dan) Cheatam, courtesy of Reginald H. Barrett.

David D. Ackerly is a professor in the Department of Integrative Biology at the University of California, Berkeley, where he held the Gill Chair in Natural History from 2010 to 2015, and he is an associate curator with the Jepson Herbarium. His current research focuses on physiology, population and community ecology of California woodlands, and the implications of climate change impacts on biodiversity for the future of conservation and land management. Ackerly coleads the Terrestrial Biodiversity and Climate Change Initiative (tbc3.org), a group focused on climate change impacts and climate adaptation strategies in the San Francisco Bay Area, and he is codirector of the Berkeley Initiative in Global Change Biology.

Ruth M. Alexander is professor of history and council chair of the Public Lands History Center at Colorado State University. Her current research in environmental history examines the rise in backcountry recreation in public lands over the course of the 20th century, the links between backcountry recreation and the tourism industry, and public land managers' efforts to mitigate the impact of backcountry visitors on natural resources. She is committed to creating synergies between scientists, environmental historians, and public land managers in order to identify management practices that will support sustainability in the face of climate change. Her research has been funded by the National Park Service, the National Endowment for the Humanities, and the Colorado Water Conservation Board.

Allison K. Barner is a PhD candidate in the Integrative Biology Department at Oregon State University completing her PhD in zoology with a minor in statistics. She uses complementary experimental and mathematical approaches to understand how diverse ecological communities of species are formed, persist, and change in space and time, with special attention to the role of interactions among species. Using the rocky intertidal ecosystem of the US Pacific Northwest coast as a model ecosystem, her work explicitly examines if and how ecological theory can be leveraged to make predictions about the impact of climate change on ecosystems.

Jill S. Baron is an ecosystem ecologist with the US Geological Survey and a senior research ecologist with the Natural Resource Ecology Laboratory at Colorado State University. Her interests include applying ecosystem concepts to management of human-dominated regions, and understanding the biogeochemical and ecological effects of climate change and atmospheric nitrogen deposition to mountain ecosystems. She is founder and codirector of the John Wesley Powell Center for Earth System Science Analysis and Synthesis. Baron was president of the Ecological Society of America in 2013 and is a fellow of this scientific society.

Steven R. Beissinger is a professor of conservation biology at the University of California, Berkeley, where he held the A. Starker Leopold Chair in Wildlife Biology from 2003 to 2013, and he is a research associate of the Museum of Vertebrate Zoology. His professional career has been devoted to producing ecological knowledge that can be used to both conserve biodiversity and to uncover basic processes in behavioral and population ecology that govern how nature works. His current research centers on two of the biggest challenges facing wildlife conservation and society—wildlife responses to global change and species extinctions—with work carried out in national parks and working landscapes in California and Latin America. He is a fellow of the American Association for the Advancement of Science. In 2010, he received the William Brewster Memorial Award for his studies of birds of the Western Hemisphere from the American Ornithologists' Union, and he became president of this scientific society in 2016.

Joel Berger is the Cox-Anthony Chair of Conservation Biology at Colorado State University and is a senior scientist at the New York–based Wildlife Conservation Society. He has worked primarily with mammals larger than a bread box including moose, muskoxen, and black rhinos, as well as in more than a dozen US national parks; he has also worked with wild yaks and elusive takin, chiru, and saiga in Asia. Using conservation and science as diplomacy, Berger has built bridges between protected area management by uniting fieldwork in US parks with those in Bhutan, on the Tibetan Plateau, and in the Russian Arctic (Beringia). Berger has been the recipient of several Guggenheim awards, is an elected fellow of the American Association for the Advancement of Science, and has received lifetime achievement awards from the Society of Conservation Biology and American Society of Mammalogists.

Edwin Bernbaum is a senior fellow at The Mountain Institute and cochair of the IUCN Specialist Group on the Cultural and Spiritual Values of Protected Areas. His research and work focus on the relationship between culture and the environment; he is currently working on integrating the cultural and spiritual significance of nature into protected area management and governance. His book *Sacred Mountains of the World* won the Commonwealth Club of California's gold medal for best work of nonfiction and an Italian award for literature of mountaineering, exploration, and the environment. At The Mountain Institute, he initiated and directed a project to develop interpretive materials with various national parks based on the evocative cultural and spiritual associations of different features of the environment in American and other cultures around the world.

Tamara Blett is an ecologist with the Air Resources Division of the US National Park Service. She works to enhance clean air and foster healthy ecosystems in all national parks across the country by developing critical loads and other ecosystem thresholds, collaborating with researchers to initiate environmental research and monitoring projects, and synthesizing scientific results for use in management, policy, and regulatory forums. She has worked on issues related to air pollution effects on natural ecosystems for over 25 years and is currently exploring how air pollution may be altering ecosystem services.

Justin S. Brashares is the G. R. & W. M. Goertz Distinguished Professor in the Department of Environmental Science, Policy, and Management at the University of California, Berkeley. His current work attempts to understand how the consumption of wild animals and conversion of natural habitats affects the dynamics of animal communities and the persistence of populations. Work in his group extends beyond traditional ecology and conservation to consider the economic, political, and social factors that drive and, in turn, are driven by, changes in biodiversity.

Stephanie M. Carlson is associate professor of freshwater fish ecology at the University of California, Berkeley. Her primary research interest centers on understanding the dynamics of freshwater fish populations, particularly the factors that shape these populations and influence their persistence. Much of her current research focuses on the ecology of salmon and trout at the southern end of their range, including research carried out in national and state parks.

Jamais A. Cascio is a distinguished fellow at the Institute for the Future in Palo Alto, California, and author of *Hacking the Earth: Understanding the Consequences of Geoengineering* (2009). Selected by *Foreign Policy* magazine as one of the top 100 Global Thinkers, Cascio has explored the intersection of environmental dilemmas, emerging technologies, and cultural evolution for nearly 20 years. Cascio's written work has appeared in the *Atlantic Monthly*, *New Scientist*, and the *New York Times*, among many others, and he has been featured in a variety of films and television programs on future issues, including National Geographic's 2008 documentary on global warming *Six Degrees*, the 2010 CBC documentary *Surviving the Future*, and the 2013 documentary *Fixed: The Science/Fiction of Human Augmentation*. Cascio speaks about future possibilities around the world at venues including the Conference on World Affairs, the National Academy of Sciences in Washington, DC, and TED.

Ruth DeFries is the Denning Family Professor of Sustainable Development in the Department of Ecology, Evolution, and Environmental Biology at Columbia University. Her research focuses on land-use change in the tropics and its implications for climate, biodiversity, and other ecosystem services. She is a member of the US National

Academy of Sciences, among other honors. She has published over a hundred scientific papers and two science books for popular audiences.

Thomas Dietz is a professor of environmental science and policy, sociology, and animal studies at Michigan State University, where he was founding director of the Environmental Science and Policy Program. His research focuses on the drivers of anthropogenic environmental stress and of human well-being, environmental values, and the relationship between science and values in environmental decision making. He has published 13 books and over 140 journal papers and book chapters. In 2005, he received the Sustainability Science Award from the Ecological Society of America.

Daniel C. Donato is a research scientist at the Washington State Department of Natural Resources and affiliate faculty at the University of Washington. He has done extensive research on forest and fire ecology in the Pacific Northwest and in Greater Yellowstone. Donato is currently studying structural development of early-seral and old-growth forests following major disturbances.

C. Josh Donlan is the director of Advanced Conservation Strategies and a visiting fellow at Cornell University. He leads the organization by building interdisciplinary teams to tackle problems in novel ways. His current research focuses on incentive program design, entrepreneurship, and how the process of innovation can contribute to biodiversity conservation. Donlan has published over 100 papers in scientific journals, and his recent book *Proactive Strategies for Protecting Species* outlines a new policy and market framework on incentivizing voluntary conservation action prior to regulatory triggers.

Holly Doremus is the James H. House and Hiram H. Hurd Professor of Environmental Regulation at the University of California, Berkeley, associate dean for faculty development and research, and co–faculty director of the Center for Law, Energy, and the Environment. She is a member scholar of the Center for Progressive Reform and serves on the board of directors of Defenders of Wildlife. She received her BS in biology from Trinity College (Hartford, CT), PhD in plant physiology from Cornell University, and JD from the University of California, Berkeley. She works in the areas of environmental and natural resources law, focusing on biodiversity protection, the intersection between property rights and environmental regulation, and the interrelationship of environmental law and science.

Kelly J. Easterday is a doctoral candidate in the Kelly Lab in the Department of Environmental Science, Policy, and Management at the University of California, Berkeley. Her research focuses on assessing drivers of change in the forests of California by integrating historical and modern data sets with geospatial technologies.

Ernesto C. Enkerlin-Hoeflich is former chair of the IUCN World Commission on Protected Areas and a professor at Monterrey Tech (Mexico), where he currently serves as leader of Legacy for Sustainability and director for Natural Solutions. As a prominent Mexican conservationist, environmentalist, and researcher, he specializes in parrot ecology, environmental policy, sustainability, and biodiversity stewardship. His efforts at the National Commission on Protected Areas of Mexico (CONANP), which he presided over from 2001 to 2010, resulted in an expansion of 50% in protected areas coverage. Mexico became the world leader in designations under international conventions and in establishing the first wilderness area in Latin America.

John Francis is a behavioral ecologist, turned filmmaker, turned administrator, who most recently served as vice president for research, conservation, and exploration at the National Geographic Society. His professional career is rooted in wildlife biology and a deep commitment to communication of the potent ties between humans and the rest of the natural world. He has served on a variety of boards and committees, including the Commission for Education and Communication of the IUCN, the US National Commission for UNESCO, and the National Advisory Board for the US National Park Service. His 10-year commitment to the BioBlitz concept in advance of the NPS centennial is emblematic of a belief that all citizens can contribute to science safeguarding our planet's natural resources.

Denis P. Galvin joined the US National Park Service in 1963 after serving as a civil engineer in the first Peace Corps group in East Africa. In a 38-year career, he worked in parks, regional offices, training centers, and service centers; the last 16 years were in Washington, DC, 9 of them as deputy director. He represented the National Park Service in over 200 congressional hearings. In 2013, he received the George Melendez Wright Award for his "distinguished lifetime record . . . on behalf of America's national parks."

Patrick Gonzalez is the principal climate change scientist of the US National Park Service and a visiting scholar at the University of California, Berkeley. A forest ecologist, he conducts research on climate change impacts, vulnerabilities, and ecosystem carbon, and works with managers and policymakers to adapt natural resource management. Gonzalez has published field research from Africa, South America, and the United States; been honored as a Fulbright Scholar, an American Association for the Advancement of Science Diplomacy Fellow, and a National Academy of Sciences Frontiers of Science speaker; and served as a lead author for three reports of the Intergovernmental Panel on Climate Change.

Kirsten Grorud-Colvert is senior research assistant professor in the Integrative Biology Department at Oregon State University. Her current research seeks to better understand the effects of marine reserves and other marine protected areas on ecological communities, including how the early life of fishes can inform marine reserve design. She also directs the international Science of Marine Reserves Project, working with a team of marine ecologists, graphic designers, communication specialists, and marine reserve scientists around the globe to catalyze, synthesize, and communicate scientific data about marine reserves to inform management and conservation.

Winslow D. Hansen is a PhD student studying at the University of Wisconsin–

Madison. As a MS student at the University of Alaska Fairbanks, he explored inter-actions between bark beetle outbreaks and fire on the Kenai Peninsula and how both disturbances affect home values. Hansen is currently studying effects of climate warming on postfire tree regeneration in Greater Yellowstone.

Brian J. Harvey is a Smith Postdoctoral Research Fellow at the University of Colorado Boulder. He completed his PhD in 2015 at the University of Wisconsin–Madison, where he studied interactions between bark beetle outbreaks and subsequent forest fires in Greater Yellowstone and the Northern Rocky Mountains. Harvey has also ex-plored regional patterns of forest fire severity and postfire tree regeneration.

Audrey F. Haynes is a doctoral candidate in the Sousa Lab in the Department of Inte-grative Biology at the University of California, Berkeley. Her research focuses on the physiology and ecology of parasitic plants. Beyond research, she can be found hiking to her goal of visiting every US National Park—the current tally is 45 out of 59.

Emily E. Kearny is a doctoral student studying under Claire Kremen in the Department of Environmental Science, Policy, and Management at the University of California, Berkeley. Her research focuses on the provision of pollination service to farmers by native insects in the Central Valley of California and the Amazonian Basin in Ecua-dor.

Cyril F. Kormos is vice president for policy at The WILD Foundation, where he works on wilderness law and policy and primary forest conservation, and he serves as vice-chair for world heritage for IUCN's World Commission on Protected Areas and sits on IUCN's World Heritage Panel, which makes official recommendations to the UNESCO World Heritage Committee on World Heritage nominations. Kormos is also an associate editor for the *International Journal of Wilderness* and an editorial board member for IUCN-WCPA's *Parks* journal. He has edited or coedited four books and publishes frequently in peer-reviewed journals. Kormos holds a BA in English from the University of California, Berkeley, a MSc in politics of the world economy from the London School of Economics, and a JD from the George Washington Uni-versity Law School.

Kelly A. Kulhanek is an undergraduate in the Department of Molecular Environmen-tal Biology at the University of California, Berkeley, concentrating in animal health and behavior. She has committed much of her undergraduate years to researching the effect of fire diversity on pollinator systems in Yosemite with the Kremen Lab. In her graduate studies she hopes to continue to find answers to many of the world's impending agricultural and ecological issues.

Laurel G. Larsen is assistant professor of earth systems science in the Department of Geography at the University of California, Berkeley, where she runs the Environmen-tal Systems Dynamics Laboratory. Larsen's research is aimed at understanding how the flow of water through the environment interacts with vegetation, sediment, and organisms from the scale of individual vegetation stems to the landscape scale. Her work has influenced restoration efforts in the Everglades, with ongoing work focusing on the Chesapeake Bay and the Wax Lake Delta, part of the greater Mississippi River Delta complex.

Christine S. Lehnertz is general superintendent of Golden Gate National Recreation Area. Previously, she served as regional director for the Pacific West Region of the National Park Service, and prior to that, she was deputy superintendent at Yellowstone National Park. She also spent 16 years working for the US Environmental Protection Agency. Lehnertz grew up in Colorado, where she hiked, camped, and fished her way around the Rockies, and received her bachelor's degree in environmental biology from the University of Colorado Boulder.

Carrie R. Levine is a doctoral candidate in the Battles Lab in the Department of Environmental Science, Policy, and Management at the University of California, Berkeley. Her research focuses on quantifying demographic changes in California forests in response to climate change and fire suppression policies. Her work informs restoration and management strategies for maintaining forest resilience in the face of global change.

Jane Lubchenco is Distinguished University Professor and Advisor in marine studies at Oregon State University. A marine biologist and environmental scientist, she has deep experience in the worlds of science, academia, and government, and is a champion of science and of the stronger engagement of scientists with society. From 2009 to 2013, she served as the administrator of the National Oceanic and Atmospheric Administration. She works to advance scientific and public understanding of the interactions between the environment and human well-being.

Gary E. Machlis is professor of environmental sustainability at Clemson University and science advisor to the director of the US National Park Service. He is the first scientist appointed to this position within the National Park Service, and he advises the director on a range of science policy issues and programs. Machlis also serves as coleader of the US Department of the Interior's Strategic Sciences Group, which conducts scientific assessments during major environmental crises. He has written numerous books and scientific papers on issues of conservation and sustainability, including *The State of the World's Parks* (1985), the first systematic study of threats to protected areas around the world; his most recent coauthored book, *The Baltimore School of Ecology: Space, Scale and Time for the Study of Cities*, was published by Yale University Press in 2015.

William C. Malm is a research scientist/scholar at the Cooperative Institute for Research in the Atmosphere and a recently retired research physicist in the National Park Service Air Resources Division, where he was program coordinator for the visibility/particulate research and monitoring program. Malm's expertise is in the general area of visibility and related topics. He made some of the first visibility and air quality measurements in the National Park System at the Grand Canyon in 1972, and he pioneered studies of visibility perception that elicit human responses, in terms of both psychophysical and value assessment, to changes in scenic quality as a function of aerosol optical properties. He has initiated and carried out large field campaigns to better characterize aerosol physical and optical properties, and the results from this work have been incorporated into the Interagency Monitoring of Protected Visual Environments (IMPROVE) program and the US Environmental Protection Agency Regional Haze Rule (RHR).

George Miller is a retired US congressman who represented the interests of his Northern California constituents for 40 years. During the course of his tenure on the Hill, Miller served as a member and chair of both the House Committee on the Environment and the House Committee on Education and Labor. His legislative legacy includes the Affordable Care Act ensuring health care access for all, the No Child Left Behind Act ensuring quality education for all children, and the Central Valley Projects Improvement Act ensuring adequate water for California's fragile environment.

Tierne M. Nickel is a doctoral student in the Beissinger Lab in the Department of Environmental Science, Policy, and Management at the University of California, Berkeley. As part of the Grinnell Resurvey Project, her research focuses on the distributional response of small mammals to climate change in the California deserts.

Meagan F. Oldfather is a doctoral candidate in the Ackerly Lab in the Department of Integrative Biology at the University of California, Berkeley. Her research is focused on the demographic drivers of range limits for long-lived plants, and she works in both alpine and forest communities in California. She is also committed to the importance of long-term monitoring in protected areas. She has received the National Science Foundation Graduate Research Fellowship, and collaborates with the Terrestrial Biodiversity and Climate Change Collaborative and the Global Observation Research Initiative in Alpine Environments.

Lauren C. Ponisio is a doctoral candidate in the Kremen Lab in the Department of Environmental Science, Policy, and Management at the University of California, Berkeley. Her research focuses on understanding the mechanisms underlying the maintenance of biodiversity in natural and human-dominated landscapes.

Maggie J. Raboin is a doctoral student studying under Damian Elias in the Department of Environmental Science, Policy, and Management at the University of California, Berkeley. Her research focuses on evolution of parental care and behavioral ecology of a newly discovered mound-building spider, the mason spider, in the Greater Yellowstone Ecosystem.

Nina S. Roberts, a dynamic educator and well-known vibrant speaker, is a professor in the Department of Recreation, Parks, and Tourism at San Francisco State University. She is a Fulbright Scholar and experiential educator whose social science research in cultural diversity and national parks has been vital to community engagement efforts. Roberts is also director of the Pacific Leadership Institute, an outdoor adventure program in partnership with the Golden Gate National Recreation Area. Her work provides National Park Service managers and partners with ideas and resources needed to respond more effectively to changing demographics and social trends across the United States.

Frances B. Roberts-Gregory is a doctoral student in the Department of Environmental Science, Policy, and Management at the University of California, Berkeley. Her research focuses on natural resource management within the southeastern United States and interrogates how class, race, and gender categories, as well as ethnic identity and culture, affect the unequal geographical distribution of environmental burdens and privileges. She has similarly worked with urban greenspace managers to better understand issues of accessibility. In addition to receiving support from the Ford Foundation and the National Science Foundation Graduate Research Fellowship Program, Frances is a Berkeley Chancellor's Fellow and Bill Gates Millennium Scholar.

William H. Romme is professor emeritus in the Natural Resource Ecology Laboratory at Colorado State University. An internationally recognized expert in fire ecology, he has studied fire and vegetation in western forests throughout his career. Romme led the fire-history studies that provided context for the 1988 Yellowstone fires, and he has collaborated with Monica Turner on research in Yellowstone for over 25 years.

Raymond M. Sauvajot is associate director for natural resource stewardship and science for the US National Park Service, where he provides leadership, oversight, and direction for natural resource science and policy, support and guidance on complex and controversial issues, and executive-level representation on national-level conservation initiatives. His expertise and interests include landscape-scale conservation, effects of habitat fragmentation on wildlife, climate change adaptation, science communication, and the interface between science, policy, and politics. He has held adjunct faculty positions in biology, ecology, and environmental science at the University of California, Berkeley, University of California, Los Angeles, and California State University, Northridge.

Kelsey J. Scheckel is a doctoral student in Neil Tsutsui's lab in the Department of Environmental Science, Policy, and Management at the University of California, Berkeley. Her research focuses on the evolutionary genetics of sociality across socially parasitic ant genera.

Adam C. Schneider is a doctoral candidate studying under Bruce Baldwin in the Department of Integrative Biology and the Jepson Herbarium at the University of California, Berkeley. His research uses a phylogentic framework to understand the evolutionary causes and ecological consequences of plant endemism.

Daniel Simberloff is the Nancy Gore Hunger Professor of Environmental Studies at the University of Tennessee. His research centers on the causes, impacts, and management of invasive nonnative species. He is editor-in-chief of *Biological Invasions*, senior editor of *Encyclopedia of Biological Invasions,* and author of *Invasive Species—What Everyone Needs to Know,* and has published nearly 400 papers in scientific journals.

Monica G. Turner is the Eugene P. Odum Professor of Ecology and Vilas Research Professor in the Department of Zoology at the University of Wisconsin–Madison. Her expertise is in forest ecosystem and landscape ecology, and she has conducted research in Yellowstone in collaboration with Bill Romme for more than 25 years. Turner is a member of the US National Academy of Sciences and is serving as the 2015–16 president of the Ecological Society of America.

Rachel E. Walsh is a doctoral candidate in Eileen Lacey's lab in the Department of Integrative Biology and Museum of Vertebrate Zoology at the University of California, Berkeley. Her research focuses on patterns of habitat use by chipmunks in Yosemite National Park in the context of understanding elevational range shifts observed in small mammal species over the past century.

A. LeRoy Westerling is associate professor of management in the School of Engineering at the University of California, Merced. Westerling was the first to establish an unambiguous relationship between recent climate warming and the occurrence of large fires throughout the western United States. He has worked extensively on fires in California and the Northern Rocky Mountains.

Edward O. Wilson is generally recognized as one of the leading biologists in the world. He is acknowledged as the creator of two scientific disciplines (island biogeography and sociobiology), three unifying concepts for science and the humanities jointly (biophilia, biodiversity studies, and consilience), and one major technological advance in the study of global biodiversity (the Encyclopedia of Life). Among the more than 100 awards he has received worldwide are the US National Medal of Science, the Crafoord Prize (equivalent of the Nobel, for ecology) of the Royal Swedish Academy of Sciences, and the International Prize of Biology of Japan; and in letters, two Pulitzer Prizes in nonfiction, the Nonino and Serono Prizes of Italy, and the COSMOS Prize of Japan. He is currently honorary curator in entomology and university research professor emeritus at Harvard University.

Jennifer Wolch is the William W. Wurster Dean of the College of Environmental Design and professor of city and regional planning at the University of California, Berkeley. Her past research includes studies of attitudes toward wildlife among urban national recreational area visitors, race and urban park space, and the distribution of urban park funding. Her recent research focuses on environmental justice and access to urban parks and recreational programs, multibenefit metropolitan open space development planning, and connections between city form, physical activity, and public health. Wolch served as a member of the National Park System Advisory Board's Planning Committee from 2010 to 2013, and she is currently a member of the board's Urban Committee.

INDEX

The letter *f* following a page number denotes a figure and the letter *t* a table.